Thi

Ergebnisse der Mathematik und ihrer Grenzgebiete

Band 80

F. F. Bonsall · J. Duncan

Complete
Normed Algebras

Springer-Verlag
Berlin Heidelberg New York 1973

Frank F. Bonsall
Mathematical Institute, University of Edinburgh

John Duncan
Department of Mathematics, University of Stirling

AMS Subject Classification (1970): 46 H 05

ISBN 3-540-06386-2 Springer-Verlag Berlin Heidelberg New York
ISBN 0-387-06386-2 Springer-Verlag New York Heidelberg Berlin

To
James Connor Alexander
1942–1972

Introduction

The axioms of a complex Banach algebra were very happily chosen. They are simple enough to allow wide ranging fields of application, notably in harmonic analysis, operator theory and function algebras. At the same time they are tight enough to allow the development of a rich collection of results, mainly through the interplay of the elementary parts of the theories of analytic functions, rings, and Banach spaces. Many of the theorems are things of great beauty, simple in statement, surprising in content, and elegant in proof. We believe that some of them deserve to be known by every mathematician.

The aim of this book is to give an account of the principal methods and results in the theory of Banach algebras, both commutative and non-commutative. It has been necessary to apply certain exclusion principles in order to keep our task within bounds. Certain classes of concrete Banach algebras have a very rich literature, namely C^*-algebras, function algebras, and group algebras. We have regarded these highly developed theories as falling outside our scope. We have not entirely avoided them, but have been concerned with their place in the general theory, and have stopped short of developing their special properties. For reasons of space and time we have omitted certain other topics which would quite naturally have been included, in particular the theories of multipliers and of extensions of Banach algebras, and the implications for Banach algebras of some of the standard algebraic conditions on rings. We have also omitted the theory of locally convex algebras and other generalizations.

Both of us have devoted almost the whole of our professional lives to the study of Banach algebras, and it is therefore difficult to attribute accurately the sources from which we have learnt the results and proofs that we give. We have undoubtedly leaned heavily on earlier books, above all on the highly authoritative book [321] by C. E. Rickart, which remains an indispensable source for anyone working seriously in this field. Our bibliography is very far from comprehensive, consisting mainly of the items to which we make direct reference together with some papers of particular historical importance. We are confident that we have omitted many references that would be relevant to our purpose.

We are also confident that this book will contain errors. It is our view that the integrity of mathematics depends on every author taking responsibility for the correctness of everything that he writes even when quoting another author. We have done our best to adhere to this principle, but none-the-less mistakes will have occurred. Accordingly, the reader is warned to verify the correctness of proofs for himself. We should have perpetrated even more errors but for the help of A. M. Sinclair who has read and criticised a large part of the manuscript. We are also indebted to Z. Sebestyén for detecting further errors.

The book is subdivided into sections numbered from 1 to 50, and these numbers are used for back references. Thus Theorem j. k refers to Theorem k in section j. For back references to items in the current section, the section number is omitted.

The reader is warned that the statements of many lemmas, propositions and theorems are incomplete in the sense that certain blanket assumptions may apply to some of the symbols. These are usually introduced at the beginning of a section, but occasionally signalled by 'Notation' in the midst of a section. Thus throughout § 32 A stands for a complex Banach annihilator algebra, in § 38 the significance of A is changed several times. Proofs end, as is now standard, with the Halmos tombstone ▯.

We owe a great deal to Mrs. Christine McLeod and Mrs. Joan Young for typing the manuscript swiftly, accurately, and with expert judgement. We are also indebted to the publishers for their patience in waiting for the manuscript and their habitual skill in converting it into a book.

April 1973   F. F. Bonsall
                                                              J. Duncan

Table of Contents

Chapter I. Concepts and Elementary Results

§ 1. Normed Algebras

Throughout this book the symbol $\mathbb{F}$ will be used to denote a field that is either the real field $\mathbb{R}$ or the complex field $\mathbb{C}$.

Definition 1. An *algebra over* $\mathbb{F}$ is a linear space A over $\mathbb{F}$ together with a mapping $(x, y) \to xy$ of $A \times A$ into A that satisfies the following axioms (for all $x, y, z \in A$, $\alpha \in \mathbb{F}$):
  (a) $x(yz) = (xy)z$,
  (b) $x(y+z) = xy + xz$, $(x+y)z = xz + yz$,
  (c) $(\alpha x)y = \alpha(xy) = x(\alpha y)$.

Remarks. (1) The full name for such an algebra is 'linear associative algebra', to distinguish it from other more general algebras. However, in a book where all the algebras are of this kind, the full name would be cumbersome.

(2) The field $\mathbb{F}$ is called the *scalar field* of A. If $\mathbb{F} = \mathbb{R}$, A is called a *real algebra*, and if $\mathbb{F} = \mathbb{C}$, a *complex algebra*.

(3) The mapping $(x,y) \to xy$ is called the *product* in A, and the vector xy the *product* of x and y.

(4) Axiom (a) asserts that the set A with its product is a semi-group. Axiom (c) is equivalent to

(c′) $\qquad (\alpha\beta)(xy) = (\alpha x)(\beta y) \quad (x, y \in A, \ \alpha, \beta \in \mathbb{F})$.

Definition 2. Let E be a set, X a linear space over $\mathbb{F}$, α an element of $\mathbb{F}$, f, g mappings of E into X. There is a natural definition of $f + g$, αf as the mappings of E into X given by

$$(f+g)(s) = f(s) + g(s), \quad (\alpha f)(s) = \alpha(f(s)) \quad (s \in E).$$

This is called the *pointwise* definition of *addition* and *scalar multiplication*. Likewise, when X is an algebra, we have the *pointwise* product

$$(fg)(s) = f(s)g(s) \quad (s \in E).$$

Example 3. Let E be a non-void set, and let A be the set of all mappings of E into $\mathbb{F}$. With the pointwise addition, scalar multiplication, and product, A is an algebra.

Notation. Given linear spaces X, Y over the same field $\mathbb{F}$, we denote by $L(X, Y)$ the linear space of all linear mappings of X into Y with the pointwise addition and scalar multiplication.

Example 4. Let X be a linear space over $\mathbb{F}$. $L(X, X)$ with the product defined by *composition*

$$(ST)x = S(Tx) \quad (x \in X),$$

is an algebra, which is denoted by $L(X)$.

Remark. Let A be an algebra, and given $a \in A$, let λ_a, ρ_a be the mappings of A into A given by

$$\lambda_a x = ax, \quad \rho_a x = xa \quad (x \in A).$$

The axioms (b) and (c) in Definition 1 are equivalent to the statement that λ_a, $\rho_a \in L(A)$. Axiom (a) is equivalent to each of the following identities

$$\lambda_a \lambda_b = \lambda_{ab}, \quad \rho_a \rho_b = \rho_{ba}, \quad \lambda_a \rho_b = \rho_b \lambda_a.$$

Notation. For the moment let θ denote the zero vector of an algebra A over $\mathbb{F}$, and 0 the zero element of $\mathbb{F}$. Since A is a linear space over $\mathbb{F}$, we have

$$0x = \theta.$$

Therefore

$$x\theta = x(0\,\theta) = 0(x\,\theta) = \theta, \quad \theta x = (0\,\theta)x = 0(\theta x) = \theta \quad (x \in A).$$

This shows that the product of x with θ on either side is equal to θ which is also the result of multiplying x by the scalar 0. For this reason, the usual convention will be adopted from now on of denoting both zero elements by 0.

Definition 5. Let A, B be algebras over the same scalar field $\mathbb{F}$. A *homomorphism* of A into B is a mapping $\phi \in L(A, B)$ such that

$$\phi(xy) = \phi(x)\phi(y) \quad (x, y \in A).$$

A *monomorphism* of A into B is an injective homomorphism of A into B, and an *isomorphism* of A onto B is a bijective homomorphism of A into B. Algebras A and B are *isomorphic* if there exists an isomorphism of A onto B.

A *subsemi-group* of A is a subset S such that

$$x, y \in S \implies xy \in S.$$

A *subalgebra* of A is a linear subspace of A that is also a subsemigroup of A.

Clearly, a subalgebra B of an algebra A is itself an algebra with the same scalar field and with the product in B the restriction to $B \times B$ of the product in A.

Example 6. Let A be an algebra, and given $a \in A$, let λ_a be the mapping of A into A defined by $\lambda_a x = ax$ $(x \in A)$. Then the mapping λ is a homomorphism of A into $L(A)$. It is called the *left regular representation* of A on A.

Definition 7. Let X be a linear space over $\mathbb{F}$, and let A be an algebra over $\mathbb{F}$. A *semi-norm* (or pre-norm or pseudo-norm) on X is a mapping p of X into $\mathbb{R}$ such that (for all $x, y \in A$ and $\alpha \in \mathbb{F}$):
  (a) $p(x) \geqslant 0$,
  (b) $p(\alpha x) = |\alpha| p(x)$,
  (c) $p(x + y) \leqslant p(x) + p(y)$.
A *norm* on X is a semi-norm p on X such that
  (d) $p(x) = 0 \Rightarrow x = 0$.
An *algebra-norm (algebra-semi-norm)* on A is a norm (semi-norm) p on A such that
  (e) $p(xy) \leqslant p(x) p(y)$ $(x, y \in A)$.

Definition 8. Let E be a subset of a linear space X over $\mathbb{F}$. Then E is said to be *absolutely convex* if

$$x, y \in E, \quad \alpha, \beta \in \mathbb{F}, \quad |\alpha| + |\beta| \leqslant 1 \Rightarrow \alpha x + \beta y \in E,$$

absorbent if, for every $x \in X$, $\lambda x \in E$ for some $\lambda > 0$, *radially bounded* if, for every $x \in X \setminus \{0\}$, the set $\{\lambda \in \mathbb{R} : \lambda x \in E\}$ is bounded or void.

The *Minkowski functional* (or gauge functional) p_E of an absorbent subset E of X is defined on X by

$$p_E(x) = \inf \{\lambda > 0 : \lambda^{-1} x \in E\}.$$

Given an algebra-semi-norm p on an algebra A, it is clear that the unit ball $\{x \in A : p(x) < 1\}$ and the closed unit ball $\{x \in A : p(x) \leqslant 1\}$ are both absolutely convex absorbent subsemi-groups of A, and are also radially bounded if p is a norm. Conversely we have the following proposition.

Proposition 9. *Let U be an absolutely convex absorbent subsemigroup of an algebra A. Then the Minkowski functional p_U is an algebra-semi-norm on A, and*

$$\{x \in A : p_U(x) < 1\} \subset U \subset \{x \in A : p_U(x) \leqslant 1\}. \tag{1}$$

If also U is radially bounded, then p_U is an algebra-norm on A.

Proof. Elementary normed linear space theory shows that p_U is a semi-norm satisfying (1); and is a norm if U is radially bounded. If $p_U(x) < 1$ and $p_U(y) < 1$, then $x, y \in U$ and so $xy \in U$, and $p_U(xy) \leqslant 1$. It follows easily that p_U is an algebra-semi-norm. □

Definition 10. A *normed algebra* is a pair (A, p), where A is an algebra and p is an algebra-norm on A. A *complete normed algebra* or *Banach algebra* is a normed algebra (A, p) such that the normed linear space A with norm p is complete (i. e. every Cauchy sequence converges). A normed algebra is an algebra A with a specified norm. Unless otherwise stated we shall denote the specified norm by $\| \cdot \|$, and we shall write 'the normed algebra A' as an abbreviation for 'the normed algebra $(A, \| \cdot \|)$'.

As is usual for normed linear spaces, a normed algebra A is regarded as a metric space with the distance function

$$d(x, y) = \| x - y \| \qquad (x, y \in A).$$

A normed algebra is also regarded as a topological space with the metric topology given by this distance function.

Remark. The name 'Banach algebra' is appropriate only because, as a normed linear space, a Banach algebra is a Banach space. However this name is too firmly established in the literature to be changed at this time. Given a free choice we should like to call complete normed algebras 'Gelfand algebras' in recognition of the distinguished pioneering work of I. M. Gelfand in this field.

Definition 11. Let A, B be normed algebras. A *topological isomorphism* of A onto B is an isomorphism of A onto B that is also a homeomorphism of the topological space A onto the topological space B. An *isometric isomorphism* of A onto B is an isomorphism T of the algebra A onto the algebra B that is also an isometric mapping of the metric space A onto the metric space B. This last condition states that

$$\| Tx - Ty \| = \| x - y \| \qquad (x, y \in A).$$

However, by the linearity of T, this is equivalent to

$$\| Tx \| = \| x \| \qquad (x \in A).$$

Similarly, for normed linear spaces X, Y, an *isometric linear isomorphism* of X onto Y is a linear mapping T of X onto Y such that $\| Tx \| = \| x \|$ $(x \in X)$.

Notation. Given normed linear spaces X, Y over the same scalar field $\mathbb{F}$, we denote by $BL(X, Y)$ the linear subspace of $L(X, Y)$, consisting of all bounded (i. e. continuous) linear mappings of X into Y.

As usual $BL(X, Y)$ is regarded as a normed linear space with the norm given by

$$\|T\| = \sup \{\|Tx\| : x \in X, \|x\| \leqslant 1\}.$$

The *dual space* $BL(X, \mathbb{F})$ of X is denoted by X', and its elements are called *continuous linear functionals*. We write $BL(X)$ for $BL(X, X)$.

We take it as known that $BL(X, Y)$ is complete whenever Y is, and in particular that the dual space X' of X is always complete. We also take it as known that every normed linear space X has a *completion*, i. e. a Banach space Y such that there exists an isometric linear isomorphism of X onto a dense linear subspace of Y.

Proposition 12. *Let A be a normed algebra. Then there exists an isometric isomorphism of A onto a dense subalgebra of a Banach algebra B. B is unique up to isometric isomorphism.*

Proof. There exists an isometric linear isomorphism T of A onto a dense linear subspace of a Banach space B. Given $x, y \in B$, there exist $x_n, y_n \in A$ such that $x = \lim_{n \to \infty} Tx_n$, $y = \lim_{n \to \infty} Ty_n$. Since T is an isometry, $\{x_n\}$ and $\{y_n\}$ are Cauchy sequences in A. Since

$$\|x_p y_p - x_q y_q\| \leqslant \|x_p\| \, \|y_p - y_q\| + \|x_p - x_q\| \, \|y_q\|,$$

$\{x_n y_n\}$ is a Cauchy sequence in A, $\{T(x_n y_n)\}$ is a Cauchy sequence in B, and $\lim_{n \to \infty} T(x_n y_n) = z \in B$. Moreover, z is independent of the choice of the sequences $\{x_n\}$, $\{y_n\}$, and so may be used to define a product in B by taking $xy = z$. □

Definition 13. A *completion* of a normed algebra A is a Banach algebra B as in Proposition 12. When considering the completion B of a normed algebra A, we shall find it convenient to identify A with its image in B under the isometric isomorphism. Thus we shall regard B as a Banach algebra having A as a dense subalgebra.

In the rest of this section we exhibit some important examples of normed and Banach algebras.

Example 14. Let X be a normed linear space. Then $BL(X)$ with composition for the product is a normed algebra, and is a Banach algebra if X is complete. The elements of $BL(X)$ are called *bounded linear operators* on X, and the norm on $BL(X)$ is called the *operator norm*.

Definition 15. Let E be a non-void set and X a normed linear space. A mapping f of E into X is said to be *bounded* if $\{\|f(s)\| : s \in E\}$ is a bounded set of real numbers. For a bounded mapping f the *uniform norm* $\|f\|_\infty$ is defined by

$$\|f\|_\infty = \sup\{\|f(s)\|: s\in E\}.$$

Example 16. Let E be a non-void set and X a normed linear space. We denote by $l^\infty(E,X)$ the normed linear space of all bounded mappings of E into X with pointwise addition and scalar multiplication and with the uniform norm. When X is a normed algebra, $l^\infty(E,X)$ becomes a normed algebra with the pointwise product. It is straightforward to prove that $l^\infty(E,X)$ is complete when X is complete.

An *algebra of functions* on E is an algebra of mappings of E into $\mathbb{F}$ with the pointwise definition of addition, scalar multiplication, and the product. A *uniform algebra of functions on* E is a subalgebra of the Banach algebra $l^\infty(E,\mathbb{F})$.

Given a non-void topological space E, $C(E,X)$ denotes the linear space of all continuous mappings of E into X with pointwise addition and scalar multiplication. When E is compact, $C(E,X)$ is a closed linear subspace of $l^\infty(E,X)$, and in particular, $C(E,\mathbb{F})$ is a uniform algebra of functions. The notation $C(E,\mathbb{C})$ is abbreviated to $C(E)$.

Given an open subset D of $\mathbb{C}$, we denote by $H(D)$ the set of all mappings of D into $\mathbb{C}$ that are *locally holomorphic* on D, i.e. complex analytic on each connected open subset of D. (We shall omit the word 'locally' even when D is not connected.) It is clear that $H(D)$ is an algebra of functions on D. The subalgebra $H^\infty(D)$, given by

$$H^\infty(D) = H(D)\cap l^\infty(D,\mathbb{C}),$$

is a uniform algebra of functions on D.

Let Δ denote the closed unit disc $\{\zeta\in\mathbb{C}:|\zeta|\leqslant 1\}$ and Δ^0 its interior. The *disc algebra* $\mathscr{A}(\Delta)$ is the uniform algebra of functions on Δ given by

$$\mathscr{A}(\Delta) = \{f\in C(\Delta): f|_{\Delta^0}\in H(\Delta^0)\}.$$

Example 17. Let A be a normed algebra. Let $a\cdot b$ denote the *reversed product* in A, i.e.

$$a\cdot b = ba \quad (a,b\in A).$$

With the reversed product and the given norm, A becomes a normed algebra called the *reversed algebra* of A and denoted by $\mathrm{rev}(A)$.

Example 18. Let G be a group, and let $l^1(G)$ denote the set of mappings f of G into $\mathbb{C}$ such that

$$\sum_{s\in G}|f(s)| < \infty.$$

With pointwise addition and scalar multiplication, with convolution

$$(f * g)(s) = \sum_{t \in G} f(t) g(t^{-1} s) \quad (s \in G)$$

as product and with the norm

$$\|f\| = \sum_{s \in G} |f(s)|,$$

$l^1(G)$ is a Banach algebra called the *discrete group algebra* of G.

Example 19. Let W denote the set of all functions a on the compact real interval $[0, 2\pi]$ of the form

$$a(t) = \sum_{k \in \mathbb{Z}} \alpha_k \exp(ikt) \quad (t \in [0, 2\pi]),$$

where the α_k $(k \in \mathbb{Z})$ are complex numbers satisfying

$$\sum_{k \in \mathbb{Z}} |\alpha_k| < \infty.$$

With pointwise addition, scalar multiplication, and product, and with the norm $\|a\| = \sum_{k \in \mathbb{Z}} |\alpha_k|$, W is a Banach algebra called the *Wiener algebra*.

It is easy to verify that the mapping $f \to f^\sim$, given by

$$f^\sim(t) = \sum_{k \in \mathbb{Z}} f(k) \exp(ikt) \quad (t \in [0, 2\pi]),$$

is an isometric isomorphism of the discrete group algebra $l^1(\mathbb{Z})$ onto the Wiener algebra W.

Example 20. Let $L^1(\mathbb{R})$ denote the space of equivalence classes (under equality almost everywhere) of Lebesgue integrable complex valued functions on $\mathbb{R}$. With the addition and scalar multiplication derived from the pointwise addition and scalar multiplication of the functions and with the norm given by

$$\|f\| = \int_{\mathbb{R}} |f(t)| dt$$

(where the same symbol f is used to denote both a function and its equivalence class), $L^1(\mathbb{R})$ is a Banach space. With convolution

$$(f * g)(s) = \int_{\mathbb{R}} f(t) g(s - t) dt \quad (s \in \mathbb{R})$$

as the product, it becomes a Banach algebra, the *group algebra* of $\mathbb{R}$.

More generally, given a locally compact group G, let μ be a left invariant Haar measure on G, and let $L^1(G)$ be the corresponding

Banach space of integrable 'functions'. Then $L^1(G)$ becomes a Banach algebra with the product given by convolution

$$(f*g)(s) = \int_G f(t)g(t^{-1}s)d\mu(t) \quad (s\in G).$$

This is the *group algebra* of G. For details see Rudin [324], or Hewitt and Ross [194].

Example 21. Let α be a continuous non-negative real valued function on $\mathbb{R}$ such that

$$\alpha(t_1+t_2) \leqslant \alpha(t_1)\alpha(t_2) \quad (t_1, t_2\in\mathbb{R}).$$

$L^1(\mathbb{R},\alpha)$ denotes the space of equivalence classes (under equality almost everywhere) of Lebesgue measurable complex valued functions f on $\mathbb{R}$ for which $|f|\alpha$ is Lebesgue integrable. With addition, scalar multiplication, and the product defined as in Example 20 and with the norm given by

$$\|f\| = \int_{\mathbb{R}} |f(t)|\alpha(t)dt,$$

$L^1(\mathbb{R},\alpha)$ is a Banach algebra.

Example 22. Let G be a locally compact group, and let $M(G)$ denote the set of all complex-valued bounded regular Borel measures on G. The elements of $M(G)$ are complex-valued functions on the set $\mathscr{B}$ of Borel subsets of G, and addition and scalar multiplication are defined pointwise on $\mathscr{B}$. The product in $M(G)$ is defined by convolution of measures,

$$(m*n)(E) = \int_G m(Es^{-1})n(ds) \quad (E\in\mathscr{B}),$$

and the norm $\|m\|$ of a measure m is its total variation

$$\|m\| = \sup\left\{\sum_{k=1}^{r} |m(E_k)|\right\},$$

where the supremum is taken over all finite sets $\{E_1,\dots,E_r\}$ of mutually disjoint elements of $\mathscr{B}$.

Example 23. Let S be a semi-group, let α be a positive real function on S such that

$$\alpha(st) \leqslant \alpha(s)\alpha(t) \quad (s,t\in S)$$

and let $l^1(S,\alpha)$ denote the set of mappings f of S into $\mathbb{C}$ such that

$$\sum_{s\in S} |f(s)|\alpha(s) < \infty.$$

With pointwise addition and scalar multiplication, with convolution

$$(f*g)(s) = \sum_{tu=s} f(t)g(u)$$

as product $((f*g)(s)=0$ if $tu=s$ has no solutions) and with the norm

$$\|f\| = \sum_{s\in S} |f(s)|\alpha(s),$$

$l^1(S,\alpha)$ is a Banach algebra. With $\alpha(s)=1$ $(s\in S)$ we obtain the *discrete semi-group algebra* of S, which we denote by $l^1(S)$.

Corollary 24. *Let α_n be any positive real sequence such that*

$$\alpha_{m+n} \leqslant \alpha_m \alpha_n \quad (m,n=1,2,3,...).$$

Then there exists a Banach algebra A and an element $a\in A$ such that $\|a^n\|=\alpha_n$ $(n=1,2,3,...)$.

Example 25. Let A be a normed algebra, let S be a subsemi-group of A (with respect to the product of A), and let B be the linear span of S in A, so that B is a subalgebra of A. For $x\in B$ let

$$p(x) = \inf\left\{ \sum_{k=1}^{r} |c_k|\,\|s_k\| : x = \sum_{k=1}^{r} c_k s_k, s_k\in S \right\}$$

where the infimum is taken over all representations of x as a linear combination of elements of S. Then (B,p) is a normed algebra with the following properties:

(i) $p(s)=\|s\|$ $(s\in S)$;

(ii) $p(x)\geqslant\|x\|$ $(x\in B)$;

(iii) if q is any algebra-norm on B with $q(s)=\|s\|$ $(s\in S)$, then $q(x)\leqslant p(x)$ $(x\in B)$.

Remarks. (1) Bade and Curtis [32] give an example of an algebra which cannot be normed.

(2) The algebra of polynomials in a single variable can be given a norm but not a complete norm. This fact follows easily by applying Baire's category theorem to the sequence $\{P_n\}$, where P_n is the set of all polynomials of degree at most n.

§ 2. Inverses

One of the good reasons for studying Banach algebras rather than more general algebras and rings is the simplicity of the behaviour of inverses in Banach algebras.

Definition 1. An element e of an algebra A is a *unit element* or *identity element* if and only if $e \neq 0$ and

$$ex = xe = x \quad (x \in A).$$

We say that A is an *algebra with unit* if it has a unit element.

An algebra has at most one unit element, for if e, e' are unit elements, then $e' = ee' = e$.

Let A be an algebra with unit. The unique unit element of A will be denoted by 1; moreover the symbol α will be used to denote a scalar α and the element of A obtained by multiplying the unit element of A by α.

Let $a \in A$. A *left inverse* of a is an element b of A such that $ba = 1$, a *right inverse* of a is an element b of A such that $ab = 1$. An *inverse* of a is an element of A that is both a left inverse of a and a right inverse of a. An element of A for which there exists an inverse is said to be *invertible* (some authors write *regular*); all other elements of A are called *singular* elements.

If a has a left inverse b and a right inverse c, then $b = c$ and so a is invertible, for

$$b = b1 = bac = 1c = c.$$

This shows, in particular, that an element has at most one inverse.

Notation. The unique inverse of an invertible element a is denoted by a^{-1}. The set of invertible elements of A is denoted by $\mathrm{Inv}(A)$, and the set of singular elements by $\mathrm{Sing}(A)$.

Definition 2. Given a subset E of an algebra A, the *commutant* of E is the subset E^c of A given by

$$E^c = \{a \in A : ax = xa \ (x \in E)\}.$$

$(E^c)^c$ is abbreviated to E^{cc}, and is called the *second commutant* of E.

It is easy to see that E^c is a subalgebra. In particular, given $a \in A$, $\{a\}^{cc}$ is a commutative subalgebra containing a.

Proposition 3. *Let* $a \in \mathrm{Inv}(A)$. *Then* a^{-1} *belongs to the second commutant* $\{a\}^{cc}$ *of* $\{a\}$.

Proof. Let $x \in \{a\}^c$. Then $ax = xa$, and so

$$xa^{-1} = a^{-1}(ax)a^{-1} = a^{-1}(xa)a^{-1} = a^{-1}x. \quad \square$$

It is entirely straightforward to verify that $\mathrm{Inv}(A)$ is a group with respect to the product in A, i.e. that

$$a, b \in \mathrm{Inv}(A) \Rightarrow ab, a^{-1} \in \mathrm{Inv}(A).$$

Inv(A) is for this reason called the *group of invertible elements* of A, or the *group of regular elements* of A.

Proposition 4. *The product in a normed algebra A is a continuous mapping of $A \times A$ into A.*

Proof. $\|xy-ab\| = \|(x-a)(y-b)+a(y-b)+(x-a)b\|$
$\leqslant \|x-a\|\,\|y-b\| + \|a\|\,\|y-b\| + \|x-a\|\,\|b\|.$ □

Lemma 5. *Let A be a normed algebra with unit. If $a,b \in \text{Inv}(A)$ and $\|b-a\| \leqslant \frac{1}{2}\|a^{-1}\|^{-1}$, then*
$$\|b^{-1}-a^{-1}\| \leqslant 2\|a^{-1}\|^2\,\|a-b\|.$$

Proof. For such a,b we have
$$\|b^{-1}\| - \|a^{-1}\| \leqslant \|b^{-1}-a^{-1}\| = \|b^{-1}(a-b)a^{-1}\| \leqslant \tfrac{1}{2}\|b^{-1}\|.$$

Thus $\|b^{-1}\| \leqslant 2\|a^{-1}\|$, and so
$$\|b^{-1}-a^{-1}\| \leqslant \|b^{-1}\|\,\|a-b\|\,\|a^{-1}\| \leqslant 2\|a^{-1}\|^2\|a-b\|. \quad □$$

Proposition 6. *Let A be a normed algebra with unit. Then, with the topology induced from the norm topology of A, the mapping $a \to a^{-1}$ is a homeomorphism of $\text{Inv}(A)$ onto itself, and $\text{Inv}(A)$ is a topological group.*

Proof. Proposition 4 and Lemma 5. □

Definition 7. Let a be an element of a normed algebra. The *spectral radius* of a, denoted by $r(a)$, is defined by
$$r(a) = \inf\left\{\|a^n\|^{\frac{1}{n}} : n = 1,2,\dots\right\}.$$

Proposition 8. *Let a be an element of a normed algebra. Then*
$$r(a) = \lim_{n\to\infty} \|a^n\|^{\frac{1}{n}}.$$

Proof. Let $\rho = r(a)$ and $\varepsilon > 0$, and select k such that
$$\|a^k\|^{\frac{1}{k}} < \rho + \varepsilon.$$
Every positive integer n can be written uniquely in the form $n = p(n)k + q(n)$ with $p(n), q(n)$ non-negative integers and $q(n) \leqslant k-1$. Since $\frac{1}{n}q(n) \to 0$, we have $\frac{1}{n}p(n)k \to 1$ as $n\to\infty$, and so
$$\|a^n\|^{\frac{1}{n}} \leqslant \|a^k\|^{\frac{1}{n}p(n)}\|a\|^{\frac{1}{n}q(n)} \to \|a^k\|^{\frac{1}{k}} < \rho + \varepsilon.$$

Thus $\|a^n\|^{\frac{1}{n}} < \rho + \varepsilon$ for all sufficiently large n. Also $\rho \leqslant \|a^n\|^{\frac{1}{n}}$ for all n. ☐

Theorem 9. *Let A be a Banach algebra with unit, let $a \in A$, and let $r(a) < 1$. Then $1 - a$ is invertible, and*

$$(1-a)^{-1} = 1 + \sum_{n=1}^{\infty} a^n.$$

Proof. Choose η with $r(a) < \eta < 1$. By Proposition 8, we have $\|a^n\| < \eta^n$ for all sufficiently large n, and therefore the series $\|1\| + \sum_{n=1}^{\infty} \|a^n\|$ converges. Since A is a Banach space, it follows that the series $1 + \sum_{n=1}^{\infty} a^n$ converges, with sum $s \in A$, say. Let $s_n = 1 + a + \cdots + a^{n-1}$. Then $s_n \to s$ and $\|a^n\| \to 0$ as $n \to \infty$, and we have

$$(1-a)s_n = s_n(1-a) = 1 - a^n.$$

Therefore, by continuity of multiplication (Proposition 4), we have,

$$(1-a)s = s(1-a) = 1. \quad ☐$$

Corollary 10. *Let A be a Banach algebra with unit. Then each element a of A with $\|1-a\| < 1$ is invertible.*

Proof. $r(1-a) \leqslant \|1-a\| < 1$, and so $a = 1 - (1-a)$ is invertible. ☐

Theorem 11. *Let A be a Banach algebra with unit. Then $\mathrm{Inv}(A)$ is an open subset of A.*

Proof. Let $a \in \mathrm{Inv}(A)$. Given $x \in A$ with $\|x\| < \|a^{-1}\|^{-1}$, we have

$$a - x = a(1 - a^{-1}x),$$

and $\|a^{-1}x\| \leqslant \|a^{-1}\| \|x\| < 1$. Therefore, by Theorem 9, $1 - a^{-1}x \in \mathrm{Inv}(A)$, and, $\mathrm{Inv}(A)$ being a group, $a - x \in \mathrm{Inv}(A)$. Thus $\mathrm{Inv}(A)$ contains the open ball with centre a and radius $\|a^{-1}\|^{-1}$. ☐

Definition 12. Let A be a normed algebra, and let $S(A)$ denote its unit sphere $S(A) = \{x \in A : \|x\| = 1\}$. An element a of A is called (i) a *left topological divisor of zero*, (ii) *a right topological divisor of zero*, (iii) *a joint topological divisor of zero* if, respectively, (i) $\inf\{\|ax\| : x \in S(A)\} = 0$, (ii) $\inf\{\|xa\| : x \in S(A)\} = 0$, (iii) $\inf\{\|ax\| + \|xa\| : x \in S(A)\} = 0$.

An element a of A is a joint topological divisor of zero if and only if there exists a sequence $\{x_n\}$ of elements of $S(A)$ such that

$$\lim_{n \to \infty} \|ax_n\| = \lim_{n \to \infty} \|x_n a\| = 0.$$

It is clear that a joint topological divisor of zero is both a left and a right topological divisor of zero. The fact that a left and right divisor of zero can fail to be a joint divisor of zero is shown by the following example due to P. G. Dixon.

Example 13. Let A_0 be the free complex algebra on the symbols
a, b, c subject to the relations $ba = ac = cb = 0$, and $cwb = 0$ for all
words w. Each non-zero element of A_0 has a unique representation of
the form $\sum_{k=1}^{n} \alpha_k w_k$ where the α_k are complex numbers and the w_k are
distinct non-zero words on the symbols a, b, c. A_0 is a normed algebra
with norm defined by

$$\left\| \sum_{k=1}^{n} \alpha_k w_k \right\| = \sum_{k=1}^{n} |\alpha_k|.$$

Let A be the completion of A_0, so that the elements of A are of the form
$\sum_{k=1}^{\infty} \alpha_k w_k$ with $\sum_{k=1}^{\infty} |\alpha_k| < \infty$. By definition, a is a left and right divisor
of zero in A. Given $d_n \in A$, we write

$$d_n = r_n + s_n + t_n$$

where the words involved in r_n are those words of d_n which end in b,
and the words in s_n are those which begin with c. Then

$$\|a d_n\| = \|a r_n + a t_n\| = \|r_n\| + \|t_n\|,$$
$$\|d_n a\| = \|s_n a + t_n a\| = \|s_n\| + \|t_n\|.$$

Therefore $d_n \to 0$ whenever $d_n a \to 0$ and $a d_n \to 0$, which proves that a
is not a joint topological divisor of zero.

Although Definition 12 is given in terms of the norm of A, it is easy
to verify that the sets of left, right and joint topological divisors of zero
are unchanged if we take an equivalent algebra-norm on A.

Notation. Given a subset E of a topological space, we denote by E^-
and ∂E respectively the closure of E and the topological boundary
(i.e. frontier) of E.

Theorem 14. *Let A be a Banach algebra with unit, and let $a \in \partial \operatorname{Inv}(A)$.
Then a is a joint topological divisor of zero.*

Proof. Since $a \in \partial \operatorname{Inv}(A)$, there exist $a_n \in \operatorname{Inv}(A)$ with $\lim_{n \to \infty} a_n = a$.
We prove first that $\{\|a_n^{-1}\|\}$ is unbounded. Suppose on the contrary
that

$$\|a_n^{-1}\| \leq M \quad (n = 1, 2, \ldots).$$

Then

$$\|a_m^{-1} - a_n^{-1}\| = \|a_m^{-1}(a_n - a_m) a_n^{-1}\| \leq M^2 \|a_n - a_m\|.$$

This shows that $\{a_n^{-1}\}$ is a Cauchy sequence. Let $b = \lim_{n \to \infty} a_n^{-1}$. Then,
by continuity of multiplication, $ab = ba = 1$, $a \in \operatorname{Inv}(A)$. But since $\operatorname{Inv}(A)$
is an open set, this contradicts the assumption that $a \in \partial \operatorname{Inv}(A)$.

By retaining only a suitable subsequence, we may now suppose that $\|a_n^{-1}\| \geqslant n$ $(n=1, 2, \ldots)$. Let $x_n = \|a_n^{-1}\|^{-1} a_n^{-1}$. Then $x_n \in S(A)$, and

$$a x_n = (a - a_n) x_n + a_n x_n = (a - a_n) x_n + \|a_n^{-1}\|^{-1}.$$

Thus $\lim_{n \to \infty} a x_n = 0$, and similarly $\lim_{n \to \infty} x_n a = 0$. ⬜

Definition 15. Let A be a normed algebra with unit. A normed algebra B is an *extension* of A if A is a subalgebra of B, the unit element of A is a unit element for B, and the norm on A is equivalent to the restriction to A of the norm on B. A singular element of A is said to be *permanently singular* if it is a singular element of every extension of A.

Proposition 16. *Let A be a normed algebra with unit, and let a be a left or a right topological divisor of zero in A. Then a is permanently singular.*

Proof. Suppose that a has an inverse b in an extension B of A and that $\lim_{n \to \infty} \|a x_n\| = 0$ for some sequence $\{x_n\}$ of elements of $S(A)$. With $\|\cdot\|'$ denoting the norm of B, we have

$$\|x_n\|' = \|b a x_n\|' \leqslant \|b\|' \cdot \|a x_n\|',$$

which gives a contradiction, since $\|\cdot\|$ is equivalent to $\|\cdot\|'$ on A. ⬜

Corollary 17. *Let A be a Banach algebra with unit. Then each element of $\partial\,\mathrm{Inv}(A)$ is permanently singular.*

Proof. Theorem 14 and Proposition 16. ⬜

Corollary 18. *Let B be an extension of a Banach algebra A with unit. Then*
 (i) $\mathrm{Inv}(A) \subset \mathrm{Inv}(B)$,
 (ii) $\partial\,\mathrm{Inv}(A) \subset \partial\,\mathrm{Inv}(B)$.

Proof. (i) is clear. By Corollary 17, $\partial\,\mathrm{Inv}(A) \subset \mathrm{Sing}(B)$, and with (i) this gives (ii). ⬜

Given an algebra A with unit, the left regular representation of A on A provides an embedding of A in $L(A)$. In this embedding all singular elements are preserved, as the following proposition shows.

Proposition 19. *Let A be a normed algebra with unit, and let λ be the left regular representation of A on A. Then*

$$a \in \mathrm{Sing}(A) \Leftrightarrow \lambda_a \in \mathrm{Sing}(L(A)).$$

Proof. Let $B=L(A)$, let $a\in A$, and suppose that $\lambda_a\in \mathrm{Inv}(B)$. Then there exists $T\in B$ with $\lambda_a T=T\lambda_a=I$, where I is the identity operator on A. Let $x=T1$. Then

$$ax=\lambda_a(T1)=I1=1.$$

Therefore $\lambda_a\lambda_x=I$. Since λ_a has a unique right inverse in B, it follows that $\lambda_x=T$. Therefore $\lambda_x\lambda_a=I$, $xa=\lambda_x\lambda_a 1=1$. Thus $a\in \mathrm{Inv}(A)$. ▯

Remark. If A is a normed algebra with unit, it follows that

$$a\in \mathrm{Sing}(A) \Rightarrow \lambda_a\in \mathrm{Sing}(BL(A)).$$

For further results on topological divisors of zero and permanently singular elements see Rickart [321], Arens [20].

§ 3. Quasi-Inverses

The important results on inverses proved in §2 are of course vacuous for algebras without a unit element, but have been extended to such algebras in two ways:

(1) by the adjunction of a unit element,
(2) by using the concept of quasi-inverse.

Definition 1. The *unitization* of a normed algebra A over $\mathbb{F}$, denoted by $A+\mathbb{F}$, is the normed algebra consisting of the set $A\times\mathbb{F}$ with addition, scalar multiplication and product defined (for all $x,y\in A$, $\alpha,\beta\in\mathbb{F}$) by

$$(x,\alpha)+(y,\beta)=(x+y,\alpha+\beta),$$
$$\beta(x,\alpha)=(\beta x,\beta\alpha),$$
$$(x,\alpha)(y,\beta)=(xy+\alpha y+\beta x,\alpha\beta),$$

and with the norm defined by

$$\|(x,\alpha)\|=\|x\|+|\alpha|.$$

It is a routine matter to verify that $A+\mathbb{F}$ is a normed algebra with unit element $(0,1)$, that $\|(0,1)\|=1$, and that the mapping $a\to(a,0)$ is an isometric isomorphism of A onto a subalgebra of $A+\mathbb{F}$.

Definition 2. Given elements x,y of an algebra A, the *quasi-product* of x,y is the element $x\circ y$ of A defined by

$$x\circ y=x+y-xy.$$

Proposition 3. (i) $(x \circ y) \circ z = x \circ (y \circ z)$,
(ii) $x \circ 0 = 0 \circ x = x$.

Proof. (i) $(x \circ y) \circ z = (x \circ y) + z - (x \circ y) z$
$$= x + y + z - xy - xz - yz + xyz = x \circ (y \circ z).$$
(ii) Clear. ☐

Proposition 3 shows that an algebra with its quasi-product is a semi-group with identity 0. Quasi-inverses, which we shall now define, are inverses with respect to this semi-group.

Definition 4. Let x be an element of an algebra A. Elements y, z of A are respectively *left* and *right quasi-inverses* of x if

$$y \circ x = 0, \qquad x \circ z = 0.$$

A *quasi-inverse* of x is an element of A that is both a left quasi-inverse and a right quasi-inverse of x. An element that has a quasi-inverse is said to be *quasi-invertible* (or *quasi-regular*), all other elements are said to be *quasi-singular*. Likewise x is left (right) quasi-singular if x is not left (right) quasi-invertible, i.e. if x has no left (right) quasi-inverse.

If x has a left quasi-inverse y and a right quasi-inverse z, then by Proposition 3, $y = z$, and x is quasi-invertible. This shows in particular that an element has at most one quasi-inverse.

Notation. The quasi-inverse of a quasi-invertible element x is denoted by x^0, the set of all quasi-invertible elements of A by q-Inv(A), and the set of all quasi-singular elements by q-Sing(A).

Proposition 5. (i) *An element x of A has the quasi-inverse y if and only if $(0, 1) - (x, 0)$ has the inverse $(0, 1) - (y, 0)$ in $A + \mathbb{F}$.*
(ii) *If A has a unit element, an element x of A has the quasi-inverse y if and only if $1 - x$ has the inverse $1 - y$.*

Proof. (ii) $(1 - x)(1 - y) = 1 - (x \circ y)$.
(i) Apply (ii) in the algebra $A + \mathbb{F}$ and use the fact that the mapping $x \to (x, 0)$ is a monomorphism of A into $A + \mathbb{F}$. ☐

Proposition 6. *If xy is left (right) quasi-invertible, then yx is left (right) quasi-invertible.*

Proof. $y x \circ (yzx - yx) = y(x y \circ z) x$,
$(yzx - yx) \circ yx = y(z \circ xy) x$. ☐

Theorem 7. *Let a be an element of a Banach algebra such that $r(a) < 1$. Then a is quasi-invertible and*

$$a^0 = - \sum_{n=1}^{\infty} a^n.$$

Proof. Either repeat the proof of Theorem 2.9 or apply that theorem to the algebra $A + \mathbb{F}$ and use Proposition 5 (i). $\square$

Theorem 8. *The set of quasi-invertible elements of a Banach algebra is open.*

Proof. Let A be a Banach algebra. Note that if $(-x, 1)$ has the inverse $(-y, \alpha)$, then $\alpha = 1$. It then follows from Proposition 5 (i) that

$$\text{q-Inv}(A) = \{x : (-x, 1) \in \text{Inv}(A + \mathbb{F})\} .$$

Since the mapping $x \rightarrow (-x, 1)$ is a continuous mapping of A into the Banach algebra $A + \mathbb{F}$, Theorem 2.11 applied to $A + \mathbb{F}$ now shows that q-Inv(A) is open. $\square$

Proposition 9. *Let A be a Banach algebra, and let $a \in \partial(\text{q-Inv}(A))$. Then there exist $x_n \in S(A)$ $(n \in \mathbb{N})$ such that*

$$\lim_{n \to \infty} x_n - a x_n = \lim_{n \to \infty} x_n - x_n a = 0 . \tag{1}$$

Proof. We have $(-a, 1) \in \partial \text{Inv}(A + \mathbb{F})$, and so, by Theorem 2.14, $(-a, 1)$ is a joint topological divisor of zero in $A + \mathbb{F}$. Thus there exist $x_n \in A$, $\alpha_n \in \mathbb{F}$ such that $\|x_n\| + |\alpha_n| = 1$ $(n \in \mathbb{N})$ and

$$\lim_{n \to \infty} (-a, 1)(x_n, \alpha_n) = \lim_{n \to \infty} (x_n, \alpha_n)(-a, 1) = 0 .$$

Therefore $\lim_{n \to \infty} \alpha_n = 0$ and then (1) holds, and $\lim_{n \to \infty} \|x_n\| = 1$, so that we may arrange that $\|x_n\| = 1$ $(n \in \mathbb{N})$. $\square$

Corollary 10. *Let A be a Banach algebra and let $a \in \partial(\text{q-Inv}(A))$. Then a is left and right quasi-singular.*

Proof. By Proposition 9, there exist $x_n \in S(A)$ satisfying (1). If a has a left quasi-inverse b, then $b - ba = -a$,

$$-a x_n = (b - ba) x_n = b(x_n - a x_n).$$

Therefore $\lim_{n \to \infty} a x_n = 0$, and so $\lim_{n \to \infty} x_n = 0$, which is impossible. $\square$

§ 4. Equivalent Norms

Throughout this section $(A, \|\cdot\|)$ will denote a normed algebra over $\mathbb{F}$, $\text{En}(A)$ will denote the set of all algebra-norms on A equivalent to the given algebra-norm $\|\cdot\|$, and, if A has a unit element,

$$\text{Eun}(A) = \{p \in \text{En}(A) : p(1) = 1\} .$$

We recall that norms p_1, p_2 on a linear space X are *equivalent* if they determine the same topology on X, and that this holds if and only if there exist positive constants M_1, M_2 such that

$$p_1(x) \leqslant M_1 p_2(x), \qquad p_2(x) \leqslant M_2 p_1(x) \qquad (x \in X).$$

It is an immediate consequence of Proposition 2.8 that equivalent algebra-norms give the same spectral radius, i.e. for all $p \in \text{En}(A)$ we have

$$r(a) = \lim_{n \to \infty} \|a^n\|^{\frac{1}{n}} = \lim_{n \to \infty} [p(a^n)]^{\frac{1}{n}} = \inf \{ [p(a^n)]^{\frac{1}{n}} : n = 1, 2, \ldots \}. \qquad (1)$$

Theorem 1. *Let S be a bounded subsemi-group of A (with respect to the product of A). Then there exists $p \in \text{En}(A)$ such that*

$$p(s) \leqslant 1 \qquad (s \in S).$$

If A has a unit element we may choose such p in $\text{Eun}(A)$.

Proof. Since A can be embedded in its unitization, there is no loss of generality in supposing that A has a unit element. Then $S \cup \{1\}$ is also a bounded semi-group, and so we may assume that $1 \in S$. Take

$$q(a) = \sup \{ \|s a\| : s \in S \},$$

and choose a positive constant M with $\|s\| \leqslant M$ $(s \in S)$. Then $q(a) \leqslant M \|a\|$. Since $1 \in S$, we have $\|a\| \leqslant q(a)$, and so

$$q(ab) = \sup \{ \|s a b\| : s \in S \} \leqslant q(a) \|b\| \leqslant q(a) q(b).$$

It is now clear that $q \in \text{En}(A)$. We define p on A by

$$p(a) = \sup \{ q(a x) : x \in A, q(x) \leqslant 1 \}.$$

Then $p(a)$ is the operator norm of λ_a (the left regular representation of a) regarded as a bounded linear operator on the normed space (A, q). Therefore $p \in \text{Eun}(A)$. Also, if $t \in S$, we have

$$q(t x) = \sup \{ \|s t x\| : s \in S \} \leqslant q(x),$$

since $st \in S$. Thus $p(t) \leqslant 1$ $(t \in S)$. $\square$

Corollary 2. (i) $r(a) = \inf \{ p(a) : p \in \text{En}(A) \}$ $(a \in A)$.
(ii) *If A has a unit element,*

$$r(a) = \inf \{ p(a) : p \in \text{Eun}(A) \} \qquad (a \in A).$$

Proof. By (1), we have

$$r(a) \leqslant p(a) \qquad (p \in \text{En}(A), a \in A).$$

If $r(a)<1$, then $\{a^n : n=1,2,...\}$ is a bounded subsemi-group of A, and so by Theorem 1 there exists $p\in\mathrm{En}(A)$ with $p(a)\leqslant 1$, and if A has a unit element there exists $p\in\mathrm{Eun}(A)$ with $p(a)\leqslant 1$. □

Corollary 3. *Let $a, b\in A$ with $ab=ba$. Then*

$$r(ab)\leqslant r(a)r(b), \qquad r(a+b)\leqslant r(a)+r(b).$$

Proof. Let $\varepsilon>0$, $u=(r(a)+\varepsilon)^{-1}a$, $v=(r(b)+\varepsilon)^{-1}b$. Then $r(u)<1$, $r(v)<1$, and so the sets $\{u^n : n=1,2,...\}$, $\{v^n : n=1,2,...\}$ are bounded. Since $uv=vu$, the set S of all elements u^k, v^j, $u^k v^j$ $(k,j=1,2,...)$ is a bounded subsemi-group of A. Therefore, by Theorem 1, there exists $p\in\mathrm{En}(A)$ with $p(u)\leqslant 1$, $p(v)\leqslant 1$. Then

$$p(a)\leqslant r(a)+\varepsilon, \qquad p(b)\leqslant r(b)+\varepsilon,$$

and so

$$r(ab)\leqslant p(ab)\leqslant p(a)p(b)\leqslant (r(a)+\varepsilon)(r(b)+\varepsilon),$$

$$r(a+b)\leqslant p(a+b)\leqslant p(a)+p(b)\leqslant r(a)+r(b)+2\varepsilon. □$$

Corollary 4. *If A has unit element, $\mathrm{Eun}(A)$ is non-void.*

Proof. Take $S=\{1\}$. □

Definition 5. $(A, \|\cdot\|)$ is said to be a *unital normed algebra* if A has a unit element and $\|1\|=1$.

Corollary 4 shows that every normed algebra with unit has an equivalent algebra-norm with which it is unital.

Proposition 6. *Let A be a Banach algebra with unit. Let $a_n\in\mathrm{Inv}(A)$, $a a_n=a_n a$ $(n=1,2,...)$, $a=\lim\limits_{n\to\infty} a_n$, and let $\{r(a_n^{-1})\}$ be bounded. Then $a\in\mathrm{Inv}(A)$.*

Proof. $1-a_n^{-1}a=a_n^{-1}(a_n-a)$, and so, by Corollary 3,

$$r(1-a_n^{-1}a)\leqslant r(a_n^{-1})r(a_n-a)\leqslant r(a_n^{-1})\|a_n-a\|.$$

Therefore $\lim\limits_{n\to\infty} r(1-a_n^{-1}a)=0$, and so, by Theorem 2.9, $a_n^{-1}a\in\mathrm{Inv}(A)$ for sufficiently large n. □

Remark. A similar result holds for quasi-inverses.

§ 5. The Spectrum of an Element of a Complex Normed Algebra

Throughout this section A will denote an algebra over $\mathbb{C}$.

Definition 1. The *spectrum* of an element a of A is the set $\mathrm{Sp}(A,a)$ of complex numbers defined as follows:

(i) if A has a unit element, $\operatorname{Sp}(A,a)=\{\lambda\in\mathbb{C}:\lambda-a\in\operatorname{Sing}(A)\}$;

(ii) if A does not have a unit element,

$$\operatorname{Sp}(A,a)=\{0\}\cup\left\{\lambda\in\mathbb{C}\setminus\{0\}:\frac{1}{\lambda}\,a\in\text{q-Sing}(A)\right\}. \quad\text{When no confusion can}$$

occur, we write $\operatorname{Sp}(a)$ to denote $\operatorname{Sp}(A,a)$.

The two concepts of spectrum in this definition are related through the unitization of A, as the following lemma shows.

Lemma 2. *Let A not have a unit element, and let B denote the unitization $A+\mathbb{C}$ of A. Then $\operatorname{Sp}(A,a)=\operatorname{Sp}(B,(a,0))$ $(a\in A)$.*

Proof. We have $(x,0)\circ(y,\eta)=((x\circ y)-\eta x,\eta)$. Therefore we have $\eta=0$ and $x\circ y=0$ if $(x,0)\circ(y,\eta)=0$. It follows that $x\in\text{q-Sing}(A)$ if and only if $(x,0)\in\text{q-Sing}(B)$, i.e., by Proposition 3.5, if and only if $(0,1)-(x,0)\in\operatorname{Sing}(B)$. If $\lambda\in\mathbb{C}\setminus\{0\}$, we have

$$\lambda\in\operatorname{Sp}(A,a)\Leftrightarrow 1\in\operatorname{Sp}\left(A,\frac{1}{\lambda}\,a\right)\Leftrightarrow 1\in\operatorname{Sp}\left(B,\left(\frac{1}{\lambda}\,a,0\right)\right)\Leftrightarrow\lambda\in\operatorname{Sp}(B,(a,0)).$$

Finally we have $0\in\operatorname{Sp}(B,(a,0))$, since $(a,0)\in\operatorname{Sing}(B)$. □

Before proving the important theorem that the spectrum of an element of a normed algebra is always non-void, we establish a few elementary algebraic propositions.

Proposition 3. *Let $a, b\in A$. Then*

$$\operatorname{Sp}(ab)\setminus\{0\} = \operatorname{Sp}(ba)\setminus\{0\}.$$

Proof. It is enough to prove that $1\in\operatorname{Sp}(ab)$ if and only if $1\in\operatorname{Sp}(ba)$, or equivalently that ab is quasi-regular if and only if ba is quasi-regular. But this is Proposition 3.6. □

Proposition 4. *Let A have a unit element, let π be a homomorphism of A into an algebra B with unit such that $\pi(1)$ is the unit element of B, and let λ be the left regular representation of A on A. Then*

(i) $\operatorname{Sp}(B,\pi(a))\subset\operatorname{Sp}(A,a)$ $(a\in A)$,

(ii) $\operatorname{Sp}(L(A),\lambda_a)=\operatorname{Sp}(A,a)$ $(a\in A)$.

Proof. (i) $\alpha-a\in\operatorname{Inv}(A)\Rightarrow\pi(\alpha-a)\in\operatorname{Inv}(B)\Rightarrow\alpha-\pi(a)\in\operatorname{Inv}(B)$.

(ii) Proposition 2.19. □

Notation. Given a non-void open subset D of $\mathbb{C}$, we denote by $P(D)$ the algebra of all complex polynomial functions on D, and by $R(D)$ the algebra of all rational functions on D with no poles in D. Let A

have a unit element, and let $a \in A$. Given $p \in P(D)$ we denote by $p(a)$ the element of A given by

$$p(a) = \alpha_0 + \alpha_1 a + \cdots + \alpha_n a^n,$$

where

$$p(z) = \alpha_0 + \alpha_1 z + \cdots + \alpha_n z^n \qquad (z \in D).$$

Since D is an infinite set, it is clear that $p(a)$ is well defined. It is a matter of formal manipulation to show that the mapping

$$p \to p(a): P(D) \to A$$

is a homomorphism.

We wish to extend the homomorphism $p \to p(a)$ to a homomorphism of $R(D)$ into A (and in §7 to a homomorphism of $H(D)$ into A). We note first that for this to be possible it is necessary that

$$\mathrm{Sp}(a) \subset D.$$

For, let $\lambda \in \mathbb{C} \setminus D$ and let $f(z) = (\lambda - z)^{-1}$ $(z \in D)$. Then $f \in R(D)$, and

$$f(z)(\lambda - z) = (\lambda - z) f(z) = 1 \qquad (z \in D).$$

If $f(a)$ denotes the image of f under a homomorphism of $R(D)$ into A that extends the homomorphism $p \to p(a)$, we have

$$f(a)(\lambda - a) = (\lambda - a) f(a) = 1,$$

and so $\lambda - a \in \mathrm{Inv}(A)$, $\lambda \in \mathbb{C} \setminus \mathrm{Sp}(a)$.

Proposition 5. *Let A have a unit element, let $a \in A$, let D be a non-void open neighbourhood of* $\mathrm{Sp}(a)$, *and let* $p \in P(D)$, *p non-constant. Then*

$$\mathrm{Sp}(p(a)) = \{p(z) : z \in \mathrm{Sp}(a)\}.$$

Proof. Let $\lambda \in \mathbb{C}$. Then there exist $\alpha_1, \ldots, \alpha_n$, $\beta \in \mathbb{C}$ such that

$$\lambda - p(z) = \beta(\alpha_1 - z)(\alpha_2 - z)\ldots(\alpha_n - z) \qquad (z \in D).$$

Therefore, using the homomorphism of $P(D)$ into A, we have

$$\lambda - p(a) = \beta(\alpha_1 - a)(\alpha_2 - a)\ldots(\alpha_n - a).$$

Since p is not the constant polynomial, we have $\beta \neq 0$. Thus $\lambda - p(a)$ is singular if and only if $\alpha_k - a$ is singular for some k, i.e. if and only if $\alpha_k \in \mathrm{Sp}(a)$. Thus $\lambda \in \mathrm{Sp}(p(a))$ if and only if $\lambda - p(z) = 0$ for some $z \in \mathrm{Sp}(a)$. ☐

Notation. Let A have a unit element, let $a \in A$, let D be a non-void open neighbourhood of $\mathrm{Sp}(a)$, and let $f \in R(D)$. Then $f = p/q$, where

$p, q \in P(D)$ and q has no zeros in D. By Proposition 5, $0 \notin \mathrm{Sp}(q(a))$, so $q(a) \in \mathrm{Inv}(A)$, and we define $f(a)$ by

$$f(a) = p(a)(q(a))^{-1}.$$

Since D is a non-void open set, the representation p/q for f is unique apart from common factors of numerator and denominator. Moreover polynomials in a and the inverses of such polynomials commute with each other (by Proposition 2.3) and so the element $f(a)$ is independent of the choice of p, q.

Proposition 6. *Let A have a unit element, let $a \in A$, and let D be a non-void open neighbourhood of $\mathrm{Sp}(a)$. The mapping $f \to f(a)$ is a homomorphism of $R(D)$ into A which extends the homomorphism of $P(D)$ into A, and satisfies*

$$\mathrm{Sp}(f(a)) = \{f(z) : z \in \mathrm{Sp}(a)\} \qquad (f \in R(D),\ f \text{ non-constant}).$$

Proof. Straightforward. □

Theorem 7. *Let a be an element of a complex normed algebra. Then there exists a complex number λ in $\mathrm{Sp}(a)$ such that $|\lambda| \geqslant r(a)$.*

Proof. If the normed algebra A does not have a unit element, then, by Lemma 2, $\mathrm{Sp}(A, a) = \mathrm{Sp}(B, (a, 0))$ with $B = A + \mathbb{C}$. Also $r((a, 0)) = r(a)$, since the mapping $x \to (x, 0)$ is an isometric isomorphism of A into B. Therefore we may assume that A has a unit element.

If $0 \notin \mathrm{Sp}(a)$, then a has an inverse a^{-1}, and $a^n(a^{-1})^n = 1$. Therefore

$$0 < \|1\| \leqslant \|a^n\| \cdot \|a^{-1}\|^n,$$

from which $r(a) \geqslant \|a^{-1}\|^{-1} > 0$. Therefore $0 \in \mathrm{Sp}(a)$ if $r(a) = 0$, and the theorem is proved in this case.

We now assume without real loss of generality, that $r(a) = 1$, and we suppose that there exists $\eta > 1$ such that the annulus $E = \{z \in \mathbb{C} : 1 \leqslant |z| \leqslant \eta\}$ has void intersection with $\mathrm{Sp}(a)$. We have

$$z - a \in \mathrm{Inv}(A) \qquad (z \in E),$$

and so, by Proposition 2.6, the mapping $z \to z(z - a)^{-1}$ is a continuous mapping of E into A. Since E is compact, this mapping is uniformly continuous; and therefore, given $\varepsilon > 0$, there exists $\delta > 0$ such that

$$\|\lambda(\lambda - a)^{-1} - \mu(\mu - a)^{-1}\| < \varepsilon \qquad (\lambda, \mu \in E, |\lambda - \mu| \leqslant \delta). \tag{1}$$

Given a positive integer n, let $\omega_1, \ldots, \omega_n$ denote the zeros of the polynomial $z^n - 1$, and consider the elementary identity

$$\frac{1}{n} \sum_{k=1}^{n} \frac{1}{1 - \omega_k^{-1} z} = \frac{1}{1 - z^n}.$$

Let $D = \mathbb{C} \setminus E$. Then D is a non-void open neighbourhood of $\text{Sp}(a)$; and, when $\lambda \in E$, we have $\lambda \omega_k \in E$, and so

$$\frac{1}{n} \sum_{k=1}^{n} \frac{1}{1 - \omega_k^{-1} \lambda^{-1} z} = \frac{1}{1 - \lambda^{-n} z^n} \qquad (z \in D)$$

relates rational functions belonging to $R(D)$. Therefore, by Proposition 6,

$$\frac{1}{n} \sum_{k=1}^{n} (1 - \omega_k^{-1} \lambda^{-1} a)^{-1} = (1 - \lambda^{-n} a^n)^{-1} \qquad (\lambda \in E). \tag{2}$$

Let $c_n(\lambda) = \frac{1}{n} \sum_{k=1}^{n} (1 - \omega_k^{-1} \lambda^{-1} a)^{-1}$ $(\lambda \in E)$. Given $\lambda, \mu \in E$ with $|\lambda - \mu| < \delta$, we have $\lambda \omega_k, \mu \omega_k \in E$ and $|\lambda \omega_k - \mu \omega_k| < \delta$. Therefore, by (1),

$$\| \lambda \omega_k (\lambda \omega_k - a)^{-1} - \mu \omega_k (\mu \omega_k - a)^{-1} \| < \varepsilon \qquad (k = 1, \ldots, n),$$

and so

$$\| c_n(\lambda) - c_n(\mu) \| < \varepsilon \qquad (\lambda, \mu \in E, |\lambda - \mu| < \delta). \tag{3}$$

If $|\lambda| > 1$, $\lim_{n \to \infty} 1 - \lambda^{-n} a^n = 1$, and so, by Proposition 2.6 and (2), $\lim_{n \to \infty} c_n(\lambda) = 1$. Now choose λ in E with $|\lambda| > 1$ and $|\lambda - 1| < \delta$. Then, by (3),

$$\| c_n(\lambda) - c_n(1) \| < \varepsilon \qquad (n = 1, 2, \ldots),$$

and so $\| 1 - c_n(1) \| \leqslant \varepsilon$ for all sufficiently large n. Therefore, as $n \to \infty$, we have in turn $c_n(1) \to 1$, $(1 - a^n)^{-1} \to 1$, $1 - a^n \to 1$, $a^n \to 0$. But this is impossible, since

$$\| a^n \| \geqslant (r(a))^n = 1 \qquad (n = 1, 2, \ldots).$$

This contradiction shows that $E \cap \text{Sp}(a)$ is not void. ☐

Theorem 8. *Let a be an element of a complex Banach algebra. Then $\text{Sp}(a)$ is a non-void compact subset of $\mathbb{C}$ and*

$$\max \{ |\lambda| : \lambda \in \text{Sp}(a) \} = r(a). \tag{4}$$

Proof. As in the proof of Theorem 7, we may assume that the Banach algebra A has a unit element. Let $|\lambda| > \rho = r(a)$. Then $r(\lambda^{-1} a) < 1$, and so, by Theorem 2.9, $1 - \lambda^{-1} a$ is invertible, i. e. $\lambda \notin \text{Sp}(a)$. This proves that $\text{Sp}(a)$ is a subset of the disc $\{\lambda \in \mathbb{C} : |\lambda| \leqslant \rho\}$, and, together with Theorem 7, this proves (4). It only remains to prove that $\text{Sp}(a)$ is closed. However, $\mathbb{C} \setminus \text{Sp}(a)$ is the inverse image with respect to the continuous mapping $z \to z - a$ of the open set $\text{Inv}(A)$. Therefore $\mathbb{C} \setminus \text{Sp}(a)$ is open. ☐

Definition 9. Let X be a Banach space over $\mathbb{C}$, and let D be an open subset of $\mathbb{C}$. A mapping f of D into X is said to be *holomorphic* on D if, for every continuous linear functional $\phi \in X'$, the complex valued function $\phi \circ f$ is holomorphic on D in the usual sense.

Remarks. (1) Note that we do not require D to be connected.

(2) For a full discussion of holomorphic vector valued functions see (Hille and Phillips [197] p. 92—108).

Notation. For the rest of this section A denotes a complex Banach algebra with unit, and a denotes an element of A.

Theorem 10. *The mapping* $z \to (z-a)^{-1}$ *is a holomorphic mapping of* $\mathbb{C} \backslash \mathrm{Sp}(a)$ *into* A.

Proof. Let $\zeta \in \mathbb{C} \backslash \mathrm{Sp}(a)$ and $\phi \in A'$. Then

$$z - a = (\zeta - a)(1 - (\zeta - z)(\zeta - a)^{-1}).$$

Therefore, if $|z - \zeta| < \|(\zeta - a)^{-1}\|^{-1}$, we have $z \in \mathbb{C} \backslash \mathrm{Sp}(a)$ and

$$(z - a)^{-1} = (\zeta - a)^{-1} + (\zeta - z)(\zeta - a)^{-2} + (\zeta - z)^2 (\zeta - a)^{-3} + \cdots,$$

with the series convergent in norm. Thus, for z in this neighbourhood of ζ, we have

$$\phi((z - a)^{-1}) = \alpha_0 + \alpha_1 (\zeta - z) + \alpha_2 (\zeta - z)^2 + \cdots,$$

with $\alpha_k = \phi((\zeta - a)^{-(k+1)})$ $(k = 0, 1, 2, \ldots)$. $\quad\square$

We know (by Theorem 2.9) that when $|z| > r(a)$, $(z - a)^{-1}$ belongs to the least closed subalgebra of A containing 1 and a. How large a subalgebra is needed to contain $(z - a)^{-1}$ for all $z \in \mathbb{C} \backslash \mathrm{Sp}(a)$? An answer is given by the following theorem, which will also give a proof (in § 6) of the classical theorem of Runge on the approximation of holomorphic functions by rational functions with preassigned poles.

Theorem 11. (An abstract Runge theorem.) *Let* Λ *be a subset of* $\mathbb{C} \backslash \mathrm{Sp}(a)$ *that has non-void intersection with each bounded component of* $\mathbb{C} \backslash \mathrm{Sp}(a)$, *and let* B *denote the least closed subalgebra of* A *containing* $1, a$ *and* $(\lambda - a)^{-1}$ *for each* $\lambda \in \Lambda$. *Then*

$$(z - a)^{-1} \in B \qquad (z \in \mathbb{C} \backslash \mathrm{Sp}(a)).$$

Proof. Let $\phi \in A'$ with $\phi(b) = 0$ $(b \in B)$, and define h on $\mathbb{C} \backslash \mathrm{Sp}(a)$ by

$$h(z) = \phi((z - a)^{-1}) \qquad (z \in \mathbb{C} \backslash \mathrm{Sp}(a)).$$

By Theorem 10, h is a holomorphic function on $\mathbb{C} \backslash \mathrm{Sp}(a)$.

If $|z| > r(a)$, we have

$$(z - a)^{-1} = \frac{1}{z} + \frac{1}{z^2} a + \frac{1}{z^3} a^2 + \cdots$$

with the series convergent in norm, and so $(z-a)^{-1} \in B$. Thus

$$h(z) = 0 \qquad (|z| > r(a)).$$

Therefore $h(z)=0$ for all z in the unbounded component of $\mathbb{C} \backslash \mathrm{Sp}(a)$.
    Let $\lambda \in \Lambda$. Then, as in the proof of Theorem 10

$$(z-a)^{-1} = (\lambda-a)^{-1} + (\lambda-z)(\lambda-a)^{-2} + (\lambda-z)^2(\lambda-a)^{-3} + \cdots$$

for all z such that $|z-\lambda| < \|(\lambda-a)^{-1}\|^{-1}$. Thus

$$(z-a)^{-1} \in B \qquad (|z-\lambda| < \|(\lambda-a)^{-1}\|^{-1}).$$

Therefore $h(z)=0$ on a neighbourhood of λ, and hence on the component of $\mathbb{C} \backslash \mathrm{Sp}(a)$ containing λ.
    We have now proved that if $z \in \mathbb{C} \backslash \mathrm{Sp}(a)$, then

$$\phi((z-a)^{-1}) = 0$$

whenever $\phi \in A'$ with $\phi(b)=0$ $(b \in B)$. By the Hahn-Banach theorem, we have $(z-a)^{-1} \in B$. ☐

Proposition 12. *Let B be a closed subalgebra of A and let $b \in B$. Then $\mathrm{Sp}(A, b) \subset \mathrm{Sp}(B, b) \cup \{0\}$, and $\partial \mathrm{Sp}(B, b) \subset \partial \mathrm{Sp}(A, b)$.*

Proof. Apply Corollary 2.18 to the subalgebra $B + \mathbb{C}1$ and use Lemma 2. ☐

Example 13. Let $A = C(\Gamma)$, where $\Gamma = \{z : |z|=1\}$, let B be the image of the disc algebra $\mathscr{A}(\Delta)$ under the isometric isomorphism $f \to f|_\Gamma$, and let $u(z)=z$. Then $\mathrm{Sp}(A, u) = \Gamma$, $\mathrm{Sp}(B, u) = \Delta$.

Proposition 14. *Let $a \in A$ be such that $\mathbb{C} \backslash \mathrm{Sp}(A, a)$ is connected, and let B denote the closed subalgebra of A generated by 1, a. Then $\mathrm{Sp}(B, a) = \mathrm{Sp}(A, a)$.*

Proof. Clear from Theorem 11. ☐

Definition 15. Let X, Y be topological spaces, and for each $x \in X$ let $\phi(x)$ be a subset of Y. The mapping ϕ is *upper semi-continuous* if and only if for each $x_0 \in X$ and each neighbourhood V of $\phi(x_0)$, there exists a neighbourhood U of x_0 such that

$$\phi(x) \subset V \qquad (x \in U).$$

Lemma 16. *Let X, Y be metric spaces, let Y be compact, and let ϕ be a mapping of X into the closed subsets of Y. Then ϕ is upper semi-continuous if and only if the following condition holds:*

$$x_n \in X, \ y_n \in \phi(x_n) \ (n=1, 2, \ldots), \ x = \lim_{n \to \infty} x_n, \ y = \lim_{n \to \infty} y_n \Rightarrow y \in \phi(x).$$

Proof. Straightforward. ☐

Proposition 17. *The mapping* $a \to \mathrm{Sp}(a)$ *of* A *into the compact subsets of* $\mathbb{C}$ *is upper semi-continuous.*

Proof. It is enough to consider $X = \{a \in A : \|a\| \leqslant R\}$ with $R > 0$, and to take $Y = \{z \in \mathbb{C} : |z| \leqslant R\}$. For each $a \in X$, $\mathrm{Sp}(a)$ is a closed subset of the compact space Y, and Lemma 16 is applicable. Let $a_n \in X$, $\lambda_n \in \mathrm{Sp}(a_n)$, $\lim_{n \to \infty} a_n = a$, $\lim_{n \to \infty} \lambda_n = \lambda$. If $\lambda \notin \mathrm{Sp}(a)$, then $\lambda - a \in \mathrm{Inv}(A)$. Since $\lambda - a = \lim_{n \to \infty} (\lambda_n - a_n)$, and $\mathrm{Inv}(A)$ is open, we have $\lambda_n - a_n \in \mathrm{Inv}(A)$ for sufficiently large n, which contradicts $\lambda_n \in \mathrm{Sp}(a_n)$.   □

Proposition 17 shows that if $a \in A$ and $\varepsilon > 0$, then there exists $\delta > 0$ such that $d(\lambda, \mathrm{Sp}(a)) = \inf\{|\lambda - \mu| : \mu \in \mathrm{Sp}(a)\} < \varepsilon$ whenever $\lambda \in \mathrm{Sp}(x)$ and $\|x - a\| < \delta$. For commuting elements we can conclude more, as the following proposition shows.

Proposition 18. *Let* $a \in A$ *and* $\varepsilon > 0$. *Then there exists* $\delta > 0$ *such that*

$$d(\lambda, \mathrm{Sp}(x)) < \varepsilon$$

whenever $\|x - a\| < \delta$, $xa = ax$, *and* $\lambda \in \mathrm{Sp}(a)$.

Proof. We prove first that if $\lambda \in \mathrm{Sp}(a)$ there exists $\delta(\lambda) > 0$ such that

$$\|x - a\| < \delta(\lambda), \qquad xa = ax \Rightarrow d(\lambda, \mathrm{Sp}(x)) < \tfrac{1}{2}\varepsilon. \tag{5}$$

Suppose that $\lambda \in \mathrm{Sp}(a)$ with no such $\delta(\lambda)$ existing. Since we may replace a, x by $a - \lambda, x - \lambda$, we may suppose that $\lambda = 0$. Then there exist $x_n \in A$ with $\lim_{n \to \infty} x_n = a$, $x_n a = a x_n$, $d(0, \mathrm{Sp}(x_n)) \geqslant \tfrac{1}{2}\varepsilon$ $(n = 1, 2, \ldots)$. We have $x_n \in \mathrm{Inv}(A)$, $\mathrm{Sp}(x_n^{-1}) = \{z^{-1} : z \in \mathrm{Sp}(x_n)\}$, and so

$$r(x_n^{-1}) \leqslant 2\varepsilon^{-1}.$$

Then Proposition 4.6 gives $a \in \mathrm{Inv}(A)$ which contradicts $0 \in \mathrm{Sp}(a)$. This contradiction proves the existence of $\delta(\lambda) > 0$ satisfying (5) for each $\lambda \in \mathrm{Sp}(a)$.

We now cover $\mathrm{Sp}(a)$ by a finite number m of open discs $U(\lambda_j, \tfrac{1}{2}\varepsilon)$ with radius $\tfrac{1}{2}\varepsilon$ and with their centres $\lambda_j \in \mathrm{Sp}(a)$, and take

$$\delta = \min\{\delta(\lambda_j) : 1 \leqslant j \leqslant m\}.   □$$

Remarks. Theorem 8 is one of the most important results in Banach algebra theory. Beurling [51] proved the result for a special class of Banach algebras. The proof of the general case given in Gelfand [150] uses function theory, the elementary proof given here is due to Rickart [320].

§ 6. Contour Integrals

In this section we review the elementary theory of integration of Banach space valued functions that we shall need.

Let α, β be real numbers with $\alpha \leqslant \beta$, and let X be a Banach space over $\mathbb{C}$. A *dissection* of $[\alpha, \beta]$ is an ordered $(n+1)$-tuple $(\alpha_0, \alpha_1, \ldots, \alpha_n)$ of real numbers with $\alpha = \alpha_0 < \alpha_1 < \cdots < \alpha_n = \beta$. The *mesh* of the dissection is $\max \{\alpha_k - \alpha_{k-1} : 1 \leqslant k \leqslant n\}$. An *X-valued step function* on $[\alpha, \beta]$ is a mapping f of $[\alpha, \beta]$ into X for which there exists a dissection $(\alpha_0, \alpha_1, \ldots, \alpha_n)$ of $[\alpha, \beta]$ and elements $c_1, \ldots, c_n$ of X such that

$$f(t) = c_k \qquad (\alpha_{k-1} < t < \alpha_k, k = 1, \ldots, n),$$ (1)

i.e. a mapping taking constant values in the open intervals determined by some dissection. We denote by $S([\alpha, \beta], X)$ the set of all X-valued step functions on $[\alpha, \beta]$. Clearly, $S([\alpha, \beta], X)$ is a linear subspace of $l^\infty([\alpha, \beta], X)$.

Given $f \in S([\alpha, \beta], X)$ we define the integral $\int_\alpha^\beta f$ by

$$\int_\alpha^\beta f = \sum_{k=1}^n (\alpha_k - \alpha_{k-1}) c_k,$$

when $(\alpha_0, \ldots, \alpha_n)$ is a dissection of $[\alpha, \beta]$ and $c_1, \ldots, c_n$ are elements of X satisfying (1). It is easy to verify that the integral is independent of the choice of such a dissection. It is clear that

$$\left\| \int_\alpha^\beta f \right\| \leqslant (\beta - \alpha) \| f \|_\infty,$$ (2)

and so the mapping $f \rightarrow \int_\alpha^\beta f$ is a bounded linear mapping of $S = S([\alpha, \beta], X)$ into X. By the theorem on extension by continuity, this mapping has a unique extension to a bounded linear mapping of S^- into X, S^- being the closure of S in $l^\infty([\alpha, \beta], X)$; and we denote the image under this mapping of $f \in S^-$ also by $\int_\alpha^\beta f$. It will often be convenient to write $\int_\alpha^\beta f(t) dt$ to denote $\int_\alpha^\beta f$. Extension by continuity does not change norms, and so (2) holds also for all $f \in S^-$. It is easy to prove that if $T \in BL(X, Y)$ with Y a Banach space, and if $f \in (S([\alpha, \beta], X))^-$, then $T \circ f \in (S([\alpha, \beta], Y))^-$, and

$$T\left(\int_\alpha^\beta f\right) = \int_\alpha^\beta (T \circ f).$$ (3)

Proposition 1. *Let* $f \in C([\alpha, \beta], X)$ *and* $\varepsilon > 0$. *Then* $f \in (S([\alpha, \beta], X))^-$, *and there exists* $\delta > 0$ *such that, for every dissection* $(\alpha_0, ..., \alpha_n)$ *of* $[\alpha, \beta]$ *with mesh not exceeding* δ *and for all* $(t_1, ..., t_n)$ *with*

$$\alpha_{k-1} \leqslant t_k \leqslant \alpha_k \quad (k = 1, 2, ..., n),$$

the following statements hold:
   (i) *there exists* $g \in S([\alpha, \beta], X)$ *with* $g(t) = f(t_k)$ $(\alpha_{k-1} < t < \alpha_k, k = 1, ... n)$
and $\| f - g \|_\infty < \varepsilon$;
   (ii) $\| \int_\alpha^\beta f - \sum_{k=1}^n (\alpha_k - \alpha_{k-1}) f(t_k) \| \leqslant \varepsilon(\beta - \alpha)$.

Proof. Compactness of $[\alpha, \beta]$. □

Notation. Let Γ be an arc of a circle in $\mathbb{C}$ with centre ζ and radius ρ, and let $f \in C(\Gamma, X)$. We have $\theta_1, \theta_2 \in \mathbb{R}$ with $0 \leqslant \theta_2 - \theta_1 \leqslant 2\pi$ such that $\Gamma = \{\zeta + \rho e^{i\theta} : \theta_1 \leqslant \theta \leqslant \theta_2\}$. The *integral of* f *along the positively oriented arc* Γ is defined by

$$\int_{\theta_1}^{\theta_2} \rho i e^{i\theta} f(\zeta + \rho e^{i\theta}) d\theta,$$

and is denoted by $\int_\Gamma f(z)dz$. Note that $\int_\Gamma f(z)dz$ is independent of the choice (modulo 2π) of such θ_1, θ_2. The *integral of* f *along the negatively oriented arc* Γ is defined to be $-\int_\Gamma f(z)dz$.

Given $\zeta \in \mathbb{C}$ and $\rho \geqslant 0$, we use the following notations for open and closed discs and their complements, and for circles with centre ζ and radius ρ.

$$U(\zeta, \rho) = \{z : |z - \zeta| < \rho\}, \quad E(\zeta, \rho) = \{z : |z - \zeta| \leqslant \rho\},$$
$$V(\zeta, \rho) = \{z : |z - \zeta| > \rho\}, \quad F(\zeta, \rho) = \{z : |z - \zeta| \geqslant \rho\},$$
$$C(\zeta, \rho) = \{z : |z - \zeta| = \rho\}.$$

Definition 2. A *punched disc* is a subset E of $\mathbb{C}$ of the form

$$E = E(\zeta_0, \rho_0) \cap \bigcap_{j=1}^m F(\zeta_j, \rho_j). \tag{4}$$

(We allow in particular the case $m = 0$, $E = E(\zeta_0, \rho_0)$).

Given subsets K, D of $\mathbb{C}$ with K compact, D open, and $K \subset D$, a *punched disc envelope* for (K, D) is a punched disc E such that

$$K \subset \text{int } E, \quad E \subset D.$$

The boundary ∂E of E consists of a finite number of arcs of the circles $C(\zeta_j, \rho_j)$ $(0 \leqslant j \leqslant m)$. Given $f \in C(\partial E, X)$ we define

$$\int_{\partial E} f(z)dz$$

to be the sum of the integrals of f along these arcs, with the arcs in $C(\zeta_0, \rho_0) \cap \partial E$ positively oriented and the arcs in $C(\zeta_j, \rho_j) \cap \partial E$ $(j = 1, \ldots, m)$ negatively oriented. Note that with this orientation, $\operatorname{int} E$ lies on the left of each arc of ∂E.

Since the interior of a finite intersection of sets is the intersection of their interiors, we have

$$K \subset \operatorname{int} E = U(\zeta_0, \rho_0) \cap \bigcap_{j=1}^{m} V(\zeta_j, \rho_j).$$

Proposition 3. *Let K, D be subsets of $\mathbb{C}$ with K compact, D open, and with $K \subset D$. Then there exists a punched disc envelope*

$$E = E(\zeta_0, \rho_0) \cap \bigcap_{j=1}^{m} F(\zeta_j, \rho_j)$$

for (K, D) with arbitrary $\zeta_0 \in \mathbb{C}$ and with $\zeta_j \in \mathbb{C} \setminus D$ $(j = 1, \ldots, m)$.

Proof. Let $\zeta_0 \in \mathbb{C}$, and choose $\rho_0 > 0$ such that $K \subset U(\zeta_0, \rho_0)$. Let $F = E(\zeta_0, \rho_0) \setminus D$. If $F = \emptyset$, $E = E(\zeta_0, \rho_0)$ is the required punched disc envelope. Otherwise, for $\zeta \in F$ there exists $\rho > 0$ such that $E(\zeta, \rho) \cap K = \emptyset$. The corresponding open discs $U(\zeta, \rho)$ cover the compact set F, and so there exists a finite covering $U(\zeta_1, \rho_1), \ldots, U(\zeta_m, \rho_m)$ by such discs. We have $K \subset V(\zeta_j, \rho_j)$ $(j = 1, \ldots, m)$, and so $E = E(\zeta_0, \rho_0) \cap \bigcap_{j=1}^{m} F(\zeta_j, \rho_j)$ is a punched disc envelope for (K, D) with the required properties. □

Proposition 4. *Let D be an open neighbourhood of a punched disc E, and let $f \in H(D)$. Then*
   (i) $\int_{\partial E} f(z) \, dz = 0$;
   (ii) $f(\zeta) = \dfrac{1}{2\pi i} \displaystyle\int_{\partial E} f(z)(z - \zeta)^{-1} \, dz$ $(\zeta \in \operatorname{int} E)$.

Proof. (i) Let E be as in (4), and consider the deformation of E produced by a small change of ρ_0. The resulting difference set is either an annulus or a finite number of regions each bounded by a simple closed contour consisting of four arcs of circles. By applying Cauchy's theorem to each of these contours, we see that $\int_{\partial E} f(z) \, dz$ is unchanged by a small change of ρ_0, and (i) follows on contracting E to a point.

(ii) Let $\zeta \in \operatorname{int} E$, and choose $\rho > 0$ with $E(\zeta, \rho) \subset \operatorname{int} E$. Then $E' = E \cap F(\zeta, \rho)$ is a punched disc, and $f(z)(z - \zeta)^{-1}$ is holomorphic in an open neighbourhood of E'. Therefore, by (i), $\int_{\partial E'} f(z)(z - \zeta)^{-1} \, dz = 0$,

$$\frac{1}{2\pi i} \int_{\partial E} f(z)(z - \zeta)^{-1} \, dz = \frac{1}{2\pi i} \int_{C(\zeta, \rho)} f(z)(z - \zeta)^{-1} \, dz = f(\zeta). \quad □$$

The next proposition may be regarded as a generalization of the Taylor series and Laurent series expansions of a holomorphic function.

Proposition 5. *Let* $E = E(\zeta_0, \rho_0) \cap \bigcap_{j=1}^{m} F(\zeta_j, \rho_j)$ *be a punched disc, let* $\Gamma_k = C(\zeta_k, \rho_k) \cap \partial E$ *with the positive orientation of* $C(\zeta_k, \rho_k)$, *let* D *be an open neighbourhood of* E, *and let* $f \in H(D)$. *Then*

$$f(\zeta) = \sum_{n=0}^{\infty} \alpha_n (\zeta - \zeta_0)^n + \sum_{j=1}^{m} \sum_{n=1}^{\infty} \beta_{jn} (\zeta - \zeta_j)^{-n} \qquad (\zeta \in \text{int } E),$$

where

$$\alpha_n = \frac{1}{2\pi i} \int_{\Gamma_0} f(z)(z - \zeta_0)^{-n-1} dz, \qquad \beta_{jn} = \frac{1}{2\pi i} \int_{\Gamma_j} f(z)(z - \zeta_j)^{n-1} dz,$$

and the series converge uniformly on compact subsets of int E.

Proof. By Proposition 4 (ii), for all $\zeta \in \text{int } E$ we have

$$f(\zeta) = \frac{1}{2\pi i} \int_{\Gamma_0} f(z)(z - \zeta)^{-1} dz - \sum_{j=1}^{m} \frac{1}{2\pi i} \int_{\Gamma_j} f(z)(z - \zeta)^{-1} dz.$$

For $z \in \Gamma_0$ and $\zeta \in U(\zeta_0, \rho_0)$, we have

$$(z - \zeta)^{-1} = (z - \zeta_0)^{-1} \left\{ 1 - \frac{\zeta - \zeta_0}{z - \zeta_0} \right\}^{-1} = \sum_{n=0}^{\infty} (\zeta - \zeta_0)^n (z - \zeta_0)^{-n-1},$$

and so

$$\frac{1}{2\pi i} \int_{\Gamma_0} f(z)(z - \zeta)^{-1} dz = \sum_{n=0}^{\infty} \alpha_n (\zeta - \zeta_0)^n,$$

with the series converging uniformly on compact subsets of $U(\zeta_0, \rho_0)$. Likewise, for $j \in \{1, 2, \ldots, m\}$, $z \in \Gamma_j$, $\zeta \in V(\zeta_j, \rho_j)$, we have

$$-(z - \zeta)^{-1} = (\zeta - \zeta_j)^{-1} \left\{ 1 - \frac{z - \zeta_j}{\zeta - \zeta_j} \right\}^{-1} = \sum_{n=1}^{\infty} (\zeta - \zeta_j)^{-n} (z - \zeta_j)^{n-1},$$

and the result follows. □

Corollary 6. *Let* D *be an open subset of* $\mathbb{C}$. *Then* $R(D)$ *is dense in* $H(D)$ *in the topology of uniform convergence on compact subsets of* D.

Proof. Let K be a compact subset of D and let $f \in H(D)$. Let E be a punched disc envelope for (K, D) as in Proposition 3 with $\zeta_j \in \mathbb{C} \setminus D$ $(j = 1, \ldots, m)$. By Proposition 5

$$f(\zeta) = \sum_{n=0}^{\infty} \alpha_n (\zeta - \zeta_0)^n + \sum_{j=1}^{m} \sum_{n=1}^{\infty} \beta_{jn} (\zeta - \zeta_j)^{-n}$$

with the series converging uniformly on K. The nth partial sum belongs to $R(D)$.   ▯

Remark. It is easy to see that Propositions 4 and 5 hold also for holomorphic mappings f of D into a complex Banach space (Definition 5.9). For example, to prove the generalization of Proposition 4 (ii), we note that for every $\phi \in X'$, we have

$$(\phi \circ f)(\zeta) = \frac{1}{2\pi i} \int_{\partial E} (\phi \circ f)(z)(z-\zeta)^{-1} dz \qquad (\zeta \in \text{int } E).$$

By equation (3), we therefore have

$$\phi(f(\zeta)) = \phi \left\{ \frac{1}{2\pi i} \int_{\partial E} f(z)(z-\zeta)^{-1} dz \right\} \qquad (\zeta \in \text{int } E),$$

and the Hahn-Banach theorem completes the proof.

For use in § 38, we note the following special case of this generalization.

Proposition 7. *Let D be an open neighbourhood of the closed unit disc $E(0, 1)$, and let f be a holomorphic mapping of D into a complex Banach space X. Then*

$$f(0) = \frac{1}{2\pi} \int_0^{2\pi} f(e^{it}) dt.$$

Proof. Above remark.   ▯

§ 7. A Functional Calculus for a Single Banach Algebra Element

Let A be a complex Banach algebra with unit, let $a \in A$, and let D be an open neighbourhood of $\text{Sp}(a)$. We construct a homomorphism $f \to \mathbf{f}(a)$ of $H(D)$ into A which extends the natural homomorphism of $R(D)$ into A.

Lemma 1. *Let E be a punched disc envelope for $(\text{Sp}(a), \mathbb{C})$. Then*

$$\frac{1}{2\pi i} \int_{\partial E} (z-a)^{-1} dz = 1.$$

Proof. Let $E = E(\zeta_0, \rho_0) \cap \bigcap_{j=1}^m F(\zeta_j, \rho_j)$, let $\phi \in A'$, and let $h(z) = \phi((z-a)^{-1})$ $(z \in \mathbb{C} \setminus \text{Sp}(a))$. By Theorem 5.10, h is holomorphic on

$\mathbb{C}\setminus \text{Sp}(a)$, and therefore, as in the proof of Proposition 6.4, $\int_{\partial E} h(z)dz$ is unchanged by a small change in ρ_j. Since $\text{Sp}(a) \subset V(\zeta_j, \rho_j)$, we may let $\rho_j \to 0$ without changing the integral. Likewise we may replace $E(\zeta_0, \rho_0)$ by a disc $E(0, R)$ with $R > r(a)$. Thus

$$\int_{\partial E} h(z)dz = \int_{|z|=R} h(z)dz .$$

With $|z| = R$, we have $(z-a)^{-1} = \dfrac{1}{z} + \dfrac{1}{z^2}a + \dfrac{1}{z^3}a^2 + \cdots$, and so

$$\phi\left(\frac{1}{2\pi i}\int_{\partial E} (z-a)^{-1}dz\right) = \frac{1}{2\pi i}\int_{\partial E} h(z)dz = \frac{1}{2\pi i}\int_{|z|=R} \frac{\phi(1)}{z}dz = \phi(1).$$

Now apply the Hahn-Banach theorem. □

Lemma 2. *Let E be a punched disc envelope for* $(\text{Sp}(a), D)$, *and let $f \in R(D)$. Then*

$$\frac{1}{2\pi i}\int_{\partial E} f(z)(z-a)^{-1}dz = f(a).$$

Proof. We have $f = p/q$ with $p, q \in P(D)$ and with the zeros of q outside D. Then $\{p(z)q(a) - q(z)p(a)\}(z-a)^{-1}$ is a polynomial in z, a; and therefore there exist $a_k \in A$, $g_k \in R(D)$ $(k=1,\dots,n)$ such that

$$\{f(z) - f(a)\}(z-a)^{-1} = \sum_{k=1}^{n} g_k(z)a_k \qquad (z \in D).$$

By Proposition 6.4,

$$\int_{\partial E} g_k(z)dz = 0 \qquad (k=1,\dots,n),$$

and therefore, by Lemma 1,

$$\frac{1}{2\pi i}\int_{\partial E} f(z)(z-a)^{-1}dz = \frac{1}{2\pi i}\int_{\partial E} f(a)(z-a)^{-1}dz = f(a). \quad □$$

Definition 3. Given $f \in H(D)$, we denote by $\mathbf{f}(a)$ the element of A defined by

$$\mathbf{f}(a) = \frac{1}{2\pi i}\int_{\partial E} f(z)(z-a)^{-1}dz,$$

where E is a punched disc envelope for $(\text{Sp}(a), D)$.

That $\mathbf{f}(a)$ does not depend on the choice of the punched disc envelope E is proved in the next theorem, which is the main result of this section.

Theorem 4. (i) *Given* $f \in H(D)$, $\mathbf{f}(a)$ *is independent of the choice of the punched disc envelope* E *for* $(\mathrm{Sp}(a), D)$.

(ii) *The mapping* $f \to \mathbf{f}(a)$ *is a homomorphism of* $H(D)$ *into* A *that extends the natural homomorphism* $f \to f(a)$ *of* $R(D)$ *into* A.

(iii) *Given a compact neighbourhood* K *of* $\mathrm{Sp}(a)$ *contained in* D, *the mapping* $f \to \mathbf{f}(a)$ *is continuous with respect to uniform convergence on* K.

(iv) $\mathrm{Sp}(\mathbf{f}(a)) = \{f(\lambda) : \lambda \in \mathrm{Sp}(a)\} \quad (f \in H(D))$.

Proof. Let E be a punched disc envelope for $(\mathrm{Sp}(a), D)$, and let $\mathbf{f}(a) = \dfrac{1}{2\pi i} \displaystyle\int_{\partial E} f(z)(z-a)^{-1} dz$. The mapping $f \to \mathbf{f}(a)$ is obviously linear on $H(D)$. Since the mapping $z \to \|(z-a)^{-1}\|$ is continuous, and therefore bounded, on ∂E, there exists a positive constant M such that

$$\|\mathbf{f}(a)\| \leqslant M |f|_{\partial E} \quad (f \in H(D)), \tag{1}$$

where

$$|f|_{\partial E} = \sup \{|f(z)| : z \in \partial E\}.$$

By Corollary 6.6, given a compact subset K of D containing E, there exists a sequence $\{f_n\}$ of elements of $R(D)$ that converges uniformly to f on K. Therefore, by (1) and Lemma 2,

$$\mathbf{f}(a) = \lim_{n \to \infty} \mathbf{f}_n(a) = \lim_{n \to \infty} f_n(a), \tag{2}$$

and (ii) is now clear.

Given another punched disc envelope E' for $(\mathrm{Sp}(a), D)$, let $\mathbf{f}(a)' = \dfrac{1}{2\pi i} \displaystyle\int_{\partial E'} f(z)(z-a)^{-1} dz$, and take $K = E \cup E'$. Then (2) holds also for $\mathbf{f}(a)'$, and so

$$\mathbf{f}(a)' = \lim_{n \to \infty} f_n(a) = \mathbf{f}(a),$$

which proves (i).

Given any compact neighbourhood K of $\mathrm{Sp}(a)$ contained in D, we may now take E to be a punched disc envelope for $(\mathrm{Sp}(a), \mathrm{int}\, K)$, and (iii) follows from (1).

Finally, we prove (iv). Fix a punched disc envelope E for $(\mathrm{Sp}(a), D)$, let $f \in H(D)$, and let $\lambda \in \mathrm{Sp}(a)$. There exist non-constant elements f_n of $R(D)$ that converge uniformly to f on E, and we have $\lim_{n \to \infty} f_n(a) = \mathbf{f}(a)$, $\lim_{n \to \infty} f_n(\lambda) = f(\lambda)$, and $f_n(\lambda) \in \mathrm{Sp}(f_n(a))$. Thus, by the upper semi-continuity of the mapping $x \to \mathrm{Sp}(x)$ (Proposition 5.17), $f(\lambda) \in \mathrm{Sp}(\mathbf{f}(a))$. Moreover, $f_n(a)$ commutes with $\mathbf{f}(a)$, and therefore Proposition 5.18 is applicable. Let $\mu \in \mathrm{Sp}(\mathbf{f}(a))$ and $\varepsilon > 0$. Then $d(\mu, \mathrm{Sp}(f_n(a))) < \varepsilon$, and so there exist $\zeta_n \in \mathrm{Sp}(a)$ such that $|\mu - f_n(\zeta_n)| < \varepsilon$ for all sufficiently large n. By replacing the sequences $\{\zeta_n\}$ and $\{f_n\}$ by subsequences, we may assume that $\lim_{n\to\infty} \zeta_n = \zeta \in \mathrm{Sp}(a)$. Then $\lim_{n\to\infty} f_n(\zeta_n) = f(\zeta)$, $|\mu - f(\zeta)| \leq \varepsilon$. Since ε is arbitrary and $f(\mathrm{Sp}(a))$ is closed, this gives $\mu \in f(\mathrm{Sp}(a))$. $\quad\square$

Remarks. (1) Theorem 4(iv) is called the '*spectral mapping theorem*'.
(2) An alternative proof that the mapping $f \to \mathbf{f}(a)$ preserves products turns on calculating

$$\mathbf{f}(a)\mathbf{g}(a) = \left(\frac{1}{2\pi i}\right)^2 \int_{\partial E} f(z)(z-a)^{-1} dz \int_{\partial F} g(\zeta)(\zeta-a)^{-1} d\zeta ,$$

where F is a punched disc envelope for (E, D), using Cauchy's integral theorem, a change of order of integration, and the identity

$$(z-a)^{-1}(\zeta-a)^{-1} = (\zeta-z)^{-1}[(z-a)^{-1} - (\zeta-a)^{-1}].$$

The generalized Laurent series is valid also for $\mathbf{f}(a)$ as the next Theorem states.

Theorem 5. *Let* $E = E(\zeta_0, \rho_0) \cap \bigcap_{j=1}^m F(\zeta_j, \rho_j)$ *be a punched disc envelope for* $(\mathrm{Sp}(a), D)$, *let* $f \in H(D)$, *and let* α_n, β_{jn} *be as in Proposition 6.5. Then*

$$\mathbf{f}(a) = \sum_{n=0}^\infty \alpha_n (a-\zeta_0)^n + \sum_{j=1}^m \sum_{n=1}^\infty \beta_{jn}(a-\zeta_j)^{-n}.$$

Proof. Let Γ_k $(k=0,1,\ldots,m)$ be as in Proposition 6.5, so that

$$\mathbf{f}(a) = \frac{1}{2\pi i} \int_{\Gamma_0} f(z)(z-a)^{-1} dz - \sum_{j=1}^m \frac{1}{2\pi i} \int_{\Gamma_j} f(z)(z-a)^{-1} dz.$$

Since $\mathrm{Sp}(a) \subset U(\zeta_0, \rho_0)$, we have $r(\alpha - \zeta_0) < \rho_0$, the series $\sum_{n=0}^\infty \alpha_n(a-\zeta_0)^n$ converges in norm, and

$$\frac{1}{2\pi i} \int_{\Gamma_0} f(z)(z-a)^{-1} dz = \frac{1}{2\pi i} \int_{\Gamma_0} f(z)(z-\zeta_0)^{-1}\{1-(z-\zeta_0)^{-1}(a-\zeta_0)\}^{-1} dz$$

$$= \sum_{n=0}^\infty \frac{1}{2\pi i} \int_{\Gamma_0} f(z)(z-\zeta_0)^{-n-1}(a-\zeta_0)^n dz$$

$$= \sum_{n=0}^\infty \alpha_n(a-\zeta_0)^n.$$

With $1 \leq j \leq m$, we have $\mathrm{Sp}(a) \subset V(\zeta_j, \rho_j)$, and so, by Proposition 5.6, $\mathrm{Sp}((a-\zeta_j)^{-1}) \subset U(0, \rho_j^{-1})$, Therefore $r((a-\zeta_j)^{-1}) < \rho_j^{-1}$, the series $\sum_{n=1}^\infty \beta_{jn}(a-\zeta_j)^{-n}$ converges and

$$-\frac{1}{2\pi i}\int_{\Gamma_j} f(z)(z-a)^{-1}dz = \frac{1}{2\pi i}\int_{\Gamma_j} f(z)(a-\zeta_j)^{-1}\{1-(z-\zeta_j)(a-\zeta_j)^{-1}\}^{-1}dz$$

$$= \sum_{n=1}^{\infty}\frac{1}{2\pi i}\int_{\Gamma_j} f(z)(z-\zeta_j)^{n-1}(a-\zeta_j)^{-n}dz$$

$$= \sum_{n=1}^{\infty}\beta_{jn}(a-\zeta_j)^{-n}. \quad \square$$

Theorem 6. *Let* D_1, D_2 *be open subsets of* $\mathbb{C}$, $f \in H(D_2)$, $g \in H(D_1)$, $\mathrm{Sp}(a) \subset D_1$, $g(D_1) \subset D_2$, $h = f \circ g$. *Then* $\mathbf{h}(a) = \mathbf{f}(\mathbf{g}(a))$.

Proof. We have $\mathrm{Sp}(\mathbf{g}(a)) = g(\mathrm{Sp}(a)) \subset g(D_1) \subset D_2$. Let $E = E(\zeta_0, \rho_0)$ $\cap \bigcap_{j=1}^{m} F(\zeta_j, \rho_j)$ be a punched disc envelope for $(\mathrm{Sp}(\mathbf{g}(a)), D_2)$. Then

$$f(z) = \sum_{k=0}^{\infty}\alpha_k(z-\zeta_0)^k + \sum_{j=1}^{m}\sum_{k=1}^{\infty}\beta_{jk}(z-\zeta_j)^{-k} \qquad (z \in \mathrm{int}\, E),$$

converging uniformly on a neighbourhood of $\mathrm{Sp}(\mathbf{g}(a))$, and

$$\mathbf{f}(\mathbf{g}(a)) = \sum_{k=0}^{\infty}\alpha_k(\mathbf{g}(a)-\zeta_0)^k + \sum_{j=1}^{m}\sum_{k=1}^{\infty}\beta_{jk}(\mathbf{g}(a)-\zeta_j)^{-k}.$$

Let

$$f_n(z) = \sum_{k=0}^{n}\alpha_k(z-\zeta_0)^k + \sum_{j=1}^{m}\sum_{k=1}^{n}\beta_{jk}(z-\zeta_j)^{-k} \qquad (z \in \mathrm{int}\, E),$$

and let $\phi_n(z) = f_n(g(z))$ $(z \in g^{-1}(\mathrm{int}\, E))$. By Theorem 4,

$$\boldsymbol{\phi}_n(a) = \sum_{k=0}^{n}\alpha_k(\mathbf{g}(a)-\zeta_0)^k + \sum_{j=1}^{m}\sum_{k=1}^{n}\beta_{jk}(\mathbf{g}(a)-\zeta_j)^{-k},$$

and so $\mathbf{f}(\mathbf{g}(a)) = \lim_{n\to\infty}\boldsymbol{\phi}_n(a)$. Also $\phi_n(z) \to h(z)$ uniformly on a neighbourhood of $\mathrm{Sp}(a)$, and therefore $\lim_{n\to\infty}\boldsymbol{\phi}_n(a) = \mathbf{h}(a)$. $\quad \square$

In the rest of this section we consider some simple applications of the functional calculus, first a proof of Runge's theorem on the uniform approximation of holomorphic functions by rational functions with preassigned poles.

Theorem 7 (Runge's theorem [325]). *Let* K *be a compact subset of* $\mathbb{C}$, *and let* Λ *be a subset of* $\mathbb{C} \backslash K$ *having non-void intersection with each bounded component of* $\mathbb{C} \backslash K$. *Let* D *be an open neighbourhood of* K, *and let* $f \in H(D)$. *Then* f *can be uniformly approximated on* K *by rational functions that have no poles outside* Λ.

Proof. Let A denote the Banach algebra $C(K)$, and a the element of A defined by

$$a(z) = z \quad (z \in K).$$

Then $\mathrm{Sp}(a) = K$, since $\lambda - a$ is invertible if and only if $\lambda - a(z) \neq 0 \ (z \in K)$. Let B denote the least closed subalgebra of A containing $1, a$ and $\{(\lambda - a)^{-1} : \lambda \in \Lambda\}$. By Theorem 5.11, we know that

$$(z - a)^{-1} \in B \quad (z \in \mathbb{C} \setminus K). \tag{3}$$

Let E be a punched disc envelope for (K, D). Then, by (3) and Theorem 5 or the definition of $\mathbf{f}(a)$, we have $\mathbf{f}(a) \in B$. Thus, given $\varepsilon > 0$, there exists a rational function g with no poles outside Λ such that $\|\mathbf{f}(a) - g(a)\| < \varepsilon$. It remains to show that $\mathbf{f}(a) = f|_K$ and that $g(a) = g|_K$.

Given $\zeta \in K$, let ϕ denote the evaluation functional at ζ, i.e.

$$\phi(x) = x(\zeta) \quad (x \in A).$$

Since ϕ is a continuous homomorphism on A and $\phi(a) = \zeta$,

$$\phi(\mathbf{f}(a)) = \frac{1}{2\pi i} \int_{\partial E} f(z) \phi((z - a)^{-1}) \, dz$$

$$= \frac{1}{2\pi i} \int_{\partial E} f(z) (z - \zeta)^{-1} \, dz = f(\zeta).$$

Thus $\mathbf{f}(a)(\zeta) = f(\zeta)$, and it is also easy to see that $g(a)(\zeta) = g(\zeta)$. $\quad\square$

Definition 8. An *idempotent* is an element x of an algebra with $x^2 = x$.

Proposition 9. *Let* $\mathrm{Sp}(a) = \bigcup_{j=1}^{m} \sigma_j$ *with* $\sigma_1, \ldots, \sigma_m$ *mutually disjoint non-void compact sets. Then there exist non-zero idempotents* $e_1, \ldots, e_m$ *belonging to the closed linear span of* $\{(z - a)^{-1} : z \in \mathbb{C} \setminus \mathrm{Sp}(a)\}$, *such that*

$$1 = e_1 + e_2 + \cdots + e_m, \quad e_k e_j = 0 \ (k \neq j).$$

Moreover, if $\sigma_j = \{\alpha_j\}$, *then* $\mathrm{Sp}(a e_j) = \{\alpha_j\}$, *and* $a e_j - \alpha_j e_j$ *is quasi-nilpotent, i. e.*

$$r(a e_j - \alpha_j e_j) = 0.$$

Proof. Let $D_1, \ldots, D_m$ be mutually disjoint open neighbourhoods of $\sigma_1, \ldots, \sigma_m$ respectively. Let $D = \bigcup_{j=1}^{m} D_j$, and define f_k on D by

$$f_k(z) = \begin{cases} 1 & (z \in D_k), \\ 0 & (z \in D \setminus D_k). \end{cases}$$

Then $f_k \in H(D)$, and $f_k^2 = f_k$. Let $e_k = f_k(a)$. Then $e_k^2 = e_k$, and e_k belongs to the closed linear span of $\{(z-a)^{-1} : z \in \mathbb{C} \backslash \mathrm{Sp}(a)\}$. Since $\mathrm{Sp}(e_k) = f_k(\mathrm{Sp}(a)) = \{0,1\}$, $e_k \neq 0$. Since

$$f_1(z) + f_2(z) + \cdots + f_m(z) = 1 \qquad (z \in D),$$

and

$$f_k(z) f_j(z) = 0 \qquad (z \in D, k \neq j),$$

we have $1 = e_1 + \cdots + e_m$ and $e_k e_j = 0 \; (k \neq j)$.

Let $\sigma_j = \{\alpha_j\}$, and consider the functions g, h defined on D by

$$g(z) = (z - \alpha_j) f_j(z), \qquad h(z) = z f_j(z).$$

Clearly we have $g(z) = 0 \; (z \in \mathrm{Sp}(a))$. Therefore

$$\mathrm{Sp}(\mathbf{g}(a)) = g(\mathrm{Sp}(a)) = \{0\}.$$

But $\mathbf{g}(a) = a e_j - \alpha_j e_j$. Also $\mathrm{Sp}(a e_j) = \mathrm{Sp}(\mathbf{h}(a)) = h(\mathrm{Sp}(a)) = \{\alpha_j\}$. ☐

Remarks. (1) We have the explicit formula

$$e_k = \frac{1}{2\pi i} \int_{\Gamma_k} (z - a)^{-1} dz,$$

where $\Gamma_k = \partial E \cap D_k$, E being a punched disc envelope for $(\mathrm{Sp}(a), D)$.

(2) We have $e_k \in \{a\}^{cc}$.

(3) It is an immediate consequence of Proposition 9 that A contains an idempotent e with $0 \neq e \neq 1$ whenever, for some $a \in A$, $\mathrm{Sp}(a)$ is not connected.

Proposition 10. *Suppose that* $\mathrm{Sp}(a) \subset U(0,1) \cup V(0,1)$. *Then the sequence* $\{(1-a^n)^{-1}\}$ *converges to an idempotent* e. *Also* $e = 1$ *if and only if* $\mathrm{Sp}(a) \subset U(0,1)$, *and* $e = 0$ *if and only if* $\mathrm{Sp}(a) \subset V(0,1)$.

Proof. Let $D = U(0,1) \cup V(0,1)$, let f be defined on D by $f(z) = 1$ $(z \in U(0,1))$, $f(z) = 0$ $(z \in V(0,1))$, and let E be a punched disc envelope for $(\mathrm{Sp}(a), D)$. Then $(1-z^n)^{-1}$ converges uniformly to $f(z)$ on E as $n \to \infty$, and $f \in H(D)$. Therefore $\lim_{n \to \infty} (1-a^n)^{-1} = \mathbf{f}(a) = e$, say. We have $e^2 = e$ since $f^2 = f$; and $\mathrm{Sp}(e) = f(\mathrm{Sp}(a))$ shows that $\mathrm{Sp}(e) = \{1\}$ if and only if $\mathrm{Sp}(a) \subset U(0,1)$, and $\mathrm{Sp}(e) = \{0\}$ if and only if $\mathrm{Sp}(a) \subset V(0,1)$. Finally, for an idempotent e, it is clear that $\mathrm{Sp}(e) = \{0\}$ if and only if $e = 0$, and consequently also $\mathrm{Sp}(e) = \{1\}$ if and only if $e = 1$. ☐

The Cauchy integral formula and the generalized Laurent series expansion are also valid for Banach space valued holomorphic functions. For applications of such ideas to resolvent equations for closed linear operators on Banach spaces see Hille and Phillips [197, Chapter V], which also contains a history of the development of the functional calculus.

We have finally the Cauchy integral formulae for the mapping $f \to \mathbf{f}(a)$.

Theorem 11. *Let D be an open neighbourhood of $\mathrm{Sp}(a)$, let E be a punched disc envelope for $(\mathrm{Sp}(a), D)$, and let $f \in H(D)$. Then, for $k = 1, 2, 3, \ldots$,*

$$\mathbf{f}^{(k)}(a) = \frac{k!}{2\pi i} \int_{\partial E} f(z)(z-a)^{-k-1} \, dz \, .$$

Proof. Let E_1 be a punched disc envelope for (E, D). Then

$$\mathbf{f}^{(k)}(a) = \frac{1}{2\pi i} \int_{\partial E} f^{(k)}(z)(z-a)^{-1} \, dz$$

$$= \frac{1}{2\pi i} \int_{\partial E} \left[\frac{k!}{2\pi i} \int_{\partial E_1} f(w)(w-z)^{-k-1} \, dw \right] (z-a)^{-1} \, dz$$

$$= \frac{k!}{2\pi i} \int_{\partial E_1} \left[\frac{1}{2\pi i} \int_{\partial E} (w-z)^{-k-1}(z-a)^{-1} \, dz \right] f(w) \, dw$$

$$= \frac{k!}{2\pi i} \int_{\partial E_1} f(w)(w-a)^{-k-1} \, dw$$

$$= \frac{k!}{2\pi i} \int_{\partial E} f(z)(z-a)^{-k-1} \, dz \, ,$$

the last step following from the obvious analogue of Theorem 7.4 (i).   □

§ 8. Elementary Functions

Throughout this section A will denote a complex Banach algebra with unit.

Definition 1. Given $a \in A$, $\exp(a)$ is defined by

$$\exp(a) = 1 + \sum_{n=1}^{\infty} \frac{1}{n!} a^n \, .$$

Since the complex function $\exp(z)$ is an entire function, Theorem 7.5 shows that $\exp(a) = \mathbf{exp}(a)$.

Proposition 2. *Let $a, b \in A$ and $ab = ba$. Then*
  (i) $\exp(a+b) = \exp(a)\exp(b)$,
  (ii) $\exp(a) \in \mathrm{Inv}(a)$ *and* $(\exp(a))^{-1} = \exp(-a)$,
  (iii) $\mathrm{Sp}(\exp(a)) = \exp(\mathrm{Sp}(a))$,
  (iv) $\exp(a) = \lim\limits_{n \to \infty} \left(1 + \dfrac{1}{n}a\right)^n$.

Proof. (i) Let $x_n, y_n, z_n, \xi_n, \eta_n, \zeta_n$ be defined by

$$x_n = 1 + \sum_{k=1}^{n} \frac{1}{k!}a^k, \qquad y_n = 1 + \sum_{k=1}^{n} \frac{1}{k!}b^k, \qquad z_n = 1 + \sum_{k=1}^{n} \frac{1}{k!}(a+b)^k,$$

$$\xi_n = 1 + \sum_{k=1}^{n} \frac{1}{k!}\|a\|^k, \quad \eta_n = 1 + \sum_{k=1}^{n} \frac{1}{k!}\|b\|^k, \quad \zeta_n = 1 + \sum_{k=1}^{n} \frac{1}{k!}(\|a\| + \|b\|)^k.$$

We have

$$x_n y_n - z_n = \sum_{j,k=1}^{n} \alpha_{jk} a^j b^k$$

with $\alpha_{jk} \geq 0$ for all j, k. Therefore

$$\|x_n y_n - z_n\| \leq \sum_{j,k=1}^{n} \alpha_{jk}\|a\|^j\|b\|^k = \xi_n \eta_n - \zeta_n.$$

But $\lim\limits_{n \to \infty} (\xi_n \eta_n - \zeta_n) = \exp(\|a\|)\exp(\|b\|) - \exp(\|a\| + \|b\|) = 0$.
  (ii) Take $b = -a$ in (i).
  (iii) $\exp(a) = \mathbf{exp}(a)$. Apply Theorem 7.4 (iv).
  (iv) Define x_n, ξ_n as in (i) and take $y_n = \left(1 + \dfrac{1}{n}a\right)^n$, $\eta_n = \left(1 + \dfrac{1}{n}\|a\|\right)^n$.
We have

$$x_n - y_n = \sum_{k=2}^{n} \alpha_k a^k$$

with $\alpha_k \geq 0$ for all k. Therefore

$$\|x_n - y_n\| \leq \sum_{k=2}^{n} \alpha_k\|a\|^k = \xi_n - \eta_n. \quad \square$$

Alternatively, (iv) can be proved using the continuity of the functional calculus, Theorem 7.3 (iii).

Notation. We denote by $\exp(A)$ the range of the exponential function in A, i.e.

$$\exp(A) = \{\exp(a) : a \in A\} .$$

Let $\mathbb{R}_- = \{t \in \mathbb{R} : t \leqslant 0\}$. We define $\log z$ for $z \in \mathbb{C} \backslash \mathbb{R}_-$ as the principal branch of the logarithm, i.e. each $z \in \mathbb{C} \backslash \mathbb{R}_-$ has a unique representation in the form $z = r e^{i\theta}$ with $r > 0$ and $-\pi < \theta < \pi$, and we take $\log z = \log r + i\theta$. This function is holomorphic on $\mathbb{C} \backslash \mathbb{R}_-$, satisfies

$$\exp(\log z) = z \qquad (z \in \mathbb{C} \backslash \mathbb{R}_-)$$

and

$$\log(\exp z) = z \qquad (-\pi < \operatorname{Im} z < \pi).$$

If $\operatorname{Sp}(a) \subset \mathbb{C} \backslash \mathbb{R}_-$, we define $\log(a)$ by

$$\log a = \mathbf{f}(a),$$

where $f(z) = \log z$ $(z \in \mathbb{C} \backslash \mathbb{R}_-)$.

Proposition 3. *Let* $\operatorname{Sp}(a) \subset \mathbb{C} \backslash \mathbb{R}_-$, *and* $\operatorname{Sp}(b) \subset \{z : -\pi < \operatorname{Im} z < \pi\}$. *Then*

(i) $\exp(\log a) = a$, *and there exists a sequence* $\{p_n\}$ *of complex polynomials such that* $\lim_{n \to \infty} p_n(z) = \log z$ *uniformly on a neighbourhood of* $\operatorname{Sp}(a)$ *and* $\lim_{n \to \infty} p_n(a) = \log a$;

(ii) $\operatorname{Sp}(\exp(b)) \subset \mathbb{C} \backslash \mathbb{R}_-$, *and* $\log(\exp(b)) = b$.

Proof. That $\exp(\log a) = a$ follows from Theorem 7.6, as does also (ii). Let K be a compact neighbourhood of $\operatorname{Sp}(a)$ such that $\mathbb{C} \backslash K$ is connected and $K \subset \mathbb{C} \backslash \mathbb{R}_-$. By Runge's theorem there exists a sequence $\{p_n\}$ of polynomials converging uniformly to $\log z$ on K; and then by the continuity of the functional calculus, $\lim_{n \to \infty} p_n(a) = \log a$. ▯

Corollary 4. *If* $r(1-a) < 1$, *then* $a \in \exp(A)$.

Proof. If $r(1-a) < 1$, $\operatorname{Sp}(a) \subset U(1, 1) \subset \mathbb{C} \backslash \mathbb{R}_-$. Then

$$a = \exp(\log a) \in \exp(A). \quad ▯$$

Definition 5. Let $G = \operatorname{Inv}(A)$. Since G is an open subset of the normed space A, G is a disjoint union of open connected subsets, the components of G. The component containing 1 is called the *principal component* of G.

Proposition 6. *The principal component of* $\operatorname{Inv}(A)$ *is a normal subgroup.*

Proof. Let $G = \operatorname{Inv}(A)$, and let G_1 be the principal component of G. It is easy to see that a homeomorphism of G onto G permutes the components of G. Given $h \in G_1$, the homeomorphism $x \to hx$ maps 1 onto h, and therefore maps G_1 onto G_1. The homeomorphism $x \to x^{-1}$ maps

1 onto 1, and therefore also maps G_1 onto G_1. Thus G_1 is a subgroup of G. Finally, given $g \in G$, the homeomorphism $x \to g^{-1} x g$ maps 1 onto 1, and therefore G_1 onto G_1. □

Proposition 7. *The least subgroup of* $\mathrm{Inv}(A)$ *containing* $\exp(A)$ *is the principal component of* $\mathrm{Inv}(A)$.

Proof. Let $G = \mathrm{Inv}(A)$, let G_1 be the principal component of G, and let G_0 be the least subgroup of G containing $\exp(A)$. Given $b \in \exp(A)$, we have $b = \exp(a)$ for some $a \in A$, and we define f on $[0, 1]$ by

$$f(t) = \exp(t a) \qquad (0 \leqslant t \leqslant 1).$$

Since $f(t) = 1 + \sum_{n=1}^{\infty} \dfrac{1}{n!} t^n a^n$, f is continuous. Also $f(t) \in G$ $(0 \leqslant t \leqslant 1)$, and $f(0) = 1$, $f(1) = b$. Therefore $b \in G_1$. Since G_1 is a subgroup, it follows that $G_0 \subset G_1$.

We prove next that G_0 is open. Given $x \in G_0$ and $y \in A$ with $\|y - x\| < \|x^{-1}\|^{-1}$, we have

$$\|1 - x^{-1} y\| \leqslant \|x^{-1}\| \, \|x - y\| < 1 ,$$

and so, by Corollary 4, $x^{-1} y \in \exp(A) \subset G_0$. Therefore $y = x(x^{-1} y) \in G_0$, and G_0 is open.

Finally, we prove that G_0 is closed relative to G. Let $a_n \in G_0$ and $\lim_{n \to \infty} a_n = a \in G$. By continuity of the mapping $x \to x^{-1}$ on G, we have $\lim_{n \to \infty} a_n^{-1} a = 1$. Therefore, by Corollary 4, $a_n^{-1} a \in \exp(A)$ for large n, $a = a_n (a_n^{-1} a) \in G_0$. □

Corollary 8. (i) *The principal component of* $\mathrm{Inv}(A)$ *is* $\exp(A)$ *if and only if* $\exp(A)$ *is a group.*

(ii) *If* A *is commutative, the principal component of* $\mathrm{Inv}(A)$ *is* $\exp(A)$.

Proof. Clear. □

Our next result gives a useful formula for the spectral radius of an element of a Banach algebra in terms of the exponential function. We need a preliminary lemma on *subadditive functions*, i.e. real functions f such that $f(s + t) \leqslant f(s) + f(t)$.

Lemma 9. *Let* f *be a continuous subadditive mapping of* $\mathbb{R}^+ \backslash \{0\}$ *into* $\mathbb{R}$. *Then*

$$\lim_{t \to \infty} \frac{1}{t} f(t) = \inf_{t > 0} \frac{1}{t} f(t).$$

Proof. Given $\alpha > \inf\limits_{t>0} \dfrac{1}{t} f(t)$, choose $s > 0$ such that $\dfrac{1}{s} f(s) < \alpha$. By continuity of f,

$$\sup\{f(t): s \leqslant t \leqslant 2s\} = m < \infty .$$

Therefore, for all positive integers n and real numbers t such that

$$(n+1)s \leqslant t \leqslant (n+2)s , \tag{1}$$

we have

$$f(t) \leqslant f(ns) + f(t-ns) \leqslant n f(s) + m ,$$

and so

$$\frac{1}{t} f(t) < \frac{ns}{t} \alpha + \frac{m}{t} .$$

Letting $t \to \infty$ and $n \to \infty$ together so that (1) holds, we obtain

$$\limsup_{t \to \infty} \frac{1}{t} f(t) \leqslant \alpha . \quad \square$$

Proposition 10. $\max\{\operatorname{Re}\zeta: \zeta \in \operatorname{Sp}(a)\} = \inf\limits_{t>0} \dfrac{1}{t} \log \|\exp(ta)\|$.

Proof. Define f on $\mathbb{R}^+ \backslash \{0\}$ by

$$f(t) = \log \|\exp(ta)\| .$$

Then f is a subadditive mapping of $\mathbb{R}^+ \backslash \{0\}$ into $\mathbb{R}$. Therefore, by Lemma 9,

$$\inf_{t>0} \frac{1}{t} \log \|\exp(ta)\| = \lim_{t \to \infty} \frac{1}{t} \log \|\exp(ta)\|$$

$$= \lim_{n \to \infty} \log \|(\exp(a))^n\|^{\frac{1}{n}} = \log(r(\exp(a)).$$

By the spectral mapping theorem (Theorem 7.4 (iv)),

$$r(\exp(a)) = \exp \max \{\operatorname{Re}\zeta: \zeta \in \operatorname{Sp}(a)\} . \quad \square$$

Proposition 11. *Let $a \in A$ and let $\mathbf{f}(a) = 0$ for some entire function f with simple zeros. Then $\operatorname{Sp}(a) = \{\lambda_1, \dots, \lambda_r\}$, $f(\lambda_j) = 0$ $(j = 1, \dots, r)$,*

$$(a - \lambda_1)(a - \lambda_2) \dots (a - \lambda_r) = 0 ,$$

and there exist non-zero idempotents $e_1, \dots, e_r$ belonging to the closed linear span of $\{(z-a)^{-1}: z \in \mathbb{C} \backslash \operatorname{Sp}(a)\}$ such that $e_j e_k = 0$ $(j \neq k)$, $1 = e_1 + \dots + e_r$, and $a e_j = \lambda_j e_j$ $(j = 1, \dots, r)$.

Proof. The spectral mapping theorem gives $f(\mathrm{Sp}(a)) = \mathrm{Sp}(\mathbf{f}(a)) = \{0\}$. Since $\mathrm{Sp}(a)$ is compact and f is entire, $\mathrm{Sp}(a)$ is finite, say $\mathrm{Sp}(a) = \{\lambda_1, ..., \lambda_r\}$ and $f(\lambda_j) = 0$ $(j = 1, ..., r)$. By Proposition 7.9, there exist non-zero idempotents $e_1, ..., e_r$ as stated with $\mathrm{Sp}(q_j) = \{0\}$ $(j = 1, ..., r)$, where $q_j = a e_j - \lambda_j e_j$.

Define g on $\mathbb{C}$ by

$$g(z) = \begin{cases} (z - \lambda_1)^{-1} ... (z - \lambda_r)^{-1} f(z) & (z \notin \{\lambda_1, ..., \lambda_r\}), \\ f'(\lambda_j) \prod_{k \neq j} (\lambda_j - \lambda_k)^{-1} & (z = \lambda_j, j = 1, ..., r). \end{cases}$$

Then g is an entire function, $g(z) \neq 0$ $(z \in \mathrm{Sp}(a))$, and so $0 \notin \mathrm{Sp}(\mathbf{g}(a))$. Since

$$g(z)(z - \lambda_1)...(z - \lambda_r) = f(z) \qquad (z \in \mathbb{C}),$$

the functional calculus gives

$$\mathbf{g}(a)(a - \lambda_1)...(a - \lambda_r) = \mathbf{f}(a) = 0.$$

But $\mathbf{g}(a) \in \mathrm{Inv}(A)$, and so

$$(a - \lambda_1)...(a - \lambda_r) = 0.$$

Let $p(z) = (z - \lambda_1)...(z - \lambda_r)$ $(z \in \mathbb{C})$. Then

$$e_1 p(a e_1) = e_1(a e_1 - \lambda_1)...(a e_1 - \lambda_r)$$
$$= (a e_1 - \lambda_1 e_1)...(a e_1 - \lambda_r e_1)$$
$$= e_1 p(a) = 0.$$

Also $e_1(a e_1 - \lambda_1) = a e_1 - \lambda_1 e_1 = q_1$. So

$$q_1(a e_1 - \lambda_2)...(a e_1 - \lambda_r) = 0.$$

By Proposition 7.9, $\mathrm{Sp}(a e_1) = \{\lambda_1\}$. So $a e_1 - \lambda_j \in \mathrm{Inv}(A)$ $(j = 2, ..., r)$. Therefore $q_1 = 0$. Similarly $q_2 = \cdots = q_r = 0$. $\square$

Proposition 11 has an obvious generalization to the situation in which $\mathbf{f}(a) = 0$ for some function f which is holomorphic on a neighbourhood of $\mathrm{Sp}(a)$. For the special case $f(z) = z^m - 1$, or $f(z) = \exp(z) - 1$, we obtain Hille's result [196] on roots and logarithms of the unit element. Further results on roots and logarithms are given in § 18, and we end this section with some elementary existence and uniqueness theorems for quasi-square roots in Banach algebras over $\mathbb{F}$.

Notation. Let B denote a Banach algebra over $\mathbb{F}$. We recall that the quasi-product $x \circ y$ is defined by

$$x \circ y = x + y - xy.$$

Given $a \in B$, a *quasi-square root* of a is an element $x \in B$ with

$$x \circ x = a.$$

We denote by $B(a)$ the least closed subalgebra of B containing a.

Lemma 12. *Let* $x, y \in B$ *and suppose that* $xy = yx$, $x \circ x = y \circ y$, *and* $r(x+y) < 2$. *Then* $x = y$.

Proof. Let $u = \frac{1}{2}(x+y)$, $v = x - y$. Since $2x - x^2 = 2y - y^2$, and $xy = yx$, we have $uv = v$, i.e. $u \circ v = u$. Also $r(u) < 1$, and so u has a quasi-inverse w, $w \circ u = 0$. Thus $x - y = v = 0 \circ v = w \circ u \circ v = w \circ u = 0$. □

Proposition 13. *Let* $a \in B$ *and* $r(a) < 1$. *Then there exists a unique quasi-square root* x *of* a *with* $r(x) < 1$. *Moreover* $x \in B(a)$.

Proof. By Corollary 4.2, we may suppose that the norm $\|\cdot\|$ is chosen so that $\|a\| < 1$. Let $\|a\| < \eta < 1$, let $E = \{x \in B(a): \|x\| \leqslant \eta\}$, and let T be the mapping defined on E by

$$Tx = \tfrac{1}{2}(a + x^2).$$

Since elements of $B(a)$ commute with each other, we have

$$\|Tx - Ty\| = \tfrac{1}{2}\|x^2 - y^2\| \leqslant \tfrac{1}{2}\|x - y\|\,\|x + y\| \leqslant \eta \|x - y\| \quad (x, y \in E).$$

Therefore T is a contradiction mapping of E into E. By the contraction mapping principle, there exists $x \in E$ with $Tx = x$, i.e. with $x \circ x = a$, and we have $r(x) \leqslant \|x\| < 1$.

Suppose now that $y \in B$, $y \circ y = a$ and $r(y) < 1$. We have

$$ay = (2y - y^2)y = y(2y - y^2) = ya.$$

Since $x \in B(a)$, it follows that $xy = yx$. Therefore, by Corollary 4.3,

$$r(x + y) \leqslant r(x) + r(y) < 2,$$

and so $y = x$ by Lemma 12. □

Theorem 14. *Let* $a \in B$ *and* $r(a^2 - a) = 0$. *Then there exists a unique element* $z \in B$ *with*

$$r(z) = 0, \qquad az = za, \qquad (a + z)^2 = a + z.$$

Proof. Let $R = \{x \in B(a): r(x) = 0\}$. Since $B(a)$ is commutative, Corollary 4.3 shows that R is a closed subalgebra of $B(a)$ that also satisfies

$$x \in R, \quad b \in B(a) \Rightarrow bx \in R, \tag{2}$$

(i. e. R is a closed ideal of $B(a)$). Let $f = a - a^2$. Since $r(4f) = 0$, Theorem 3.7 shows that $4f \in \text{q-Inv}(R)$, i.e. $4f$ has a quasi-inverse g in R.

By Proposition 13 applied to the algebra R, there exists $x \in R$ with $x \circ x = g$. Then, since x and a commute with each other and the quasi-product is associative,

$$((2a) \circ x) \circ ((2a) \circ x) = (2a) \circ (2a) \circ (x \circ x) = (4f) \circ g = 0. \tag{3}$$

We have $(2a) \circ x = 2(a + z)$, where $z = \frac{1}{2} x - ax$. Then by (2), $z \in R$, and by (3), $4[(a + z) - (a + z)^2] = (2(a + z)) \circ (2(a + z)) = 0$.

Suppose that $w \in B$ with $r(w) = 0$, $aw = wa$, and $(a + w)^2 = a + w$. Since $z \in B(a)$, we have $zw = wz$. Let $h = x \circ (2a) \circ x$. Then by (3), $(2a) \circ h = 0$. Let $y = 2w - h(2w)$, so that

$$y - 2ay = 2w - [2a + h - 2ah]2w = 2w,$$

and therefore $(2a) \circ y = 2(a + w)$. Then

$$(4f) \circ (y \circ y) = (2a) \circ (2a) \circ (y \circ y) = ((2a) \circ y) \circ ((2a) \circ y)$$
$$= (2a + 2w) \circ (2a + 2w) = 0.$$

Since quasi-inverses are unique it follows that $y \circ y = g = x \circ x$. Since $h \in B(a)$, we have $hw = wh$, and so $r(y) = 0$. Finally, by the uniqueness assertion in Proposition 13, $y = x$, $w = \frac{1}{2} x - ax = z$. ☐

§ 9. Ideals and Modules

In this section we develop only the most elementary properties of ideals and modules.

Notation. Let A denote an algebra with or without unit. Given $a \in A$ and subsets E, F of A we denote by EF, $E(1 - a)$, $(1 - a)E$ respectively the sets $\{xy : x \in E, y \in F\}$, $\{x - xa : x \in E\}$, $\{x - ax : x \in E\}$.

Definition 1. A *left ideal* of A is a linear subspace J of A such that $AJ \subset J$. An element u of A is a *right modular unit* for a linear subspace E of A if $A(1 - u) \subset E$. A *modular left ideal* is a left ideal for which there exists a right modular unit.

A left ideal J of A is *proper* if $J \neq A$, *maximal* if it is proper and not contained in any other proper left ideal, *maximal modular* if it is proper and modular and not contained in any other such left ideal.

Similar definitions apply to *right ideals*, *left modular units*, etc., in terms of the inclusions $JA \subset J$, $(1 - u)A \subset E$.

A *two-sided ideal*, or *bi-ideal*, is a linear subspace that is both a left ideal and a right ideal.

It is clear that a left ideal of A is a subalgebra of A, and that if A has a unit element 1, then 1 is a right modular unit for every linear subspace of A. To avoid tedious repetition, we shall state results only for left ideals and right modular units. We collect together some related elementary facts in the following Proposition.

Proposition 2. (i) *If u is a right modular unit for a linear subspace E, it is a right modular unit for any subspace F containing E.*

(ii) *If u is a right modular unit for a proper left ideal J, then $u \notin J$.*

(iii) *u is left quasi-regular if and only if $A(1-u) = A$.*

(iv) *If J is a proper modular left ideal of A, then J is contained in a maximal left ideal.*

(v) *Maximal modular left ideals are maximal left ideals.*

(vi) *If A has a unit element, every proper left ideal is contained in a maximal left ideal.*

Proof. (i) Clear.

(ii) Let u be a right modular unit for a left ideal J. Then

$$a = (a - au) + au \in J \qquad (a \in A)$$

if $u \in J$.

(iii) Let u be left quasi-regular. Then $u \in A(1-u)$, and, since u is a right modular unit for the left ideal $A(1-u)$, (ii) gives $A(1-u) = A$. The converse is clear.

(iv) Let J be a proper modular left ideal with right modular unit u, and consider the set Γ of all proper left ideals K with $J \subset K$. By (i), u is a right modular unit for each $K \in \Gamma$, and so, by (ii),

$$u \notin K \qquad (K \in \Gamma).$$

Therefore the set Γ is inductively ordered by the relation of inclusion, and so, by Zorn's lemma, it has maximal elements.

(v) and (vi). Clear. □

Theorem 3. *Let A be a Banach algebra, let J be a proper modular left ideal of A, and let u be a right modular unit for J. Then*

(i) *$\|u - x\| \geqslant 1 \ (x \in J)$;*

(ii) *the closure of J is a proper modular left ideal of A.*

Proof. (i) Suppose that J contains an element x with $\|u - x\| < 1$. Then by Theorem 3.7 $u - x$ has a quasi-inverse y. Therefore

$$u = x - y + y(u - x) = x - yx - (y - yu) \in J,$$

since $x, yx, y - yu \in J$. This contradicts Proposition 2 (ii).

(ii) By (i), J^- is proper, and the rest is routine. □

Corollary 4. *Each maximal modular left ideal of a Banach algebra is closed.*

Proof. Theorem 3 (ii). □

Corollary 5. *Let A be a Banach algebra with unit. Then every proper left ideal of A is contained in a maximal left ideal of A, and every maximal left ideal of A is closed.*

Proof. All left ideals of A are modular. ☐

Remark. A bi-ideal J is *modular* if it is modular both as a left and as a right ideal. Let u be a right modular unit and v be a left modular unit. Then $v - vu$, $u - vu \in J$ and so $u - v \in J$. Thus $a - ua = a - va + (u - v)a \in J$ $(a \in A)$ and so u is also a left modular unit.

Definition 6. Let L be a linear subspace of a linear space X. The relation $\sim$ on X obtained by writing $x \sim y$ when $x - y \in L$ is an equivalence relation on X. The equivalence class to which x belongs is called the L-coset of x and is denoted by x'. The set of all L-cosets becomes a linear space with addition and scalar multiplication defined by

$$z_1 + z_2 = (x_1 + x_2)', \qquad \lambda z_1 = (\lambda x_1)',$$

where x_1, x_2 are arbitrary elements of cosets z_1, z_2 respectively. This linear space is called the *difference space of X modulo L*, and is denoted by $X - L$. The mapping $x \to x'$ of X onto $X - L$ is called the *canonical mapping* (or *quotient mapping*). When L is a closed linear subspace of X, a norm is defined on $X - L$ by

$$\|z\| = \inf\{\|x\|: x \in z\} \qquad (z \in X - L).$$

This norm is called the *canonical norm* on $X - L$. Unless otherwise stated, $X - L$ will always be equipped with the canonical norm.

Proposition 7. *Let L be a closed linear subspace of a normed linear space X. Then* (i) *the canonical mapping is a bounded linear mapping of X onto X − L of norm 1, unless L = X when its norm is 0;*
(ii) *if F is a complete subset of X such that F + L = F, then the canonical image F′ = {f′: f ∈ F} is a complete subset of X − L;*
(iii) *if X is complete, so is X − L.*

Proof. (ii) Let $\{y_n\}$ be a sequence of elements of F' with $\|y_{n+1} - y_n\| < 2^{-n}$ $(n = 1, 2, \ldots)$. We show that we can choose f_n in F such that $f_n' = y_n$ and $\|f_{n+1} - f_n\| < 2^{-n}$ $(n = 1, 2, \ldots)$. For each n, there exists $g_n \in F$ with $g_n' = y_n$. Take $f_1 = g_1$ and choose $v_2 \in (y_2 - y_1)$ with $\|v_2\| < 2^{-1}$. Then $v_2 - (g_2 - f_1) \in L$, therefore $v_2 + f_1 \in F + L = F$, and so $v_2 + f_1 = f_2 \in F$, $\|f_2 - f_1\| < 2^{-1}$. Continue the construction in this way.
(i) and (iii). Clear. ☐

Definition 8. Given a bi-ideal J of an algebra A, a product is defined on $A - J$ by

$$z_1 z_2 = (x_1 x_2)',$$

where x_1, x_2 are arbitrary elements of z_1, z_2 respectively. The coset $(x_1 x_2)'$ is independent of the choice of x_1, x_2 since

$$(x_1 + j_1)(x_2 + j_2) - x_1 x_2 = x_1 j_2 + j_1 x_2 + j_1 j_2 \in J$$

for all j_1, j_2 in J. With this product $A - J$ is an algebra, which is called the *quotient algebra* (or *factor algebra*) of A *modulo* J, and is denoted by A/J. If J is a closed bi-ideal of a normed algebra A, the canonical norm is an algebra-norm on A/J.

Obviously, the canonical mapping is a homomorphism of A onto A/J.

Proposition 9. *Let J be a bi-ideal of an algebra A, and let a' denote the J-coset of an element $a \in A$. Then*

$$\mathrm{Sp}(A/J, a') \subset \mathrm{Sp}(A, a).$$

Proof. Proposition 5.4 applied to the homomorphism $a \to a'$. ☐

Proposition 10. *Let p be an algebra-semi-norm on a complex Banach algebra A. Then*

$$\lim_{n \to \infty} (p(a^n))^{\frac{1}{n}} \leqslant r(a) \qquad (a \in A).$$

Proof. Let $N = \{x \in A : p(x) = 0\}$. Then N is a bi-ideal of A. Let $B = A/N$ and define a norm $|\cdot|$ on B by

$$|z| = p(x) \qquad (x \in z \in B).$$

Then $(B, |\cdot|)$ is a normed algebra. Let a' denote the N-coset of an element $a \in A$. By Proposition 9, we have

$$\mathrm{Sp}(B, a') \subset \mathrm{Sp}(A, a).$$

Applying Theorem 5.7 to the normed algebra B and Theorem 5.8 to the Banach algebra A, we conclude from this that

$$\lim_{n \to \infty} |(a')^n|^{\frac{1}{n}} \leqslant r(a),$$

i. e.

$$\lim_{n \to \infty} (p(a^n))^{\frac{1}{n}} \leqslant r(a). \quad ☐$$

We now consider the definition and some examples of A-modules. We are not concerned in this book with the theory of A-modules but regard them as convenient auxiliary concepts for the study of Banach algebras. In particular they are useful in connection with approximate identities and central to representation theory for non-commutative Banach algebras.

Definition 11. Let A be an algebra over $\mathbb{F}$, and M a linear space over $\mathbb{F}$. M is said to be a *left A-module* if a mapping $(a,m) \to am$ of $A \times M$ into M is specified which satisfies the following axioms:

$LM\,1$ for each fixed $a \in A$, the mapping $m \to am$ is linear on M;
$LM\,2$ for each fixed $m \in M$, the mapping $a \to am$ is linear on A;
$LM\,3$ $a_1(a_2 m) = (a_1 a_2)m$ $(a_1, a_2 \in A, m \in M)$.

The specified mapping $(a,m) \to am$ is called the *module multiplication*.

Likewise M is a *right A-module* if a mapping $(a,m) \to ma$ of $A \times M$ into M is specified which satisfies linearity axioms analogous to $LM\,1$ and $LM\,2$ and also

$RM\,3$ $(ma_1)a_2 = m(a_1 a_2)$ $(a_1, a_2 \in A, m \in M)$.

Recall that $\mathrm{rev}(A)$ denotes the algebra consisting of A with its product reversed. Then a right A-module is a left $\mathrm{rev}(A)$-module.

M is said to be an *A-bimodule* if it is both a left A-module and a right A-module and the module multiplications are related by the axiom:

BM $a(mb) = (am)b$ $(a, b \in A, m \in M)$.

A left A-module M is said to be *unit linked* if A has a unit element e and $em = m$ $(m \in M)$. Similar definitions apply to *unit linked* right A-modules and A-bimodules; for the latter we require of course that $em = me = m$ $(m \in M)$.

If A has a unit element and M is a left A-module that is not unit linked confusion can occur if the unit element of A is denoted by 1, since $1m$ may have two different meanings corresponding to the two different meanings of 1.

Definition 12. Let A be a normed algebra over $\mathbb{F}$, and let M be a normed linear space over $\mathbb{F}$. M is said to be a *normed left A-module* if M is a left A-module and also satisfies the axiom:

NLM there exists a positive constant K such that

$$\|am\| \leqslant K \|a\| \|m\| (a \in A, m \in M).$$

A similar definition applies to *normed right A-modules*, and a *normed A-bimodule* is an A-bimodule that is both a normed left A-module and a normed right A-module. A normed left A-module is called a *Banach left A-module* if it is complete as a normed linear space. Similarly for *Banach right A-modules* and *Banach A-bimodules*.

Example 13. Let A be a normed algebra.

(i) A is a normed A-bimodule with the product of A giving the two module multiplications. More generally each bi-ideal of A is a normed A-bimodule.

(ii) Each left ideal of A is a normed left A-module and each right ideal is a normed right A-module, again with the product of A giving the module multiplications.

(iii) Let L be a closed left ideal of A, let $X = A - L$, and let $a \to a'$ denote the canonical mapping of A onto X. Then the normed linear space X becomes a normed left A-module with the module multiplication given by

$$a x = (a b)' \qquad (b \in x \in X, a \in A).$$

Similarly a closed right ideal J gives a normed right A-module $A - J$.

(iv) Let X be a normed left A-module with dual space X'. Then X', with the module multiplication given by

$$(f a)(x) = f(a x) \qquad (a \in A, f \in X', x \in X)$$

is a Banach right A-module called the *dual* Banach right A-module X'.

Similarly each normed right A-module X has a dual Banach left A-module X', and each normed A-bimodule X has a dual Banach A-bimodule X'.

(v) Let $A'' = (A')'$ be the second dual space of A, and given $a \in A$, let $a^\wedge$ denote its canonical image in A'', i. e.

$$a^\wedge(f) = f(a) \qquad (f \in A').$$

Then A'' is a Banach space and $a \to a^\wedge$ is an isometric linear monomorphism of A into A''. We construct a product in A'', called the *Arens product* (Arens [22, 23]), with which A'' becomes a Banach algebra and $a \to a^\wedge$ an isometric monomorphism of A into A''.

A is a normed left A-module as in (ii), with dual Banach right A-module A' as in (iv). Given $F \in A''$, $f \in A'$, let $F f \in A'$ be defined by

$$(F f)(a) = F(f a) \qquad (a \in A). \tag{1}$$

Given $F, G \in A''$, let $F G$ (the *Arens product* of F, G) be defined by

$$(F G)(f) = F(G f) \qquad (f \in A').$$

It is straightforward to verify that the Arens product has the stated properties. Observe that A' is then a Banach left A''-module with module multiplication given by (1).

Let $\text{rev}(A)$ be the reversed algebra of A, as in Example 1.17, with reversed product $a \cdot b = b a$. We can construct the corresponding Arens product $F \cdot G$ in $(\text{rev}(A))''$. In general $F \cdot G$ is not the reversed product of FG, i. e. we do not have in general $F \cdot G = GF$. The Arens product is said to be *regular* if $F \cdot G = GF$ for all G, F in A'' (Civin and Yood [100]).

(vi) Let $\{M_\alpha : \alpha \in \Lambda\}$ be a set of Banach left A-modules for which the axiom NLM is satisfied with a constant K independent of α. We denote by $c_0(M_\alpha : \alpha \in \Lambda)$ the set of functions f belonging to the Cartesian product $\prod_{\alpha \in \Lambda} M_\alpha$ such that, for each $\varepsilon > 0$, the set $\{\alpha \in \Lambda : \|f(\alpha)\| > \varepsilon\}$ is finite. With pointwise addition and scalar multiplication and the uniform norm, and with module multiplication given, for $a \in A$ and $f \in c_0(M_\alpha : \alpha \in \Lambda)$, by

$$(af)(\alpha) = af(\alpha) \qquad (\alpha \in \Lambda), \tag{2}$$

$c_0(M_\alpha : \alpha \in \Lambda)$ becomes a Banach left A-module. In particular, given a Banach left A-module M, we denote by $c_0(M)$ the Banach left A-module $c_0(M_\alpha : \alpha \in \Lambda)$ obtained by taking $\Lambda = \mathbb{N}$ and $M_\alpha = M$ $(\alpha \in \Lambda)$.

Similarly we define the Banach left A-module $l^1(M_\alpha : \alpha \in \Lambda)$ by taking the linear space of functions $f \in \prod_{\alpha \in \Lambda} M_\alpha$ such that

$$\sum_\alpha \|f(\alpha)\| < \infty,$$

with the norm defined by $\|f\| = \sum_\alpha \|f(\alpha)\|$ and with the module multiplication given by (2).

(vii) Let M be a closed A-submodule of a normed left A-module X, let $Y = X - M$ and let π be the canonical mapping of X onto Y. Then Y is a normed left A-module with the module multiplication given by

$$ay = \pi(ax) \qquad (x \in y \in Y, a \in A).$$

Moreover π is a *module homomorphism*, i. e.

$$a(\pi x) = \pi(ax) \qquad (a \in A, x \in X).$$

Y' is a Banach right A-module and π^* is a module monomorphism of Y' into X':

$$\pi^*(\phi a) = (\pi^* \phi)a \qquad (\phi \in Y', a \in A).$$

(viii) Let M be a normed left A-module and let B be the unitization $A + \mathbb{F}$ of A. Then M becomes a unit linked normed left B-module with the module multiplication given by $(a, \alpha)m = am + \alpha m$ $(a \in A, \alpha \in \mathbb{F}, m \in M)$.

§ 10. The Numerical Range of an Element of a Complex Normed Algebra

Throughout this section $(A, \|\cdot\|)$ will denote a complex unital normed algebra. In §§ 1—9 the concepts and results are independent of the algebra-norm in the sense that they are unchanged if the norm is replaced by an equivalent algebra-norm. The concept of numerical range, with which the present section is concerned, depends on the choice of algebra-norm.

We recall that given a normed linear space X, we denote by $S(X)$, X_1, X' respectively the unit sphere, the closed unit ball, and the dual space of X.

Definition 1. We define sets $D(A;1)$, $V(A;a)$ by

$$D(A;1) = \{f \in S(A'): f(1) = 1\},$$
$$V(A;a) = \{f(a): f \in D(A;1)\} \qquad (a \in A).$$

The elements of $D(A;1)$ are called *normalized states* on A, $V(A;a)$ is called the *numerical range* (or the *algebra numerical range*) of a. We write $D(1)$, $V(a)$ for $D(A;1)$, $V(A;a)$ when no confusion can occur. The next proposition shows that $V(A;a)$ is independent of the choice of the algebra A.

Proposition 2. *Let B be a subalgebra of A such that $1 \in B$. Then*

$$V(B;b) = V(A;b) \qquad (b \in B).$$

Proof. By the Hahn-Banach theorem, the restriction mapping $f \to f|_B$ maps $D(A;1)$ onto $D(B;1)$. □

Remark. Proposition 2 shows, in particular, that $V(A;a)$ is unchanged when A is replaced by its completion.

Lemma 3. $D(1)$ *is a non-void weak* compact convex subset of A'.*

Proof. If $\|f\| \leqslant 1$, and $f(1) = 1$, we have $\|f\| = 1$. Therefore

$$D(1) = \{f \in A_1' : f(1) = 1\}.$$

This exhibits $D(1)$ as the intersection of the weak* compact convex set A_1' with the weak* closed convex set $\{f \in A' : f(1) = 1\}$. The Hahn-Banach theorem shows that $D(1)$ is not void. □

Proposition 4. (i) $V(a)$ *is a non-void compact convex subset of* $\mathbb{C}$.
(ii) $V(\alpha + \beta a) = \alpha + \beta V(a)$, $\quad V(a+b) \subset V(a) + V(b)$ $\quad (a, b \in A, \alpha, \beta \in \mathbb{C})$.
(iii) $|z| \leqslant \|a\|$ $(z \in V(a))$.

Proof. (i) This follows at once from Lemma 3 and the linearity and weak* continuity of the mapping $f \to f(a)$.
(ii) and (iii). Clear. □

The following Lemma expresses $V(a)$ as an intersection of closed discs.

Lemma 5. $V(a) = \bigcap_{z \in \mathbb{C}} E(z, \|z - a\|)$.

Proof. If $\lambda \in V(a)$, then $\lambda = f(a)$ for some $f \in D(1)$, and for all $z \in \mathbb{C}$ we have

$$|\lambda - z| = |f(a-z)| \leqslant \|a-z\|,$$

i. e.

$$\lambda \in E(z, \|z-a\|) \quad (z \in \mathbb{C}). \tag{1}$$

Suppose on the other hand that λ satisfies (1). If $a = \zeta$ for some $\zeta \in \mathbb{C}$, then $\|z-a\| = |z-\zeta|$, and taking $z = \zeta$ we have $\lambda \in E(\zeta, 0)$, i. e. $\lambda = \zeta$. But also when $a = \zeta$, $V(a) = \{\zeta\}$. Suppose then that $1, a$ are linearly independent, and define f_0 on their linear span by

$$f_0(\alpha + \beta a) = \alpha + \beta \lambda \quad (\alpha, \beta \in \mathbb{C}).$$

Since λ satisfies (1), we have $|f_0(\alpha + \beta a)| \leqslant \|\alpha + \beta a\|$, $\|f_0\| \leqslant 1$. By the Hahn-Banach theorem, f_0 can be extended to $f \in A'$ with $\|f\| \leqslant 1$. Then $f \in D(1)$ and $f(a) = f_0(a) = \lambda$. □

Proposition 6. *Let A be complete. Then* $\mathrm{Sp}(a) \subset V(a)$.

Proof. Let $\lambda \in \mathbb{C} \backslash V(a)$. Then, by Lemma 6, there exists $z \in \mathbb{C}$ such that $|z - \lambda| > \|z - a\|$. Therefore $\|(z-\lambda)^{-1}(z-a)\| < 1$, and so by Theorem 2.9,

$$1 - (z-\lambda)^{-1}(z-a) \in \mathrm{Inv}(A).$$

It follows that $\lambda - a \in \mathrm{Inv}(A)$, and so $\lambda \in \mathbb{C} \backslash \mathrm{Sp}(a)$. □

It is entirely straightforward to verify that if ϕ is a norm decreasing homomorphism of A into a complex unital normed algebra B and $\phi(1) = 1$, then $V(B; \phi(a)) \subset V(A; a)$ for all $a \in A$. When ϕ is the canonical mapping onto the quotient of A by a closed bi-ideal we have a more precise result given in the following proposition.

Proposition 7. *Let J be a closed bi-ideal of A, let $a \in A$ and let a' denote the J-coset of a. Then*

$$V(A/J; a') = \bigcap \{V(a+j) : j \in J\}.$$

Proof. By definition of the canonical norm, we have $|z - \lambda| \leqslant \|z - a'\|$ if and only if

$$|z - \lambda| \leqslant \|z - (a+j)\| \quad (j \in J).$$

Thus $E(z, \|z-a'\|) = \bigcap_{j \in J} E(z, \|z-(a+j)\|)$, and Lemma 5 completes the proof. □

Remark. Proposition 7 is the foundation of the theory of the *essential range* of an operator. For this one takes $A = BL(X)$ and J to be the closed bi-ideal of compact linear operators on X. See Stampfli and Williams [368], Bonsall and Duncan [78].

Lemma 8. $V(a) = \bigcup \{V(a,x) : x \in S(A)\}$,
where $V(a,x) = \bigcap_{z \in \mathbb{C}} E(z, \|(z-a)x\|)$.

Proof. Let $x \in S(A)$. Then $\|(z-a)x\| \leqslant \|z-a\| = \|(z-a)1\|$, and so Lemma 5 gives $V(a,x) \subset V(a) = V(a,1)$. □

Remark. An alternative expression for $V(a,x)$ is given by

$$V(a,x) = \{f(ax) : f \in D(x)\},$$

where $D(x) = \{f \in A' : f(x) = \|f\| = 1\}$.

Lemma 9. $\inf\{\operatorname{Re}\lambda : \lambda \in V(a)\} \leqslant \inf\{\|ax\| : x \in S(A)\}$.

Proof. Let $x \in S(A)$. We have $V(a,x) \subset E(0, \|ax\|)$, and so, by Lemma 8, $\inf\{\operatorname{Re}\lambda : \lambda \in V(a)\} \leqslant \inf\{\operatorname{Re}\lambda : \lambda \in V(a,x)\} \leqslant \|ax\|$. □

Theorem 10. $\sup\{\operatorname{Re}\lambda : \lambda \in V(a)\} = \inf\left\{\dfrac{1}{\alpha}\left[\|1 + \alpha a\| - 1\right] : \alpha > 0\right\}$

$$= \lim_{\alpha \to 0+} \frac{1}{\alpha}\left[\|1 + \alpha a\| - 1\right].$$

Proof. Let $\mu = \sup\{\operatorname{Re}\lambda : \lambda \in V(a)\}$. By Lemma 5,

$$V(a) \subset E\left(-\frac{1}{\alpha}, \left\|\frac{1}{\alpha} + a\right\|\right) \qquad (\alpha > 0),$$

and so

$$\mu \leqslant \frac{1}{\alpha}\left[\|1 + \alpha a\| - 1\right] \qquad (\alpha > 0). \tag{2}$$

Let $\alpha > 0$ with $1 - \alpha\mu > 0$. We have

$$\inf\{\operatorname{Re}\lambda : \lambda \in V(1 - \alpha a)\} = \inf\{\operatorname{Re}(1 - \alpha\lambda) : \lambda \in V(a)\} = 1 - \alpha\mu.$$

By Lemma 9, it follows that

$$\|(1 - \alpha a)x\| \geqslant 1 - \alpha\mu \qquad (x \in S(A)),$$

and therefore

$$\|(1 - \alpha a)x\| \geqslant (1 - \alpha\mu)\|x\| \qquad (x \in A).$$

Taking $x = 1 + \alpha a$, we have

$$\|1 + \alpha a\| \leqslant (1 - \alpha\mu)^{-1}\|1 - \alpha^2 a^2\|$$
$$\leqslant (1 - \alpha\mu)^{-1}(1 + \alpha^2 \|a^2\|).$$

Therefore $\dfrac{1}{\alpha}\left[\|1 + \alpha a\| - 1\right] \leqslant (1 - \alpha\mu)^{-1}(\mu + \alpha\|a^2\|)$ for all sufficiently small positive α, and with (2) this completes the proof. □

Theorem 11. $\sup\{\mathrm{Re}\,\lambda: \lambda \in V(a)\} = \sup\left\{\frac{1}{\alpha}\log\|\exp(\alpha a)\| : \alpha > 0\right\}$

$$= \lim_{\alpha \to 0+} \frac{1}{\alpha}\log\|\exp(\alpha a)\|.$$

Proof. Let $\mu = \sup\{\mathrm{Re}\,\lambda: \lambda \in V(a)\}$. If $\alpha > 0$ with $1 - \alpha\mu > 0$, we have, as in the proof of Theorem 10,

$$\|(1 - \alpha a)x\| \geqslant (1 - \alpha\mu)\|x\| \qquad (x \in A),$$

and so, by induction,

$$\|(1 - \alpha a)^n x\| \geqslant (1 - \alpha\mu)^n\|x\| \qquad (x \in A, n = 1, 2, \ldots). \tag{3}$$

Given $\alpha > 0$, we have $1 - \dfrac{\alpha}{n}\mu > 0$ for all sufficiently large n. Therefore we may replace α by $\dfrac{\alpha}{n}$ in (3) and let $n \to \infty$, to obtain, by Proposition 8.2 (iv),

$$\|\exp(-\alpha a)x\| \geqslant \exp(-\alpha\mu)\|x\| \qquad (\alpha > 0, x \in A).$$

Taking $x = \exp(\alpha a)$, we have

$$\|\exp(\alpha a)\| \leqslant \exp(\alpha\mu) \qquad (\alpha > 0),$$

$$\sup\left\{\frac{1}{\alpha}\log\|\exp(\alpha a)\| : \alpha > 0\right\} \leqslant \mu.$$

On the other hand, we have

$$\|\exp(\alpha a)\| = \|1 + \alpha a\| + \lambda(\alpha),$$

where, for some $M > 0$, $|\lambda(\alpha)| \leqslant M\alpha^2 \, (0 \leqslant \alpha \leqslant 1)$. Using the inequality $\log t \geqslant (t-1)t^{-1} \, (t > 0)$, we therefore have

$$\frac{1}{\alpha}\log\|\exp(\alpha a)\| \geqslant \frac{\dfrac{1}{\alpha}\left[\|1 + \alpha a\| - 1\right] + \dfrac{1}{\alpha}\lambda(\alpha)}{\|1 + \alpha a\| + \lambda(\alpha)}.$$

Now apply Theorem 10.   □

Definition 12. An element a of A is *dissipative* if $\mathrm{Re}\,\lambda \leqslant 0 \, (\lambda \in V(a))$, *Hermitian* if $V(a) \subset \mathbb{R}$.

Corollary 13. *An element a of A is dissipative if and only if*

$$\|\exp(\alpha a)\| \leqslant 1 \qquad (\alpha \geqslant 0),$$

and is Hermitian if and only if

$$\|\exp(i\alpha a)\| = 1 \qquad (\alpha \in \mathbb{R}).$$

Proof. Clear.   □

Theorem 14. $\dfrac{1}{e}\,\|a\| \leqslant \sup\{|\lambda|: \lambda \in V(a)\} \leqslant \|a\|.$

Proof. We may assume, without loss of generality, that A is complete. Let $a \in A$ and $\sup\{|\lambda|: \lambda \in V(a)\} \leqslant 1$. If $|z| = 1$, we have $\|\exp(za)\| \leqslant e$ by Theorem 11. Since

$$a = \frac{1}{2\pi i}\int\limits_{C(0,\,1)} \exp(za)\frac{dz}{z^2},$$

it follows that $\|a\| \leqslant e$. □

Corollary 15. *The set $D(1)$ of normalized states separates the points of A.*

Proof. If $a \neq 0$, $\sup\{|\lambda|: \lambda \in V(a)\} \neq 0$, and so there exists $f \in D(1)$ with $f(a) \neq 0$. □

Remark. Corollary 15 is due to Bohnenblust and Karlin [60]. R. T. Moore has proved the stronger result that A' is the linear span of $D(1)$. See Moore [282] or Bonsall and Duncan [78].

Let $\arcsin z$ be defined for $|z| \leqslant 1$ by

$$\arcsin z = z + \tfrac{1}{2}\cdot\tfrac{1}{3}z^3 + \tfrac{1}{2}\cdot\tfrac{3}{4}\cdot\tfrac{1}{5}z^5 + \cdots$$

$$= \sum_{n=1}^{\infty} \alpha_n z^n.$$

If A is complete, $a \in A$, and $\mathrm{Sp}(a) \subset U(0, 1)$, we have

$$\arcsin a = \sum_{n=1}^{\infty} \alpha_n a^n.$$

Also, for all $a \in A$, we have $\sin a = a - \dfrac{1}{3!}a^3 + \cdots = \dfrac{1}{2i}\{\exp(ia) - \exp(-ia)\}$.

Lemma 16. *Let A be complete, let $h \in A$, and let $\mathrm{Sp}(h)$ be contained in the open interval $]-\pi/2, \pi/2[$. Then $\mathrm{Sp}(\sin h) \subset U(0, 1)$, and*

$$\arcsin(\sin h) = h.$$

Proof. We have $\mathrm{Sp}(\sin h) = \{\sin \lambda: \lambda \in \mathrm{Sp}(h)\}$, and so $\mathrm{Sp}(\sin h) \subset]-1, 1[\subset U(0, 1)$. Therefore

$$\arcsin(\sin h) = \sum_{n=1}^{\infty} \alpha_n(\sin h)^n.$$

Let D be an open neighbourhood of $\mathrm{Sp}(h)$ such that $\sin z \in U(0, 1)\,(z \in D)$. Then $\arcsin(\sin z)$ is a holomorphic function of z on D, and, since $\arcsin(\sin t) = t\;(t \in [-\pi/2, \pi/2])$, we have

$$\arcsin(\sin z) = z \qquad (z \in D).$$

Let $s_n(z) = \sum_{k=1}^{n} \alpha_k (\sin z)^k$. Then $s_n(z) \to z$ uniformly on a neighbourhood of $Sp(h)$. Therefore $\lim_{n \to \infty} s_n(h) = h$ (by the continuity of the functional calculus), i.e. $h = \arc \sin(\sin h)$. ☐

Theorem 17. *Let h be a Hermitian element of A. Then $r(h) = \|h\|$.*

Proof. We may assume that A is complete, and that $r(h) < \pi/2$. Since $Sp(h) \subset V(h) \subset \mathbb{R}$, we have $Sp(h) \subset]-\pi/2, \pi/2[$, and so

$$h = \arc \sin(\sin h) = \sum_{n=1}^{\infty} \alpha_n (\sin h)^n .$$

By Corollary 13, $\|\exp(\pm ih)\| = 1$, and so $\|\sin h\| \leqslant 1$. Since $\alpha_n \geqslant 0$ $(n = 1, 2, \ldots)$, it follows that

$$\|h\| \leqslant \sum_{n=1}^{\infty} \alpha_n = \pi/2 . ☐$$

Remarks. (1) Theorem 17 is due to A. M. Sinclair who proved in [356] the stronger result

$$r(h + i\gamma) = \|h + i\gamma\| \qquad (\gamma \in \mathbb{R}) .$$

(2) Theorems 10—17 remain valid if $V(a)$ is replaced by certain smaller numerical ranges, notably the numerical ranges $W(a)$ defined in terms of semi-inner products by G. Lumer, to whom many of the basic ideas are due. It suffices to replace V by a mapping $a \to \Phi(a)$ of A into the non-void subsets of $\mathbb{C}$ such that
   (i) $\Phi(\alpha a + \beta) = \alpha \Phi(a) + \beta$,
   (ii) $\sup\{|\lambda| : \lambda \in \Phi(a)\} \leqslant \|a\|$,
   (iii) $\inf\{\mathrm{Re}\,\lambda : \lambda \in \Phi(a)\} \leqslant \inf\{\|ax\| : x \in S(A)\}$.
For further details see Lumer [271], and Bonsall and Duncan [78].

§ 11. Approximate Identities

Let A be a normed algebra over $\mathbb{F}$. We do not assume that A has a unit element. We recall that a *directed set* is a partially ordered set Λ such that, given $\lambda_1, \lambda_2 \in \Lambda$, there exists $\lambda \in \Lambda$ with $\lambda_k \leqslant \lambda$ $(k = 1, 2)$. A *net* in a topological space E is a mapping of a directed set into E, and a net $\{x(\lambda)\}_{\lambda \in \Lambda}$ in E *converges* to $x \in E$, and we write $x(\lambda) \to x$, if, given a neighbourhood U of x, there exists $\lambda_0 \in \Lambda$ such that

$$x(\lambda) \in U \qquad (\lambda \geqslant \lambda_0) .$$

Definition 1. A *left approximate identity* for A is a net $\{e(\lambda)\}_{\lambda \in \Lambda}$ in A such that

$$e(\lambda) x \to x \qquad (x \in A). \tag{1}$$

A net $\{e(\lambda)\}_{\lambda\in\Lambda}$ in A is *bounded* if there exists a positive constant k such that

$$\|e(\lambda)\| \leq k \quad (\lambda\in\Lambda).$$

A *bounded left approximate identity* is a left approximate identity which is also a bounded net. *Right approximate identities* are similarly defined by replacing $e(\lambda)x$ in (1) by $xe(\lambda)$. A *two-sided approximate identity* is a net which is both a left and a right approximate identity.

The following result of Altman [17] shows that A has a bounded left approximate identity whenever it has a bounded left approximating set.

Proposition 2. *Let A contain a bounded set U such that, given $x\in A$, $\varepsilon>0$, there exists $u\in U$ with $\|x-ux\|<\varepsilon$. Then A has a bounded left approximate identity.*

Proof. Let $\varepsilon>0$. We show that for each finite subset F of A there exists $w\in W=\{v\circ u: v, u\in U\}$ such that

$$\|x-wx\|<\varepsilon \quad (x\in F).$$

The result is then clear by consideration of the product directed set $\mathbb{N}\times F(A)$, where $F(A)$ denotes the directed set of all finite subsets of A.

Let $\|u\|\leq M$ $(u\in U)$. Given $F=\{x_1,x_2\}$ choose $u,v\in U$ such that

$$\|x_1-ux_1\|<\varepsilon/(1+M), \quad \|(x_2-ux_2)-v(x_2-ux_2)\|<\varepsilon.$$

With $w=v\circ u$ we then have $\|x_j-wx_j\|<\varepsilon$ $(j=1,2)$. Suppose the result has been established for sets of n elements. Let $F=\{x_1,...,x_{n+1}\}$, and let $\alpha=\max\{\|x_j\|:j=1,...,n\}$. There exists $y\in W$ with

$$\|x_j-yx_j\|<\varepsilon/3(1+M)^2 \quad (j=1,...,n).$$

Now choose $w\in W$ such that

$$\|y-wy\|<\varepsilon/3\alpha, \quad \|x_{n+1}-wx_{n+1}\|<\varepsilon.$$

Then for $j=1,...,n$ we have

$$\|x_j-wx_j\| \leq \|x_j-yx_j\|+\|y-wy\|\|x_j\|+\|w\|\|x_j-yx_j\|<\varepsilon. \quad \square$$

Definition 3. A *weak left approximate identity* for A is a net $\{e(\lambda)\}_{\lambda\in\Lambda}$ in A such that

$$f(e(\lambda)x)\to f(x) \quad (x\in A, f\in A').$$

Each concept of approximate identity in Definition 1 has a 'weak' counterpart.

Proposition 4. *Let A have a bounded weak left approximate identity. Then A has a bounded left approximate identity.*

Proof. Let $\{e(\lambda)\}_{\lambda \in \Lambda}$ be a weak left approximate identity for A, and let U be the convex hull of $\{e(\lambda): \lambda \in \Lambda\}$. Since $e(\lambda)x \to x$ in the weak topology of A there exists $u \in U$ such that $\|x - ux\| < \varepsilon$. Apply Proposition 2. $\square$

Remark. The above techniques may be applied to show that A has a bounded left approximate identity if it has a bounded weak left approximating set.

Lemma 5. *Let* $\{e(\lambda)\}_{\lambda \in \Lambda}$ *be a left approximate identity for A and let* $\{f(\mu)\}_{\mu \in M}$ *be a bounded net in A. Then*

$$\{f(\mu) \circ e(\lambda)\}_{(\lambda, \mu) \in \Lambda \times M}$$

is a left approximate identity for A, and is bounded if $\{e(\lambda)\}_{\lambda \in \Lambda}$ *is bounded.*

Proof. Here $\Lambda \times M$ is the directed set obtained by writing $(\lambda, \mu) \leqslant (\lambda', \mu')$ if $\lambda \leqslant \lambda'$ and $\mu \leqslant \mu'$. We have

$$\|(f(\mu) \circ e(\lambda))x - x\| = \|e(\lambda)x - x + f(\mu)x - f(\mu)e(\lambda)x\|$$
$$\leqslant \|e(\lambda)x - x\| + \|f(\mu)\| \|e(\lambda)x - x\|. \quad \square$$

The following result is due to Dixon [121].

Proposition 6. *Let* $\{e(\lambda)\}_{\lambda \in \Lambda}$ *and* $\{f(\mu)\}_{\mu \in M}$ *be respectively bounded left and right approximate identities for A. Then*

$$\{f(\mu) \circ e(\lambda)\}_{(\lambda, \mu) \in \Lambda \times M}$$

is a bounded two-sided approximate identity for A.

Proof. By Lemma 5, $\{f(\mu) \circ e(\lambda)\}_{(\lambda, \mu) \in \Lambda \times M}$ is a bounded left approximate identity for A. Similarly, it is a bounded right approximate identity for A. $\square$

Example 7. The group algebra $L^1(G)$ of a locally compact group G has a bounded two-sided approximate identity. To see this, let Λ be the directed set of compact neighbourhoods of the identity element of G with $\lambda \geqslant \lambda'$ if $\lambda \subset \lambda'$. For each $\lambda \in \Lambda$, let $e(\lambda)$ be a non-negative function in $L^1(G)$ with $\|e(\lambda)\| = 1$ and with its support contained in λ. (See, for example, Rudin [324] Theorem 1.18.)

The main results of this section are concerned with the factorization of elements of a Banach algebra with a bounded left approximate identity. For their proof it will be convenient to work in terms of a Banach left A-module.

Notation. Let A be a Banach algebra over $\mathbb{F}$, and let X be a Banach left A-module. We denote by B the unitization $A + \mathbb{F}$ of A, regard A

as a subalgebra of B, and regard X as a Banach left B-module, as in Example 9.13 (viii).

Definition 8. A *bounded approximate identity* in A *for* X is a bounded net $\{e(\lambda)\}_{\lambda \in \Lambda}$ in A such that

$$e(\lambda)x \to x \qquad (x \in X).$$

We need a technical lemma.

Lemma 9. *Let* $e \in A$, $C \in \mathbb{R}$, $C > 1$, $\|e\| \leqslant C$, $\gamma = (4C)^{-1}$. *Then* $1 - \gamma + \gamma e \in \mathrm{Inv}(B)$, *and* $f = (1 - \gamma + \gamma e)^{-1}$ *satisfies*
  (i) $\|f\| \leqslant 2$,
  (ii) *given* $\varepsilon > 0$, *there exists* $\eta > 0$, *such that*

$$\|f x - x\| \leqslant \varepsilon \|x\|$$

whenever $x \in X$ *and* $\|e x - x\| \leqslant \eta \|x\|$.

Proof. We have $0 < \gamma < \frac{1}{4}$, $1 - \gamma > \frac{3}{4}$, $\gamma C/(1 - \gamma) < \frac{1}{3}$, and so $\|\gamma(1-\gamma)^{-1} e\| < \frac{1}{3}$. Therefore $1 + \gamma(1-\gamma)^{-1} e \in \mathrm{Inv}(B)$, and

$$\|f\| = \|(1 - \gamma + \gamma e)^{-1}\| \leqslant (1-\gamma)^{-1}\left[1 + \sum_{k=1}^{\infty} \|\gamma(1-\gamma)^{-1} e\|^k\right] \leqslant 2.$$

Let $\varepsilon > 0$, and choose a positive integer N such that

$$(1-\gamma)^{-1} \sum_{k=N+1}^{\infty} \gamma^k (1-\gamma)^{-k} C^k < \varepsilon/4. \tag{2}$$

Then choose $\eta > 0$ such that

$$\eta(1-\gamma)^{-1} \sum_{k=1}^{N} \gamma^k (1-\gamma)^{-k}(1 + C + \cdots + C^{k-1}) < \varepsilon/2. \tag{3}$$

Since

$$f = (1-\gamma)^{-1}\left\{1 + \sum_{k=1}^{\infty} \gamma^k (\gamma - 1)^{-k} e^k\right\},$$

and

$$1 = (1-\gamma)^{-1}\left\{1 + \sum_{k=1}^{\infty} \gamma^k (\gamma - 1)^{-k}\right\},$$

we have

$$f x - x = (1-\gamma)^{-1} \sum_{k=1}^{\infty} \gamma^k (\gamma - 1)^{-k}(e^k x - x) \qquad (x \in X).$$

With (2), this gives

$$\|f x - x\| \leqslant (1-\gamma)^{-1} \sum_{k=1}^{N} \gamma^k (1-\gamma)^{-k} \|e^k x - x\| + \frac{\varepsilon}{2} \|x\|. \tag{4}$$

Since

$$\|e^k x - x\| = \|(1 + e + \cdots + e^{k-1})(ex - x)\|$$
$$\leqslant (1 + C + \cdots + C^{k-1})\|ex - x\|,$$

(3) and (4) give

$$\|fx - x\| \leqslant \varepsilon \|x\|$$

whenever $\|ex - x\| \leqslant \eta \|x\|$. $\quad\square$

Theorem 10. *Let A have a bounded approximate identity for X, let $z \in X$ and $\delta > 0$. Then there exists $a \in A$, $y \in X$ such that $z = ay$ and $\|z - y\| \leqslant \delta$.*

Proof. Let $\{e(\lambda)\}_{\lambda \in \Lambda}$ be a bounded approximate identity in A for X, let C be a real constant with $C > 1$ such that

$$\|e(\lambda)\| \leqslant C \quad (\lambda \in \Lambda),$$

let $\gamma = (4C)^{-1}$, and let $f(\lambda) = (1 - \gamma + \gamma e(\lambda))^{-1}$, which exists in B by Lemma 9.

We construct a sequence $\{\lambda_n\}$ in Λ such that, with $e_n = e(\lambda_n)$, we have $b_n = (1 - \gamma)^n + \sum_{k=1}^n \gamma (1 - \gamma)^{k-1} e_k \in \mathrm{Inv}(B)$, and

$$\|t_n z - t_{n-1} z\| \leqslant \delta 2^{-n} \quad (n = 1, 2, \ldots), \tag{5}$$

where $t_0 = 1$ and $t_n = b_n^{-1}$ $(n = 1, 2, \ldots)$.

The existence of λ_1 satisfying (5) with $n = 1$ is immediate from Lemma 9 with $x = z$. Suppose that $\lambda_1, \ldots, \lambda_m$ have been chosen to satisfy (5) for $n = 1, \ldots, m$. Let

$$u(\lambda) = (1 - \gamma)^m + \sum_{k=1}^m \gamma (1 - \gamma)^{k-1} f(\lambda) e_k.$$

We have

$$u(\lambda) - b_m = \sum_{k=1}^m \gamma (1 - \gamma)^{k-1} (f(\lambda) e_k - e_k).$$

Therefore, by Lemma 9 with $X = A$, $\|u(\lambda) - b_m\|$ is arbitrarily small provided that

$$\|e(\lambda) e_k - e_k\| \leqslant \kappa \quad (k = 1, \ldots, m), \tag{6}$$

with κ sufficiently small. Since $\mathrm{Inv}(B)$ is open and the mapping $x \to x^{-1}$ is continuous on $\mathrm{Inv}(B)$, it follows that $u(\lambda) \in \mathrm{Inv}(B)$ and $\|(u(\lambda))^{-1} - t_m\|$ is arbitrarily small provided that (6) holds with κ sufficiently small. We choose λ_{m+1} so that $\|e(\lambda_{m+1})z - z\| \leqslant \kappa$ with κ so small that $u(\lambda_{m+1}) \in \mathrm{Inv}(B)$ and

$$2\|(u(\lambda_{m+1}))^{-1} - t_m\|\,\|z\| + \|t_m\|\,\|f(\lambda_{m+1})z - z\| \leqslant \frac{\delta}{2^{m+1}}. \tag{7}$$

Since $(1-\gamma+\gamma e(\lambda))f(\lambda)=1$, we have

$$(1-\gamma+\gamma e(\lambda))u(\lambda) = (1-\gamma)^{m+1} + \gamma(1-\gamma)^m e(\lambda) + \sum_{k=1}^{m} \gamma(1-\gamma)^{k-1} e_k,$$

and so

$$(1-\gamma+\gamma e_{m+1})u(\lambda_{m+1}) = b_{m+1}.$$

Therefore $b_{m+1} \in \mathrm{Inv}(B)$, and its inverse t_{m+1} is given by

$$t_{m+1} = (u(\lambda_{m+1}))^{-1} f(\lambda_{m+1}).$$

Therefore, by (7) and Lemma 9,

$$
\begin{aligned}
\|t_{m+1}z - t_m z\| &= \|(u(\lambda_{m+1}))^{-1} f(\lambda_{m+1})z - t_m z\| \\
&\leqslant \|(u(\lambda_{m+1}))^{-1} - t_m\|\,\|f(\lambda_{m+1})z\| + \|t_m\|\,\|f(\lambda_{m+1})z - z\| \\
&\leqslant 2\|(u(\lambda_{m+1}))^{-1} - t_m\|\,\|z\| + \|t_m\|\,\|f(\lambda_{m+1})z - z\| \\
&\leqslant \delta/2^{m+1}.
\end{aligned}
$$

Thus λ_{m+1} satisfies (5), and we have a recursive construction for $\{\lambda_n\}$.

Let $y_n = t_n z$. Then $z = b_n y_n$, and $\{y_n\}$ is a Cauchy sequence in X which therefore converges to $y \in X$ satisfying $\|z-y\| \leqslant \delta$. We have

$$\lim_{n\to\infty} b_n = \sum_{k=1}^{\infty} \gamma(1-\gamma)^{k-1} e_k = a \in A,$$

and $z = \lim\limits_{n\to\infty} b_n y_n = a y$. □

Corollary 11. *Let A be a Banach algebra with a bounded left approximate identity, let $z \in A$ and $\delta > 0$. Then there exist $x, y \in A$ such that $z = xy$, $\|z-y\| \leqslant \delta$, and y belongs to the least closed left ideal of A containing z.*

Proof. Let X denote the least closed left ideal of A containing z. Then X is a Banach left A-module, and A has a bounded approximate identity for X. □

Corollary 12. *Let A be a Banach algebra with a bounded left approximate identity and let $z_n \in A$ with $\lim\limits_{n\to\infty} z_n = 0$. Then there exist $a, y_n \in A$ with $z_n = a y_n$ $(n=1,2,\ldots)$ and $\lim\limits_{n\to\infty} y_n = 0$.*

Proof. Let X denote the space of all sequences $\{a_n\}$ with $a_n \in A$ and $\lim\limits_{n\to\infty} a_n = 0$, with pointwise addition and scalar multiplication and with the norm

$$\|\{a_n\}\| = \sup\{\|a_n\| : n = 1,2,\ldots\}.$$

Then X is a Banach left A-module with $a\{a_n\} = \{a a_n\}$. Let $\{e(\lambda)\}_{\lambda \in \Lambda}$ be the given bounded left approximate identity in A. It is straightforward, using the boundedness of $\{e(\lambda)\}_{\lambda \in \Lambda}$ and the fact that $\lim_{n \to \infty} a_n = 0$, to verify that $e(\lambda)\{a_n\} - \{a_n\} \to 0$ for all $\{a_n\} \in X$. Therefore Theorem 10 shows that there exist $a \in A$ and $\{y_n\} \in X$ with

$$\{z_n\} = a\{y_n\},$$

i. e. $z_n = a y_n \ (n=1,2,\ldots)$. □

Theorem 10 is given in Hewitt and Ross [195, Theorem 32.22] and Pták [466]. Corollary 11 is due to Cohen [103] the proof being essentially that given for Theorem 10 here. Corollary 12 is due to Varopoulos [394].

§ 12. Involutions

Throughout this section X denotes a complex linear space and A a complex algebra.

Definition 1. A *linear involution* on X is a mapping $x \to x^*$ of X into X that satisfies the following axioms (for all $x, y \in X$, $\alpha \in \mathbb{C}$):
  (i) $(x+y)^* = x^* + y^*$,
  (ii) $(\alpha x)^* = \alpha^* x^*$,
  (iii) $(x^*)^* = x$.
In axiom (ii) (and throughout) α^* denotes the complex conjugate of α. Axiom (iii) implies that a linear involution is bijective.

Definition 2. An element $x \in X$ is *self-adjoint* with respect to the involution $*$ if $x^* = x$. The set of all self-adjoint elements of X is denoted by $\mathrm{Sym}(X)$.

Lemma 3. *Let* $*$ *be a linear involution on* X. *Then* $\mathrm{Sym}(X)$ *is a real linear subspace of* X, *and* $X = \mathrm{Sym}(X) \oplus i \, \mathrm{Sym}(X)$.

Proof. The first part is clear. Let $x \in \mathrm{Sym}(X) \cap i \, \mathrm{Sym}(X)$. Then $x = iy$, where $y^* = y$, and we have $x = x^* = -iy^* = -iy = -x$, $x = 0$. Given $x \in X$, we have $\dfrac{1}{2}(x + x^*)$, $\dfrac{1}{2i}(x - x^*) \in \mathrm{Sym}(X)$ and

$$x = \frac{1}{2}(x + x^*) + i \frac{1}{2i}(x - x^*).$$ □

Lemma 4. *Let* V *be a real linear subspace of* X *such that* $X = V \oplus iV$. *Then the mapping* $u + iv \to u - iv$ $(u, v \in V)$ *is a linear involution on* X *with* $\mathrm{Sym}(X) = V$.

Proof. Clear. □

Definition 5. An *algebra involution,* or more simply an *involution* on A is a linear involution on A that also satisfies

(iv) $(xy)^* = y^* x^*$ $(x, y \in A)$.

An algebra A with an involution $*$ is called a *star algebra.* It follows from axiom (iv) that a star algebra with a unit element satisfies $1^* = 1$, $1 \in \text{Sym}(A)$.

Definition 6. Given $x, y \in A$ the *real* and *imaginary Jordan products* of x and y are defined respectively by

$$\frac{1}{2}(xy+yx), \quad \frac{1}{2i}(xy-yx).$$

If A is a star algebra, it is clear that $\text{Sym}(A)$ is closed under the real and imaginary Jordan products.

Lemma 7. *Let V be a real linear subspace of A such that $A = V \oplus iV$ and V is closed under the real and imaginary Jordan products. Then the mapping $u+iv \to u-iv$ $(u,v \in V)$ is an involution on A with $\text{Sym}(A) = V$.*

Proof. By Lemma 4 it is sufficient to show that $(ab)^* = b^* a^*$ $(a, b \in A)$. Let $a = p+iq$, $b = r+is$ with $p, q, r, s \in V$. Then

$$ab+b^* a^* = (p+iq)(r+is)+(r-is)(p-iq)$$
$$= (pr+rp)-(qs+sq)+i(qr-rq)+i(ps-sp)$$
$$\in V$$

and similarly $\frac{1}{i}(ab-b^* a^*) \in V$. Since

$$ab = \frac{1}{2}(ab+b^* a^*) + i\frac{1}{2i}(ab-b^* a^*)$$

it follows that

$$(ab)^* = \frac{1}{2}(ab+b^* a^*) - i\frac{1}{2i}(ab-b^* a^*) = b^* a^*. \quad \square$$

Remark. It is easy to show that V is closed under the real and imaginary Jordan products if and only if $(u+iv)(u-iv) \in V$ whenever $u, v \in V$.

An involution on A is obviously a real algebra isomorphism of A onto the reversed algebra $\text{rev}(A)$. It therefore serves to interchange 'left' and 'right' properties of A.

Definition 8. Let $(A, *)$, $(B, \times)$ be star algebras. A *star homomorphism* ϕ is a homomorphism ϕ of A into B such that

$$\phi(a^*) = (\phi(a))^\times \qquad (a \in A).$$

The kernel of a star homomorphism is clearly a *star bi-ideal*, i. e. a bi-ideal K such that $a \in K$ implies $a^* \in K$.

Definition 9. A *normed star algebra* is a complex normed algebra with an involution. A *Banach star algebra* is a complete normed star algebra.

Lemma 10. *Let A be a Banach star algebra.*
(i) q-Inv$(A)^* =$ q-Inv(A); *and, if A has a unit,* Inv$(A)^* =$ Inv(A).
(ii) Sp$(a^*) =$ Sp$(a)^*$ $(a \in A)$.
(iii) $r(a^*) = r(a)$ $(a \in A)$.

Proof. Clear. ☐

The result below is known as Ford's square root lemma (see [137]).

Proposition 11. *Let A be a Banach star algebra, let $a \in$ Sym(A), and let $r(a) < 1$. Then there exists a unique $x \in$ Sym(A) with $2x - x^2 = a$ and $r(x) < 1$.*

Proof. By Proposition 8.13, there exists a unique $x \in A$ with $2x - x^2 = a$ and $r(x) < 1$. But $a = a^* = (2x - x^2)^* = 2x^* - x^{*2}$, and $r(x^*) = r(x) < 1$. Therefore by the uniqueness of x, $x = x^*$. ☐

If A is a Banach star algebra with unit and continuous involution, it is clear that

$$\exp(a^*) = (\exp(a))^* \qquad (a \in A). \tag{1}$$

We do not know if (1) holds without some continuity assumption on the involution, but we have the following partial result.

Proposition 12. *Let A be a Banach star algebra, let $a \in A$ and let $\exp(a) = 1$. Then $\exp(a^*) = 1$.*

Proof. It follows from Proposition 8.11 that $a = \sum_{j=1}^n \lambda_j e_j$ where $\exp(\lambda_j) = 1$, $e_j^2 = e_j$, $e_j e_k = 0$ $(j \neq k)$. Then

$$\exp(a^*) = \prod_{j=1}^n \exp(\lambda_j^* e_j^*)$$

$$= \prod_{j=1}^n \{1 + (\exp(\lambda_j^*) - 1)e_j^*\}$$

$$= 1. \quad ☐$$

Definition 13. Let A be a star algebra with unit. An element $u \in A$ is *unitary* if $uu^* = u^*u = 1$.

Proposition 14. *Let A be a Banach star algebra with unit. Then A is the linear span of its unitary elements.*

Proof. Let $h \in \operatorname{Sym}(A)$ with $r(h) < 1$. By Proposition 11 there exists $k \in \operatorname{Sym}(A)$ such that $2k - k^2 = h^2$ and so $(1-k)^2 = 1 - h^2$. Moreover, by Proposition 8.13, $h(1-k) = (1-k)h$. Let $u = h + i(1-k)$. Then

$$uu^* = u^*u = h^2 + (1-k)^2 = 1$$

and so u, u^* are unitary. But $h = \frac{1}{2}(u + u^*)$. □

Definition 15. A normed star algebra A is said to be *star-normed* if the involution is an isometry, i. e. $\|x^*\| = \|x\|$ ($x \in A$). A normed star algebra A with continuous involution is clearly star-normed with the equivalent norm $|\cdot|$ defined by

$$|x| = \max\{\|x\|, \|x^*\|\} \quad (x \in A).$$

Definition 16. A *B*-algebra* is a Banach star algebra such that

$$\|x^*x\| = \|x\|^2 \quad (x \in A).$$

A B^*-algebra is clearly star-normed, for we have $\|x\|^2 \leqslant \|x^*\|\,\|x\|$ ($x \in A$).

Example 17. Let E be a compact Hausdorff space and let $A = C(E)$. Then A is a commutative B^*-algebra with involution defined by

$$a^*(t) = (a(t))^* \quad (a \in A, t \in E).$$

Let H be a complex Hilbert space and let $T \in BL(H)$. Let T^* denote the *Hilbert space adjoint* of T, i. e. the unique element T^* of $BL(H)$ such that

$$(Tx, y) = (x, T^*y) \quad (x, y \in H).$$

Clearly $T \to T^*$ is an involution on $BL(H)$. Since

$$\|Tx\|^2 = (Tx, Tx) = (T^*Tx, x) \leqslant \|T^*T\|\,\|x\|^2 \quad (x \in H)$$

we have $\|T\|^2 \leqslant \|T^*T\|$, and hence $\|T^*T\| = \|T\|^2$ ($T \in BL(H)$).

Definition 18. A *C*-algebra* is a closed star subalgebra of $BL(H)$ for some Hilbert space H. Every C^*-algebra is a B^*-algebra. We shall prove later that every B^*-algebra admits an isometric star isomorphism onto a C^*-algebra.

If a B^*-algebra has a unit element then $\|1\|^2 = \|1^*1\| = \|1\|$, and so $\|1\| = 1$, i. e. the algebra is then unital. It is a very useful technical fact, due to Yood, that any B^*-algebra can be embedded in a unital B^*-algebra.

Lemma 19. *Let A be a B^*-algebra without unit. Then there exists a unital B^*-algebra B and an isometric star monomorphism of A into B.*

Proof. As usual let $a \to \lambda_a$ be the left regular representation of A on A. Let B be the subalgebra of $BL(A)$ given by

$$B = \{\lambda_a + \alpha I : a \in A, \alpha \in \mathbb{C}\}$$

where I is the identity operator on A. An involution is defined on B by

$$(\lambda_a + \alpha I)^* = \lambda_{a^*} + \alpha^* I.$$

and then $a \to \lambda_a$ is a star monomorphism of A into B. Since

$$\|a\|^2 = \|a a^*\| = \|\lambda_a a^*\| \leqslant \|\lambda_a\| \|a^*\| = \|\lambda_a\| \|a\|$$

we have $\|a\| = \|\lambda_a\|$ ($a \in A$). It remains to show that B is a B^*-algebra. Let $T = \lambda_a + \alpha I$ and let $x \in A$. Then

$$\|Tx\|^2 = \|(ax + \alpha x)^* (ax + \alpha x)\| = \|x^*(T^* Tx)\| \leqslant \|T^* T\| \|x\|^2$$

and so $\|T\|^2 \leqslant \|T^* T\|$. It follows that $\|T^* T\| = \|T\|^2$, as required. ☐

The final result in this section is a key tool in the geometric study of B^*-algebras.

Proposition 20. *An element h of a unital B^*-algebra A is self-adjoint if and only if it is Hermitian.*

Proof. Let $h \in A$, $h^* = h$ and let $\alpha \in \mathbb{R}$. We have

$$\|1 + \alpha^2 h^2\| = \|(1 + i\alpha h)(1 - i\alpha h)\| = \|1 + i\alpha h\|^2$$

and therefore

$$\lim_{\alpha \to 0} \frac{1}{\alpha} \{\|1 + i\alpha h\| - 1\} = \lim_{\alpha \to 0} \frac{1}{\alpha} \{\|1 + \alpha^2 h^2\|^{\frac{1}{2}} - 1\} = 0.$$

It follows from Theorem 10.10 that h is Hermitian.

Conversely let h be Hermitian and let $h = p + iq$ with $p^* = p$, $q^* = q$. Then

$$V(h) = \{f(p) + i f(q) : f \in D(1)\}.$$

Since h, p, q are all Hermitian we have $f(q) = 0$ for all $f \in D(1)$. Therefore $q = 0$ by Corollary 10.15, and h is self-adjoint. ☐

§ 13. The Complexification of a Real Algebra

Many of our concepts and results, particularly in §§ 5, 6, 7, 8, 10 have been established only for complex algebras. It is therefore valuable to have a procedure for embedding real algebras in complex ones.

Definition 1. Given an algebra A over $\mathbb{R}$, the *complexification* $A_{\mathbb{C}}$ of A is the set $A \times A$ with addition, scalar multiplication and product defined (for all $a, b, c, d \in A$, $\alpha, \beta \in \mathbb{R}$) by

$$(a, b) + (c, d) = (a + c, b + d),$$
$$(\alpha + i\beta)(a, b) = (\alpha a - \beta b, \alpha b + \beta a),$$
$$(a, b)(c, d) = (ac - bd, ad + bc).$$

It is a routine matter to verify that $A_{\mathbb{C}}$ is an algebra over $\mathbb{C}$, and that the mapping $a \rightarrow (a, 0)$ is a monomorphism of A into $A_{\mathbb{C}}$ ($A_{\mathbb{C}}$ being regarded, at this point, as an algebra over $\mathbb{R}$).

Definition 2. Let E be a subset of a linear space X over $\mathbb{F}$. The *absolutely convex hull* of E, denoted by $|\text{co}|(E)$, is the intersection of all absolutely convex subsets of X containing E.

The set of all linear combinations $\sum_{k=1}^{n} \alpha_k x_k$ with $x_k \in E$, $\alpha_k \in \mathbb{F}$ and $\sum_{k=1}^{n} |\alpha_k| \leqslant 1$ is absolutely convex and is contained in every absolutely convex set containing E. It is therefore $|\text{co}|(E)$.

Proposition 3. *Let $(A, \|\cdot\|)$ be a normed algebra over $\mathbb{R}$, let $A_{\mathbb{C}}$ be the complexification of A, let $U = \{a \in A : \|a\| < 1\}$, let $V = |\text{co}|(U \times \{0\})$, and let p be the Minkowski functional of V. Then*
   (i) *p is an algebra-norm on $A_{\mathbb{C}}$;*
   (ii) *$V = \{x \in A_{\mathbb{C}} : p(x) < 1\}$;*
   (iii) *$\max(\|a\|, \|b\|) \leqslant p((a, b)) \leqslant 2\max(\|a\|, \|b\|)$ $(a, b \in A)$;*
   (iv) *$p((a, 0)) = \|a\|$ $(a \in A)$;*
   (v) *$(A_{\mathbb{C}}, p)$ is complete whenever $(A, \|\cdot\|)$ is complete.*

Proof. The absolutely convex set V is absorbent, for, given $(a, b) \in A_{\mathbb{C}}$ and $\mu > \max(\|a\|, \|b\|)$, we have

$$\frac{1}{2\mu}(a, b) = \frac{1}{2}\left(\frac{1}{\mu}a, 0\right) + \frac{1}{2}i\left(\frac{1}{\mu}b, 0\right) \in V.$$

Thus the Minkowski functional p of V is defined on $A_{\mathbb{C}}$, and

$$p((a, b)) \leqslant 2\max(\|a\|, \|b\|).$$

We prove next that

$$(a, b) \in V \Rightarrow a, b \in U. \tag{1}$$

Given $(a, b) \in V$, we have $(a, b) = \sum_{k=1}^{n} (\beta_k + i\gamma_k)(u_k, 0)$ with $\beta_k, \gamma_k \in \mathbb{R}$, $\sum_{k=1}^{n} |\beta_k + i\gamma_k| \leqslant 1$, $u_k \in U$. Then $a = \sum_{k=1}^{n} \beta_k u_k$, $b = \sum_{k=1}^{n} \gamma_k u_k$, $\sum_{k=1}^{n} |\beta_k| \leqslant 1$, $\sum_{k=1}^{n} |\gamma_k| \leqslant 1$, and so $a, b \in U$. It follows that V is radially bounded, for if $\lambda \in \mathbb{R}$ and $\lambda(a, b) \in V$, (1) gives $\lambda a, \lambda b \in U$, and so $|\lambda| \max(\|a\|, \|b\|) < 1$. This inequality also gives

$$\max(\|a\|, \|b\|) \leqslant p((a, b)).$$

Straightforward verification shows that V is a subsemi-group of $A_{\mathbb{C}}$, and therefore, by Proposition 1.9, p is an algebra-semi-norm on $A_{\mathbb{C}}$ and $\{x : p(x) < 1\} \subset V$. The opposite inclusion also holds; for, given $x \in V$, we have $x = \sum_{k=1}^{n} \alpha_k (u_k, 0)$ with $\alpha_k \in \mathbb{C}$, $\sum_{k=1}^{n} |\alpha_k| \leqslant 1$, $u_k \in U$. Choose λ with

$$\max\{\|u_k\| : 1 \leqslant k \leqslant n\} < \lambda < 1.$$

Then $\dfrac{1}{\lambda} u_k \in U$, and so $\dfrac{1}{\lambda} x \in V$, $p(x) \leqslant \lambda < 1$.

We have now proved (i), (ii), (iii). To prove (iv), we note that, by (1), $(a, 0) \in V$ if and only if $a \in U$. Finally, (v) follows from (iii). ☐

Definition 4. The *complexification* of a real normed algebra $(A, \|\cdot\|)$ is the complex algebra $A_{\mathbb{C}}$ with the norm p defined in Proposition 3.

It is clear that the mapping $a \to (a, 0)$ is an isometric monomorphism of $(A, \|\cdot\|)$ into the complexification $(A_{\mathbb{C}}, p)$. It is often convenient to identify A with its image $A \times \{0\}$ in the complexification.

Notation. For the rest of this section A will denote a real normed algebra, and $A_{\mathbb{C}}$ its complexification. As in § 12, λ^* will denote the complex conjugate of $\lambda \in \mathbb{C}$.

Proposition 5. (i) *A has a unit element if and only if $A_{\mathbb{C}}$ has a unit element. If 1 is the unit element of A, then $(1, 0)$ is the unit element of $A_{\mathbb{C}}$.*
   (ii) $a \in \text{Inv}(A)$ *if and only if* $(a, 0) \in \text{Inv}(A_{\mathbb{C}})$; $(a, 0)^{-1} = (a^{-1}, 0)$.
   (iii) $a \in \text{q-Inv}(A)$ *if and only if* $(a, 0) \in \text{q-Inv}(A_{\mathbb{C}})$; $(a, 0)^\circ = (a^\circ, 0)$.
   (iv) $(x, y) \in \text{q-Inv}(A_{\mathbb{C}})$ *if and only if* $(x, -y) \in \text{q-Inv}(A_{\mathbb{C}})$.

Proof. (i) If 1 is a unit element in A, it is clear that $(1, 0)$ is a unit element in $A_{\mathbb{C}}$. On the other hand, if (u, v) is a unit element in $A_{\mathbb{C}}$, then $(au, av) = (a, 0)(u, v) = (a, 0) = (u, v)(a, 0) = (ua, va)$, so that $au = ua = a$ $(a \in A)$.

(ii) If $a \in \text{Inv}(A)$, then by the monomorphism $x \to (x, 0)$ and (i), $(a, 0) \in \text{Inv}(A_{\mathbb{C}})$ and $(a, 0)^{-1} = (a^{-1}, 0)$. On the other hand, if $(a, 0)^{-1} = (x, y)$, then

$$(ax, ay) = (a, 0)(x, y) = (1, 0) = (x, y)(a, 0) = (xa, ya),$$

$ax = 1 = xa$.

(iii) Similar to (ii).
(iv) If $(x, y)\circ(a, b)=(w, z)$, then $(x, -y)\circ(a, -b)=(w, -z)$. □

Definition 6. Let $a \in A$. The *spectrum* of a is the set $\mathrm{Sp}(A, a)$ of complex numbers given by

$$\mathrm{Sp}(A, a)=\mathrm{Sp}(A_\mathbb{C}, (a, 0)).$$

We denote $\mathrm{Sp}(A, a)$ by $\mathrm{Sp}(a)$ when no confusion is possible.

Remark. An algebra B over $\mathbb{C}$ can be regarded as an algebra, B^r say, over $\mathbb{R}$. It should be noticed that $\mathrm{Sp}(B, a)$ and $\mathrm{Sp}(B^r, a)$ do not in general coincide, $\mathrm{Sp}(B^r, a)$ being defined in terms of the complexification of B^r.

Theorem 7. *Let* $a \in A$. *Then*
(i) $(\mathrm{Sp}(a))^* = \mathrm{Sp}(a)$;
(ii) $\mathrm{Sp}(a)$ *contains a complex number* λ *with* $|\lambda| \geqslant r(a)$;
(iii) *if* A *is complete*, $\mathrm{Sp}(a)$ *is compact and* $\max\{|\lambda|: \lambda \in \mathrm{Sp}(a)\} = r(a)$.

Proof. (i) Let $\lambda \in \mathrm{Sp}(a)\setminus\{0\}$, $\lambda = \alpha + i\beta$, $\alpha, \beta \in \mathbb{R}$. Then $\frac{1}{\lambda}(a, 0) \in$ q-Sing$(A_\mathbb{C})$,

i. e. $(\alpha|\lambda|^{-2} a, -\beta|\lambda|^{-2} a) \in$ q-Sing$(A_\mathbb{C})$.   By   Proposition   5   (iv),

$(\alpha|\lambda|^{-2} a, \beta|\lambda|^{-2} a) \in$ q-Sing$(A_\mathbb{C})$; i. e. $\frac{1}{\lambda^*}(a, 0) \in$ q-Sing$(A_\mathbb{C})$, $\lambda^* \in \mathrm{Sp}(a)$.

(ii) and (iii). Proposition 3 and Theorems 5.7, 5.8 applied to $A_\mathbb{C}$. □

The following theorem describes $\mathrm{Sp}(a)$ directly in terms of A.

Theorem 8. *Let* $a \in A$. *Then*
(i) $0 \notin \mathrm{Sp}(a)$ *if and only if* A *has a unit element and* $a \in \mathrm{Inv}(A)$;
(ii) *Let* $\lambda \in \mathbb{C}\setminus\{0\}$. *Then* $\lambda \in \mathrm{Sp}(a)$ *if and only if*

$$|\lambda|^{-2}((\lambda+\lambda^*)a - a^2) \in \text{q-Sing}(A);$$

(iii) *Let* $\lambda \in \mathbb{R}\setminus\{0\}$. *Then* $\lambda \in \mathrm{Sp}(a)$ *if and only if* $\lambda^{-1} a \in$ q-Sing(A).
(iv) *If* A *has a unit element, and* $\xi, \eta \in \mathbb{R}$, *then* $\xi+i\eta \in \mathrm{Sp}(a)$ *if and only if*

$$(\xi-a)^2 + \eta^2 \in \mathrm{Sing}(A).$$

Proof. (i) Proposition 5.
(ii) Let $b = |\lambda|^{-2}((\lambda+\lambda^*)a - a^2)$. We have

$$\left(\frac{1}{\lambda^*}(a, 0)\right)\circ\left(\frac{1}{\lambda}(a, 0)\right) = \left(\frac{1}{\lambda}(a, 0)\right)\circ\left(\frac{1}{\lambda^*}(a, 0)\right)$$

$$= \frac{1}{\lambda}(a,0) + \frac{1}{\lambda^*}(a,0) - \frac{1}{|\lambda|^2}(a,0)^2 = (b,0). \tag{2}$$

If $\lambda \in \mathrm{Sp}(a)$, we have $\dfrac{1}{\lambda}(a, 0) \in \mathrm{q\text{-}Sing}(A_\mathbb{C})$, and so $(b, 0) \in \mathrm{q\text{-}Sing}(A_\mathbb{C})$, $b \in \mathrm{q\text{-}Sing}(A)$. If on the other hand $\lambda \notin \mathrm{Sp}(a)$, then, by Theorem 7, $\lambda^* \notin \mathrm{Sp}(a)$, and (2) gives $(b, 0) \in \mathrm{q\text{-}Inv}(A_\mathbb{C})$, $b \in \mathrm{q\text{-}Inv}(A)$.

(iii) Let $\lambda \in \mathbb{R} \setminus \{0\}$. Then $\lambda^{-1}(a, 0) = (\lambda^{-1}a, 0)$ and Proposition 5 completes the proof.

(iv) Follows from (i) and (ii). $\quad\square$

§ 14. Normed Division Algebras

Definition 1. *A normed division algebra is a normed algebra A with unit such that*

$$\mathrm{Inv}(A) = A \setminus \{0\} .$$

Theorem 2. *Let D be a complex normed division algebra. Then $D = \mathbb{C}1$.*

Proof. Let $a \in D$. By Theorem 5.7, there exists $\lambda \in \mathrm{Sp}(a)$. Therefore

$$\lambda - a \in \mathrm{Sing}(D) = \{0\} . \quad\square$$

For real normed division algebras the situation is necessarily more complicated. It is clear that $\mathbb{C}$ and $\mathbb{R}$ are both real normed division algebras. In fact there is one other such algebra, the real quaternion algebra $\mathbb{H}$ which we now define.

Definition 3. Let $1, i, j, k$ denote the usual basis vectors in $\mathbb{R}^4$, i.e. $1 = (1, 0, 0, 0)$, $i = (0, 1, 0, 0)$, $j = (0, 0, 1, 0)$, $k = (0, 0, 0, 1)$. The *real quaternion algebra* $\mathbb{H}$ is the linear space $\mathbb{R}^4$ with the product defined by taking

$$1^2 = 1, \quad 1i = i1 = i, \quad 1j = j1 = j, \quad 1k = k1 = k, \quad i^2 = j^2 = k^2 = -1,$$
$$ij = -ji = k, \quad jk = -kj = i, \quad ki = -ik = j,$$

extended by linearity to linear combinations of the basis elements, and with the Euclidean norm

$$|x| = \left(\sum_{k=1}^{4} \xi_k^2 \right)^{\frac{1}{2}} \quad (x = (\xi_1, \xi_2, \xi_3, \xi_4)) .$$

Routine, though tedious, calculation shows that $\mathbb{H}$ is an algebra over $\mathbb{R}$ and that

$$|xy| = |x||y| \quad (x, y \in \mathbb{H}),$$

so that $\mathbb{H}$ is a normed algebra. It has the unit element 1.

Given $x = (\xi_1, \xi_2, \xi_3, \xi_4) \in \mathbb{H}$, let

$$x^* = (\xi_1, -\xi_2, -\xi_3, -\xi_4) .$$

Then

$$x x^* = x^* x = |x|^2 ,$$

and so $\mathbb{H}$ is a division algebra.

We aim to prove that every normed division algebra over $\mathbb{R}$ is isomorphic to $\mathbb{R}, \mathbb{C}$, or $\mathbb{H}$.

Notation. Let D be a normed division algebra over $\mathbb{R}$, and let

$$Y = \{x \in D: -x^2 \in \mathbb{R}^+\}, \qquad Y_1 = \{x \in D: -x^2 = 1\} .$$

Lemma 4. *Y is a linear subspace of D, $Y = \mathbb{R}\, Y_1$, and*

$$D = \mathbb{R} \oplus Y .$$

Proof. Let $x \in D$. By Theorem 13.7, there exist $\xi, y \in \mathbb{R}$ with $\xi + i\eta \in \mathrm{Sp}(x)$. Then, by Theorem 13.8 (iv)

$$(\xi - x)^2 + \eta^2 \in \mathrm{Sing}(D) .$$

Since D is a division algebra, it follows that

$$(\xi - x)^2 + \eta^2 = 0 ,$$

and so $x - \xi \in Y$.

We have proved that every $x \in D$ can be written in the form

$$x = \xi + y \tag{1}$$

with $\xi \in \mathbb{R}$ and $y \in Y$. Suppose that also $x = \xi' + y'$ with $\xi' \in \mathbb{R}$ and $y' \in Y$. Then $y' = (\xi - \xi') + y$, and so

$$y'^2 = (\xi - \xi')^2 + 2(\xi - \xi') y + y^2 .$$

This gives $(\xi - \xi') y \in \mathbb{R}$. Therefore $\xi - \xi' = 0$ or $y \in \mathbb{R}$. If $y \in \mathbb{R}$, then $y^2 \in \mathbb{R}^+ \cap (-\mathbb{R}^+)$, $y = 0$. Thus either $\xi = \xi'$ or $y = 0$. Similarly $\xi = \xi'$ or $y' = 0$, and it follows that the expression (1) for x is unique.

It only remains to be proved that Y is a linear subspace, and it is clear that $\alpha y \in Y$ ($\alpha \in \mathbb{R}$, $y \in Y$). Let $x, y \in Y$. Then $x + y = \alpha + w$, $x - y = \beta + z$ with $\alpha, \beta \in \mathbb{R}$ and $w, z \in Y$, and so

$$2 x^2 + 2 y^2 = (x+y)^2 + (x-y)^2 = \alpha^2 + \beta^2 + 2\alpha w + 2\beta z + w^2 + z^2 .$$

Therefore $\alpha w + \beta z = \lambda \in \mathbb{R}$, i.e.

$$\lambda - \alpha w = 0 + \beta z .$$

By the uniqueness of the expression (1), this gives

$$\alpha w + \beta z = 0 .$$

Therefore

$$(a+\beta) x + (\alpha - \beta) y = \alpha(x+y) + \beta(x-y) = \alpha(\alpha+w) + \beta(\beta+z) = \alpha^2 + \beta^2 .$$

Again using the uniqueness of the expression (1), we have $\alpha^2 + \beta^2 = 0$,

$$x + y = w \in Y. \quad \square$$

Lemma 5. *Let* $u \in Y_1$, *and let* $W(u) = \{x \in Y : ux \in Y\}$. *Then*

$$Y = \mathbb{R}u \oplus W(u).$$

Proof. Given $x \in Y$, by Lemma 4, $ux = \lambda + y$ with $\lambda \in \mathbb{R}$, $y \in Y$. So

$$u(x + \lambda u) = ux - \lambda = y \in Y.$$

Since Y is a linear subspace, $x + \lambda u \in Y$, and so $x + \lambda u \in W(u)$, $x \in \mathbb{R}u + W(u)$. If $\alpha u \in W(u)$ for some $\alpha \in \mathbb{R}$, we have $-\alpha = \alpha u^2 \in Y$, $\alpha = 0$. Thus the sum is direct. $\quad \square$

Lemma 6. *Let* $u, v \in Y_1$ *and* $v \in W(u)$. *Then* $uv \in Y_1$ *and* $vu = -uv$.

Proof. Since $v \in W(u)$, we have $uv \in Y$, $uvuv = -\lambda$ with $\lambda \in \mathbb{R}^+$. Therefore

$$vu = u^2(vu)v^2 = u(uvuv)v = -\lambda uv.$$

Also $2(uv + vu) = (u + v)^2 - (u - v)^2 \in \mathbb{R}$, and so

$$(1 - \lambda)uv \in \mathbb{R}.$$

Since $Y \cap \mathbb{R} = \{0\}$ and $uv \neq 0$, $\lambda = 1$. $\quad \square$

Theorem 7. *Let* D *be a real normed division algebra. Then there exists an isomorphism* T *of* D *onto* $\mathbb{R}$, $\mathbb{C}$, *or* $\mathbb{H}$, *and*

$$|Tx| = r(x) \quad (x \in D).$$

Proof. By Lemma 4, if $Y = \{0\}$, we have $D = \mathbb{R}$. If $\dim Y \geq 1$, we take $i \in Y_1$. If $\dim Y = 1$, we have $D = \mathbb{R} \oplus \mathbb{R}i$, and $i^2 = -1$, so that D is isomorphic to $\mathbb{C}$. Suppose then that $\dim Y \geq 2$. By Lemma 5, $W(i) \neq 0$, and we may choose $j \in W(i) \cap Y_1$. Then, by Lemma 6, $ij = k \in Y_1$, and $ji = -k$. Also

$$jk = j(ij) = -j(ji) = -j^2 i = i, \quad kj = (ij)j = ij^2 = -i.$$

Similarly $ki = -ik = j$. We prove next that if $L = W(i) \cap W(j) \cap W(k)$, then

$$Y = \mathbb{R}i \oplus \mathbb{R}j \oplus \mathbb{R}k \oplus L. \tag{2}$$

In fact, by Lemma 5,

$$Y = \mathbb{R}i \oplus W(i) = \mathbb{R}j \oplus W(j) = \mathbb{R}k \oplus W(k);$$

so that, given $y \in Y$, we have in turn $y = \alpha i + a$ with $a, ia \in Y$, $a = \beta j + b$ with $b, jb \in Y$, $b = \gamma k + c$ with $c, kc \in Y$. Thus

$$y = \alpha i + \beta j + \gamma k + c,$$

and $ic=i(a-\beta j-\gamma k)=ia-\beta k+\gamma j\in Y$, $jc=j(b-\gamma k)=jb-\gamma i\in Y$. Thus $c\in L$, and $Y=\mathbb{R}\,i+\mathbb{R}\,j+\mathbb{R}\,k+L$. To prove that the sum is direct, suppose that $\lambda,\mu,\nu\in\mathbb{R}$, $x\in L$ and

$$\lambda i+\mu j+\nu k+x=0\,.$$

Premultiplication by i gives

$$-\lambda+\mu k-\nu j+ix=0\,,$$

from which $\lambda\in\mathbb{R}\cap Y$, $\lambda=0$. Similarly $\mu=\nu=0$, and (2) is proved.

We prove next that $L=\{0\}$. Otherwise, we choose $x\in L\cap Y_1$. Then by Lemma 6,

$$ix,jx,kx\in Y_1,\quad\text{and}\quad ix=-xi,\quad jx=-xj,\quad kx=-xk\,.$$

Thus

$$xk=x(ij)=(x\,i)j=-(ix)j=-i(xj)=i(jx)=(ij)x=kx\,.$$

Therefore $kx=0$, contradicting $kx\in Y_1$. Therefore $L=\{0\}$, and D is isomorphic to $\mathbb{H}$ when $\dim Y\geqslant 2$.

Let T denote the isomorphism of D onto $\mathbb{R}$, $\mathbb{C}$, $\mathbb{H}$ as the case may be, and let $\|x\|_1=|Tx|$ $(x\in D)$. Then $\|\cdot\|_1$ is an algebra-norm on D which is equivalent to the given algebra-norm since D has finite dimension. Also $\|xy\|_1=\|x\|_1\|y\|_1$, and so

$$\|x\|_1=\lim_{n\to\infty}\|x^n\|_1^{\frac{1}{n}}=r(x)\,.\quad\square$$

Remarks. Theorem 2 is known as the Gelfand-Mazur theorem. Mazur [279] announced the result and Gelfand [150] gave a proof. Subsequent elementary proofs which avoided function theory correspond essentially to Rickart's proof of Theorem 5.7 A clear account of finite-dimensional real division algebras is given in Dickson [115].

Chapter II. Commutativity

§ 15. Commutative Subsets

Definition 1. Let E be a subset of an algebra A. The *commutant* of E (in A) is the subset E^c of A given by

$$E^c = \{a \in A : xa = ax \ (x \in E)\}.$$

The *second commutant* of E is the set $(E^c)^c$, which we write E^{cc}.

A subset E of A is a *commutative subset* of A if $xy = yx$ for all $x, y \in E$. The *centre* of A is the commutant A^c of A in A.

Proposition 2. *Let E be a subset of an algebra A. Then the following statements hold.*

(i) *E^c is a subalgebra of A, which contains 1 if A has a unit element, and is closed if A is a normed algebra.*

(ii) *E is a commutative subset if and only if $E \subset E^c$.*

(iii) *$F^c \subset E^c$ if $E \subset F$.*

(iv) *If E is a commutative subset, then E^{cc} is a commutative subalgebra of A and $E \subset E^{cc} \subset E^c$.*

(v) *The centre of A is a commutative subalgebra of A.*

Proof. (i), (ii), (iii). Clear.

(iv) Let E be a commutative subset. Then $E \subset E^c$ and so $E^{cc} \subset E^c$. Obviously also $E \subset E^{cc}$. Given $x, y \in E^{cc}$, we have $x \in E^c$, $y \in E^{cc}$ and so $xy = yx$; E^{cc} is commutative.

(v) Clear. ▢

Proposition 3. *Each commutative subset of an algebra A is contained in a maximal commutative subset of A. Each maximal commutative subset is a commutative subalgebra of A, contains 1 if A has a unit element, and is closed if A is a normed algebra.*

Proof. The set of commutative subsets of A is inductively ordered by the relation of inclusion. Therefore each commutative subset is contained in a maximal commutative subset. Let M be a maximal commutative subset of A. Then $M \subset M^{cc}$ and M^{cc} is a commutative subalgebra. So $M = M^{cc}$, and the rest is clear. ▢

Theorem 4. *Let A be a complex algebra with unit, let $a \in A$, and let M be a maximal commutative subset of A containing $\{a\}$. Then*
  (i) $(z-a)^{-1} \in M$ $(z \in \mathbb{C} \setminus \mathrm{Sp}(A, a))$,
  (ii) $\mathrm{Sp}(M, a) = \mathrm{Sp}(A, a)$.

Proof. Let $z \in \mathbb{C} \setminus \mathrm{Sp}(A, a)$. Then $(z-a)^{-1} \in \{a\}^{cc}$. Given $y \in M$, we have $y \in \{a\}^{c}$ and so $(z-a)^{-1}$ commutes with y. Therefore $M \cup \{(z-a)^{-1}\}$ is a commutative subset of A. Then, by maximality of M, $(z-a)^{-1} \in M$, $z \in \mathbb{C} \setminus \mathrm{Sp}(M, a)$. □

The concept of maximal commutative subset reduces many problems to the consideration of commutative algebras.

We give next some conditions for the commutativity of complex Banach algebras. The results are derived from the following general proposition of Baker and Pym [35], which is based upon ideas of Le Page [257] and Hirschfeld and Zelazko [199].

Proposition 5. *Let A be a complex Banach algebra with unit, M a unit linked Banach left A-module, X a complex Banach space, and $h: A \times M \to X$ a continuous bilinear mapping. The following conditions are equivalent.*
  (i) $h(a, m) = h(1, am)$ $(a \in A, m \in M)$.
  (ii) *There exists* $\kappa > 0$ *with* $\|h(a, m)\| \leqslant \kappa \|am\|$ $(a \in A, m \in M)$.

Proof. (i) $\Rightarrow$ (ii). Clear.
(ii) $\Rightarrow$ (i). Let condition (ii) hold and let $f \in X'$. Given $a \in A$, $m \in M$, let F be defined on $\mathbb{C}$ by

$$F(z) = (f \circ h)(\exp(-za), (\exp(za))m).$$

Then F is an entire function and, for $z \in \mathbb{C}$,

$$|F(z)| \leqslant \|f\| \, \|h(\exp(-za), (\exp(za))m)\|$$
$$\leqslant \kappa \|f\| \, \|m\|.$$

By Liouville's theorem F is constant, and so the coefficient of z in the power series expansion of F is zero, i.e.

$$f(h(1, am)) - f(h(a, m)) = 0.$$

But $f \in X'$ was arbitrary. □

Corollary 6. *Let A be a complex Banach algebra with unit such that, for some $\kappa > 0$, $\|ab\| \leqslant \kappa \|ba\|$ $(a, b \in A)$. Then A is commutative.*

Proof. Apply Proposition 5 with $X = M = A$, $h(a, b) = ba$. □

Corollary 7. *Let A be a complex Banach algebra with unit such that, for some $\kappa > 0$, $\|a\| \leqslant \kappa r(a)$ $(a \in A)$. Then A is commutative.*

Proof. This follows from Corollary 6 since, by Proposition 5.3, $r(a\,b)=r(b\,a)$ $(a, b \in A)$.   ☐

Corollary 8. *Let A be a complex Banach algebra with unit such that, for some $\kappa > 0$, $\|a\|^2 \leqslant \kappa \|a^2\|$ $(a \in A)$. Then A is commutative.*

Proof. An induction argument gives

$$\|a\| \leqslant \kappa^{1-2^{-n}} \|a^{2^n}\|^{2^{-n}} \qquad (n = 1, 2, 3, \ldots).$$

Apply Corollary 7.   ☐

A completion argument shows that the above three corollaries hold if A is only a complex normed algebra with unit. Corollaries 7 and 8 also hold for algebras without unit; this follows from the fact that

$$r(a) + |\alpha| \leqslant 3r((a, \alpha)) \qquad ((a, \alpha) \in A + \mathbb{C}).$$

§ 16. Multiplicative Linear Functionals

Let A denote a complex Banach algebra.

Definition 1. *A multiplicative linear functional on A is a non-zero linear functional ϕ on A such that*

$$\phi(x\,y) = \phi(x)\,\phi(y) \qquad (x, y \in A),$$

i.e. a non-zero homomorphism of A into $\mathbb{C}$.

Examples 2. (i) Let A be a uniform algebra of functions on a set E (see Example 1.16), let $t \in E$ and let

$$\phi(f) = f(t) \qquad (f \in A).$$

(ii) Let $A = L^1(G)$, where G is a locally compact Abelian group with normalized Haar measure m, let γ be a character on G, i.e. γ is a continuous complex function on G with $\gamma(g\,h) = \gamma(g)\gamma(h)$ $(g, h \in G)$, and let

$$\phi(f) = \int_G \gamma f \, dm \qquad (f \in A).$$

Motivated by Example 2(ii) some authors refer to multiplicative linear functionals as characters.

Proposition 3. *Let ϕ be a multiplicative linear functional on A. Then ϕ is continuous, and $\|\phi\| \leqslant 1$. If A has a unit element, $\phi(1) = 1$.*

Proof. Suppose that there exists $x \in A$ with $\|x\| < 1$ and $\phi(x) = 1$, and let $y = \sum_{n=1}^{\infty} x^n$. Then $x + xy = y$, and so

$$1 + \phi(y) = \phi(x) + \phi(x)\phi(y) = \phi(x + xy) = \phi(y),$$

which is absurd. □

If A is a complex unital Banach algebra, every multiplicative linear functional ϕ satisfies the condition $\phi(1) = 1 = \|\phi\|$. In general a multiplicative linear functional on a complex Banach algebra A can have arbitrarily small norm. To see this, let ϕ be a multiplicative linear functional on A, let $t \geqslant 1$ and let

$$p(a) = \|a\| + t|\phi(a)| (a \in A).$$

Then p is a Banach algebra norm on A equivalent to $\|\cdot\|$ and

$$|\phi(a)| \leqslant t^{-1} p(a)   (a \in A).$$

Proposition 4. *The maximal modular bi-ideals of A of codimension 1 are the kernels of the multiplicative linear functionals on A.*

Proof. Let ϕ be a multiplicative linear functional on A and let $\ker(\phi) = \{x \in A : \phi(x) = 0\}$. Then $\ker(\phi)$ is a linear subspace of A, $\ker(\phi) \neq A$, and if $y \in A \setminus \ker(\phi)$, then the linear span of $\ker(\phi) \cup \{y\}$ is A, for

$$a - \frac{\phi(a)}{\phi(y)} y \in \ker(\phi)   (a \in A).$$

We have

$$\phi(ax) = \phi(xa) = \phi(a)\phi(x) = 0   (a \in A, x \in \ker(\phi))$$

and so $\ker(\phi)$ is a bi-ideal which is maximal. It is also modular, for there exists $u \in A$ with $\phi(u) = 1$, and then

$$\phi(a - au) = \phi(a) - \phi(a)\phi(u) = 0,$$

$a - au \in \ker(\phi) \ (a \in A)$.

Suppose that M is a maximal modular bi-ideal of A with codimension 1. Let j be a (left and right) modular unit for M. Then $A/M = \mathbb{C}j'$, and therefore, given $a \in A$, there exists a complex number $\phi(a)$ such that $a' = \phi(a)j'$. Since $a \to a'$ is a homomorphism and $\phi(j) = 1$, it follows that ϕ is a multiplicative linear functional on A. Finally,

$$\ker(\phi) = \{a \in A : a' = 0\} = M. □$$

It is easy to construct examples of non-commutative algebras which admit multiplicative linear functionals, but the real interest with multiplicative linear functionals lies in the commutative case. Moreover, an algebra is always commutative modulo the intersection of the kernels of the multiplicative linear functionals.

When A is commutative there is no distinction between left, right, and bi-ideals, and so the word 'ideal' is used without any such qualification.

Theorem 5. *Let A be commutative. Then the maximal modular ideals of A are the kernels of the multiplicative linear functionals.*

Proof. By Theorem 4 it is enough to show that each maximal modular ideal of A has codimension 1. Let M be a maximal modular ideal of A with modular unit j. Then by Corollary 9.4, M is closed. Therefore the quotient algebra $B = A/M$ is a complex Banach algebra. Since M is maximal, the only ideals of B are $\{0\}$ and B. For if J is an ideal of B with $\{0\} \neq J \neq B$, then $K = \{a : a' \in J\}$ is an ideal of A with $M \subsetneqq K \neq A$. Moreover, j' is a unit element for B, since $a - aj \in M$ $(a \in A)$. Therefore B is a complex normed division algebra, and so by Theorem 14.2, $B = \mathbb{C}j'$. ☐

Let A have a unit element and let ϕ be a multiplicative linear functional on A. Then $a - \phi(a)1 \in \ker(\phi) \subset \mathrm{Sing}(A)$ and so

$$\phi(a) \in \mathrm{Sp}(a) \qquad (a \in A). \tag{1}$$

We show below that condition (1) distinguishes the multiplicative linear functionals amongst the linear functionals on A. The preliminary result is due to Želazko [431]. A *Jordan functional* on A is a non-zero linear functional ϕ on A such that $\phi(a^2) = \phi(a)^2$ $(a \in A)$.

Proposition 6. *Every Jordan functional ϕ on A is multiplicative.*

Proof. Consideration of the identity

$$\phi((a+b)^2) = (\phi(a+b))^2$$

gives

$$\phi(ab + ba) = 2\phi(a)\phi(b) \qquad (a, b \in A)$$

and so the result is trivial if A is commutative.

If ϕ is not multiplicative there exist $a, b \in A$ such that $\phi(a) = 0$, $\phi(ab) = 1$. Then $\phi(ba) = -1$. Let $c = bab$. Then

$$0 = 2\phi(a)\phi(c) = \phi(ac + ca) = \phi((ab)^2) + \phi((ba)^2) = 2. \qquad ☐$$

Remark. The above simplification of Želazko's proof was communicated to us by A. M. Sinclair. An alternative analytic proof is available by showing that $|\phi(a)| \leqslant r(a)$ $(a \in A)$ and then applying Proposition 15.5.

Theorem 7 (Gleason [159], Kahane and Żelazko [226]). *Let A have a unit element and let φ be a linear functional on A. The following conditions are equivalent:*
   (i) $\phi(1)=1$, $\ker(\phi)\subset\mathrm{Sing}(A)$;
   (ii) $\phi(a)\in\mathrm{Sp}(a)$ $(a\in A)$;
   (iii) ϕ *is multiplicative.*

Proof. (i) ⇔ (ii), (iii) ⇒ (i). Clear.

Let condition (i) hold. Given $x\in A$, $\lambda\in\mathbb{C}$, $\|x\|\leqslant 1$, $|\lambda|>1$ we have $1-\lambda^{-1}x\in\mathrm{Inv}(A)$ and so $1-\lambda^{-1}\phi(x)\neq 0$, $\phi(x)\neq\lambda$, $|\phi(x)|\leqslant 1$. Therefore ϕ is continuous and $\|\phi\|\leqslant 1$. Given $a\in A$ let F be defined on $\mathbb{C}$ by

$$F(z)=\phi(\exp(za))$$

$$=1+\sum_{n=1}^{\infty}\frac{1}{n!}\phi(a^n)z^n. \tag{2}$$

Then F is an entire function and

$$|F(z)|\leqslant 1+\sum_{n=1}^{\infty}\frac{1}{n!}\|a\|^n|z|^n=\exp(\|a\|\,|z|)\quad(z\in\mathbb{C}),$$

so that F has order at most 1. Since $\exp(za)\in\mathrm{Inv}(A)$ we have $F(z)\neq 0$ $(z\in\mathbb{C})$. By Hadamard's theorem (see e. g. [384] Theorem 8.24) there exists $\alpha\in\mathbb{C}$ such that

$$F(z)=\exp(\alpha z)\quad(z\in\mathbb{C})$$

$$=1+\sum_{n=1}^{\infty}\frac{1}{n!}\alpha^n z^n. \tag{3}$$

Comparing coefficients in (2) and (3) we obtain

$$\phi(a)=\alpha,\qquad\phi(a^2)=\alpha^2.$$

Therefore ϕ is a Jordan functional and Theorem 6 applies. □

Remark. A unitization argument shows that conditions (ii) and (iii) are equivalent without the assumption that A has a unit element.

Proposition 8. *Let j be a modular unit for the kernel of a multiplicative linear functional φ. Then φ(j)=1.*

Proof. We have $\phi(a)-\phi(a)\phi(j)=\phi(a-aj)=0$ $(a\in A)$. Since $\phi\neq 0$, we have $\phi(a)\neq 0$ for some $a\in A$. □

We can refine (1) for commutative algebras.

Proposition 9. *Let A be commutative. Let λ be a non-zero complex number and let $a \in A$. Then $\lambda \in \mathrm{Sp}(a)$ if and only if $\lambda = \phi(a)$ for some multiplicative linear functional ϕ. If A has a unit element, this holds also for $\lambda = 0$.*

Proof. If $\lambda \in \mathrm{Sp}(a)$, then $\lambda^{-1} a$ is quasi-singular and so

$$\lambda^{-1} a \notin J = \{\lambda^{-1} a x - x : x \in A\}.$$

J is a modular ideal with modular unit $\lambda^{-1} a$, and $J \neq A$. Therefore, by Theorem 5, J is contained in the kernel $\ker(\phi)$ of a multiplicative linear functional ϕ. Then $\lambda^{-1} a$ is a modular unit for $\ker(\phi)$, and Proposition 8 gives $\phi(\lambda^{-1} a) = 1$, $\phi(a) = \lambda$.

Suppose on the other hand that $\phi(a) = \lambda$ for some multiplicative linear functional ϕ. Then $\phi(x(\lambda^{-1} a) - x) = 0$ $(x \in A)$, and so the set $J = \{x(\lambda^{-1} a) - x : x \in A\}$ is a proper modular ideal of A with modular unit $\lambda^{-1} a$. By Proposition 9.2(ii), $\lambda^{-1} a \notin J$. Therefore $\lambda^{-1} a$ is quasi-singular, $\lambda \in \mathrm{Sp}(a)$.

Finally suppose that A has a unit element. Then $0 \in \mathrm{Sp}(a)$ if and only if $a \in \mathrm{Sing}(A)$, i.e. if and only if a belongs to a proper ideal of A. Since all ideals are modular, every proper ideal is contained in the kernel of a multiplicative linear functional. $\square$

§ 17. The Gelfand Representation of a Commutative Banach Algebra

Let A denote a commutative complex Banach algebra. The modifications needed for real algebras are briefly considered at the end of this section.

Notation. The set of all multiplicative linear functionals on A is denoted by Φ_A, the identically zero functional on A by ϕ_∞, and $\Phi_A \cup \{\phi_\infty\}$ by Φ_A^∞. By Proposition 16.3, Φ_A and Φ_A^∞ are subsets of the dual space A'.

Definition 1. The *A-topology* on Φ_A (or on Φ_A^∞) is the relative topology on Φ_A (Φ_A^∞) induced by the weak* topology on A'. Thus a basis of neighbourhoods of ϕ is given by sets of the form $V(\phi; x_1, \ldots, x_n; \varepsilon)$ where

$$V(\phi; x_1, \ldots, x_n; \varepsilon) = \{\psi : |\psi(x_k) - \phi(x_k)| < \varepsilon \ (k = 1, \ldots, n)\},$$

for arbitrary positive integers n, elements $x_1, \ldots, x_n \in A$, and $\varepsilon > 0$.

The *carrier space* for A is the set Φ_A with the A-topology. Some authors use 'maximal ideal space' or 'spectrum'. We also regard Φ_A^∞ as a topological space with the A-topology.

Proposition 2. *The carrier space Φ_A is a locally compact Hausdorff space with one point compactification Φ_A^∞. If A has a unit element, Φ_A is compact.*

Proof. By Proposition 16.3, Φ_A^∞ is a subset of the closed unit ball of A'. To prove that Φ_A^∞ is compact in the A-topology it is therefore sufficient to prove that it is a weak* closed subset of A'. Suppose that $\phi \in A' \setminus \Phi_A^\infty$. Then there exist $x, y \in A$ and $\delta > 0$ such that

$$|\phi(x)\phi(y) - \phi(xy)| = \delta.$$

Let $\varepsilon = \min\left(\frac{1}{2}\delta^{\frac{1}{2}}, \frac{1}{4}(1 + |\phi(x)| + |\phi(y)|)^{-1}\delta\right)$, and let $\psi \in A'$ with

$$|\psi(x) - \phi(x)| < \varepsilon, \quad |\psi(y) - \phi(y)| < \varepsilon, \quad |\psi(xy) - \phi(xy)| < \varepsilon.$$

Then

$$|\psi(x)\psi(y) - \phi(x)\phi(y)|$$
$$\leqslant |\psi(x) - \phi(x)|\,|\psi(y) - \phi(y)| + |\phi(x)|\,|\psi(y) - \phi(y)| + |\psi(x) - \phi(x)|\,|\phi(y)|$$
$$< \varepsilon^2 + \varepsilon(|\phi(x)| + |\phi(y)|) < (\tfrac{1}{2}\delta^{\frac{1}{2}})^2 + \tfrac{1}{4}\delta = \tfrac{1}{2}\delta.$$

Since also $|\psi(xy) - \phi(xy)| < \frac{1}{4}\delta$, we have $\psi(xy) \neq \psi(x)\psi(y)$, $\psi \notin \Phi_A^\infty$. Thus ϕ has a weak* neighbourhood that does not meet Φ_A^∞, and therefore Φ_A^∞ is weak* closed.

We have now proved that Φ_A^∞ is compact in the A-topology. Given $\phi \in \Phi_A$, there exists $a \in A$ with $\phi(a) \neq 0$. Then the set

$$\{\psi \in \Phi_A^\infty : |\psi(a)| \geqslant \tfrac{1}{2}|\phi(a)|\}$$

is a compact neighbourhood of ϕ that does not contain ϕ_∞. Thus Φ_A is locally compact.

If A has a unit element, then $\Phi_A = \{\phi \in \Phi_A^\infty : \phi(1) = 1\}$. This shows that Φ_A is a closed subset of Φ_A^∞ and is therefore compact. □

Definition 3. The *Gelfand representation* of A is the mapping $a \to a^\wedge$ of A into $C(\Phi_A)$ defined by

$$a^\wedge(\phi) = \phi(a) \qquad (\phi \in A).$$

(That $a^\wedge$ is a continuous complex valued function on Φ_A is clear from the definition of the A-topology.)

Theorem 4. *Let A be a commutative complex Banach algebra with unit, with carrier space Φ_A and Gelfand representation $a \to a^\wedge$. Then*
  (i) *Φ_A is a compact Hausdorff space;*
  (ii) *the mapping $a \to a^\wedge$ is a homomorphism of A into $C(\Phi_A)$;*
  (iii) *$a^\wedge(\Phi_A) = \mathrm{Sp}(a)$;*
  (iv) *$\|a^\wedge\|_\infty = r(a)$;*
  (v) *$a \in \mathrm{Sing}(A)$ if and only if $a^\wedge(\phi) = 0$ for some $\phi \in \Phi_A$.*

Proof. Propositions 2, 16.9 and Theorem 5.8. □

Notation. Given a topological space E, we denote by $C_0(E)$ the uniform algebra of all continuous complex valued functions f on E such that for every $\varepsilon > 0$, $\{x \in E : |f(x)| \geq \varepsilon\}$ is compact. Note that if E is compact, $C_0(E) = C(E)$.

Theorem 5. *Let A be a commutative complex Banach algebra without unit, with carrier space Φ_A and Gelfand representation $a \to a^\wedge$. Then*
(i) *Φ_A is a locally compact Hausdorff space;*
(ii) *the mapping $a \to a^\wedge$ is a homomorphism of A into $C_0(\Phi_A)$;*
(iii) *$a^\wedge(\Phi_A) \setminus \{0\} = \mathrm{Sp}(a) \setminus \{0\}$;*
(iv) *$\|a^\wedge\|_\infty = r(a)$;*
(v) *$a \in \mathrm{q} - \mathrm{Sing}(A)$ if and only if $a^\wedge(\phi) = 1$ for some $\phi \in \Phi_A$.*

Proof. Given $a \in A$ and $\varepsilon > 0$, the set $\{\phi \in \Phi_A^\infty : |\phi(a)| \geq \varepsilon\}$ is a closed subset of Φ_A^∞ not containing ϕ_∞. It is therefore a compact subset of Φ_A, and
$$\{\phi \in \Phi_A : |a^\wedge(\phi)| \geq \varepsilon\} = \{\phi \in \Phi_A^\infty : |\phi(a)| \geq \varepsilon\} .$$
Therefore $a^\wedge \in C_0(\Phi_A)$.

We have $a \in \mathrm{q} - \mathrm{Sing}(A)$ if and only if $\{ax - x : x \in A\}$ is a proper modular ideal with modular unit a. The rest is now clear. ☐

Definition 6. The *radical* of A is the subset $\mathrm{rad}(A)$ of A given by
$$\mathrm{rad}(A) = \cap \{\ker(\phi) : \phi \in \Phi_A\} .$$
A is *semi-simple* if $\mathrm{rad}(A) = \{0\}$.

Corollary 7. *The radical of A is the set $\{a \in A : r(a) = 0\}$, and the following statements are equivalent:*
(i) *A is semi-simple;*
(ii) *Φ_A separates the points of A;*
(iii) *the Gelfand representation is a monomorphism of A into $C_0(\Phi_A)$;*
(iv) *the spectral radius is a norm on A.*

Proof. Clear. ☐

Theorem 8. *Let α be a homomorphism of a complex Banach algebra B into a semi-simple commutative complex Banach algebra A. Then α is continuous.*

Proof. Let $\{a_n\} \subset A$, $\lim_{n \to \infty} a_n = 0$, $\lim_{n \to \infty} \alpha(a_n) = b$. Given $\phi \in \Phi_B$ we have $\phi \circ \alpha \in \Phi_A^\infty$ and Proposition 16.3 gives
$$\phi(b) = \lim_{n \to \infty} \phi(\alpha(a_n)) = \lim_{n \to \infty} \phi \circ \alpha(a_n) = 0 \quad (\phi \in \Phi_B) .$$
Since B is semi-simple we have $b = 0$, and the closed graph theorem applies. ☐

Theorem 8 has several obvious corollaries, but we defer these since we shall prove a deeper result on the automatic continuity of homomorphisms in §25. In the opposite direction, Allan [15] has shown that the algebra of formal power series in a single indeterminate can be embedded in certain Banach algebras. As a corollary, he obtains discontinuous homomorphisms from the disc algebra into the unitization of the radical convolution algebra $L^1[0,1]$.

Proposition 9. *Let A be a semi-simple commutative complex Banach algebra, and let A have an element u such that, for some $\delta > 0$, $|\phi(u)| \geqslant \delta$ for every $\phi \in \Phi_A$. Then A has a unit element.*

Proof. Let $B = A + \mathbb{C}$, the unitization of A, and let

$$\phi_0(a + \alpha) = \alpha \quad (a + \alpha \in B).$$

We may suppose without loss that $\delta > 1$, so that $\mathrm{Sp}(B, u) \cap C(0,1) = \emptyset$. By Proposition 7.10, the sequence $\{1 - (1 - u^n)^{-1}\}$ converges to an idempotent, j say. Since $1 - (1 - u^n)^{-1} \in \ker(\phi_0)$, we have $j \in \ker(\phi_0) = A$. For $\phi \in \Phi_A$,

$$\phi(j) = 1 - \lim_{n \to \infty} \frac{1}{1 - \phi(u)^n} = 1.$$

Therefore, for $a \in A$, $\phi(a - aj) = 0$ $(\phi \in \Phi_A)$, $a = aj$. □

It is easy to see that the conclusion of Proposition 9 can fail if A is not semi-simple.

Example 10. Let E be a compact Hausdorff space and let $A = C(E)$. Given $t \in E$ let

$$\phi_t(f) = f(t) \quad (f \in A).$$

Then the mapping $t \to \phi_t$ is a homeomorphism of E onto Φ_A. To see this, let $\phi \in \Phi_A$ and let $\{f_1, \ldots, f_n\}$ be a finite subset of $\ker(\phi)$. If $f_1, \ldots, f_n$ have no common zero, then $p = \sum_{k=1}^n f_k f_k^*$ is strictly positive on E, $p \in \mathrm{Inv}(A)$ and so

$$0 \neq \phi(p) = \sum_{k=1}^n \phi(f_k)\phi(f_k^*) = 0.$$

This contradiction shows that every finite subset of $\ker(\phi)$ has a common zero. Since E is compact, all the functions in $\ker(\phi)$ have a common zero, say t, and thus $\ker(\phi) \subset \ker(\phi_t)$. Since $\phi(1) = \phi_t(1) = 1$, it follows that $\phi = \phi_t$. Therefore $t \to \phi_t$ is a bijection of E onto Φ_A, which is clearly continuous. But E and Φ_A are both compact Hausdorff spaces.

We note also that $C(E)$ is semi-simple.

Notation. Let B denote a commutative real Banach algebra. We denote by Φ_B the set of all non-zero homomorphisms of B into $\mathbb{C}$, i.e. non-zero (real) linear mappings of B into $\mathbb{C}$ such that $\phi(xy) = \phi(x)\phi(y)$ $(x, y \in B)$. We denote by $B_{\mathbb{C}}$ the complexification of B.

Definition 11. The *B-topology* on Φ_B is the topology in which a basis of open neighbourhoods of $\phi \in \Phi_B$ is given by the sets of the form

$$V(\phi; x_1, \ldots, x_n; \varepsilon) = \{\psi \in \Phi_B : |\psi(x_k) - \phi(x_k)| < \varepsilon \, (k = 1, \ldots, n)\}$$

with $n \in \mathbb{N}$, $x_1, \ldots, x_n \in B$, $\varepsilon > 0$. The *carrier space* for B is Φ_B with the B-topology. The *Gelfand representation* of B is the mapping $b \to b^\wedge$ of B into $C(\Phi_B)$, given by $b^\wedge(\phi) = \phi(b)$ $(\phi \in \Phi_B)$.

Lemma 12. *Given* $\phi \in \Phi_{B_{\mathbb{C}}}$, *let* ϕ' *be defined on B by*

$$\phi'(x) = \phi((x, 0)) \qquad (x \in B).$$

Then the mapping $\phi \to \phi'$ *is a homeomorphism of the carrier space $\Phi_{B_{\mathbb{C}}}$ onto the carrier space Φ_B.*

Proof. Given $\phi \in \Phi_B$, define ψ on $B_{\mathbb{C}}$ by $\psi(x, y) = \phi(x) + i\phi(y)$. Then $\psi \in \Phi_{B_{\mathbb{C}}}$ and $\psi' = \phi$. The rest is routine verification. □

Theorem 13. *Let B be a commutative real Banach algebra with unit, Φ_B its carrier space, and $b \to b^\wedge$ the Gelfand representation. Then*
   (i) *Φ_B is a compact Hausdorff space;*
   (ii) *the mapping $b \to b^\wedge$ is a homomorphism of B into $C(\Phi_B)$;*
   (iii) *$b^\wedge(\Phi_B) = \mathrm{Sp}(b)$;*
   (iv) *$\|b^\wedge\|_\infty = r(b)$;*
   (v) *$b \in \mathrm{Sing}(B)$ if and only if $b^\wedge(\phi) = 0$ for some $\phi \in \Phi_B$.*

Proof. Lemma 12, Theorem 4 and the definition of $\mathrm{Sp}(b)$ (Definition 13.6). □

Remarks. (1) It should be noted that the functions $b^\wedge$ are not in general real valued functions.
   (2) Analogous results hold for commutative real Banach algebras without unit.
   (3) The results in this section are due to I.M. Gelfand.

§ 18. Derivations and Automorphisms

Definition 1. A *derivation* on an algebra A (or an A-derivation) is a linear mapping D of A into A such that

$$D(ab) = aDb + (Da)b \qquad (a, b \in A).$$

Example 2. Given $c \in A$, let δ_c denote the mapping of A into A defined by

$$\delta_c(a) = ac - ca \quad (a \in A).$$

Then δ_c is a derivation on A. Each such derivation δ_c is called an *inner derivation* on A. If A is a normed algebra, each inner derivation δ_c is continuous (i.e. belongs to $BL(A)$) and

$$\|\delta_c\| \leqslant 2 \|c\| .$$

It is also obvious that A is commutative if and only if 0 is the only inner derivation.

Example 3. Let A be an algebra with unit and let a be an element of A that is not algebraic, i.e. such that the elements $1, a, a^2, \ldots$ are linearly independent. Let B be the subalgebra of A generated by $1, a$, and define a mapping D of B into B by

$$D(\alpha_0 + \alpha_1 a + \cdots + \alpha_n a^n) = \alpha_1 + 2\alpha_2 a + \cdots + n\alpha_n a^{n-1} .$$

Then D is a derivation on B which is not inner, since B is commutative and $D \neq 0$.

We first establish some elementary algebraic properties of derivations.

Proposition 4. *Let D be a derivation on an algebra A. Then the following statements hold (with the convention that $D^0 = I$):*

(i) *(Leibnitz rule)* $D^n(ab) = \sum_{r=0}^{n} \binom{n}{r} (D^{n-r} a)(D^r b)$ $(n \in \mathbb{N}, a, b \in A)$;

(ii) $D(a^n) = n a^{n-1} Da$ $(n - 1 \in \mathbb{N})$ *if and only if* $a Da = (Da) a$;

(iii) *if* $D^2 a = 0$, *then* $D^n(a^n) = n! (Da)^n$ $(n \in \mathbb{N})$.

Proof. (i) and (ii). Induction.

(iii). Let $D^2 a = 0$. Then, by induction,

$$D(Da)^k = 0 \quad (k \in \mathbb{N}). \tag{1}$$

Assume that $D^{n-1}(a^{n-1}) = (n-1)! (Da)^{n-1}$. Then, by (1),

$$D^n(a^{n-1}) = 0 .$$

Therefore, by Leibnitz rule,

$$D^n(a^n) = D^n(a \cdot a^{n-1})$$

$$= a D^n(a^{n-1}) + \binom{n}{1}(Da) D^{n-1}(a^{n-1}) + \cdots + \binom{n}{n}(D^n a) a^{n-1}$$

$$= n(Da) D^{n-1}(a^{n-1}) = n(Da)(n-1)! (Da)^{n-1} = n! (Da)^n . \quad \square$$

Remark. If D is a continuous derivation on a complex Banach algebra A with unit, $a \in A$, $aDa = (Da)a$, and f is holomorphic on a neighbourhood of $\mathrm{Sp}(a)$, then

$$D(\mathbf{f}(a)) = \mathbf{f}'(a)Da .$$

The proof is a routine exercise with the integral formula or the power series representation of $\mathbf{f}(a)$.

Proposition 5. *Let D be a derivation on an algebra A, and let e be an idempotent in A. The following statements hold:*
  (i) $e(De)e = 0$;
  (ii) *if $eDe = (De)e$, then $De = 0$;*
  (iii) *if A has a unit element, then $D1 = 0$.*

Proof. (i) $De = De^2 = eDe + (De)e$, $eDe = eDe + e(De)e$.
(ii) If $eDe = (De)e$, we have

$$De = eDe + (De)e = e^2 De + (De)e^2 = 2e(De)e = 0 .$$

(iii) Particular case of (ii) or easily proved directly. □

Definition 6. An *automorphism* on an algebra A is an isomorphism of A onto A. Given an algebra A with unit, and $c \in \mathrm{Inv}(A)$, let α_c be the mapping of A into A given by

$$\alpha_c(a) = c^{-1}ac \quad (a \in A) .$$

It is clear that α_c is an automorphism on A. An *inner automorphism* on A is an automorphism of the form α_c for some $c \in \mathrm{Inv}\, A$.

Remark. It is clear that $\alpha_c = I$ if c belongs to the centre A^c of A. In particular, if A is commutative, I is the only inner automorphism. Conversely, if A is a Banach algebra with unit and I is the only inner automorphism of A, then A is commutative. To see this note that (by Theorem 2.9) A is the linear span of $\mathrm{Inv}(A)$.

Proposition 7. *Let D be a continuous derivation on a Banach algebra A. Then $\exp D$ is a continuous automorphism on A.*

Proof. For $a, b \in A$ we have (as in the proof of Proposition 8.2),

$$(\exp D)(ab) = \sum_{n=0}^{\infty} \frac{1}{n!} D^n(ab)$$

$$= \sum_{n=0}^{\infty} \frac{1}{n!} \sum_{r=0}^{n} \binom{n}{r} (D^{n-r}a)(D^r b)$$

$$= \sum_{n=0}^{\infty} \sum_{r=0}^{n} \left(\frac{1}{(n-r)!} D^{n-r}a \right) \left(\frac{1}{r!} D^r b \right)$$

$$= \left(\sum_{n=0}^{\infty} \frac{1}{n!} D^n a \right) \left(\sum_{n=0}^{\infty} \frac{1}{n!} D^n b \right) = ((\exp D)a)((\exp D)b).$$

Also $\exp D$ is an invertible element of $BL(A)$, and is therefore a bijective mapping of A onto A. ☐

The automorphism corresponding to an inner derivation is easily identified, as follows.

Proposition 8. *Let c be an element of a Banach algebra A with unit. Then*

$$\exp \delta_c = \alpha_{\exp c}.$$

Proof. For $x \in A$, we have, by induction,

$$(\delta_c)^n(x) = x c^n - \binom{n}{1} c x c^{n-1} + \cdots + (-1)^n \binom{n}{n} c^n x.$$

Therefore

$$\begin{aligned}
(\exp \delta_c)(x) &= \sum_{n=0}^{\infty} \frac{1}{n!} \sum_{r=0}^{n} (-1)^r \binom{n}{r} c^r x c^{n-r} \\
&= \sum_{n=0}^{\infty} \sum_{r=0}^{n} \left(\frac{(-1)^r}{r!} c^r \right) x \left(\frac{1}{(n-r)!} c^{n-r} \right) \\
&= (\exp(-c)) x \exp c \\
&= (\exp c)^{-1} x \exp c = (\alpha_{\exp c})(x). ☐
\end{aligned}$$

Notation. Until Theorem 16, A will denote a complex Banach algebra with unit.

Proposition 9. *Let $b \in A$ and $c \in \mathrm{Inv}(A)$. Then*
(i) $\mathrm{Sp}(\delta_b) \subset \{z - w : z, w \in \mathrm{Sp}(b)\}$;
(ii) $\mathrm{Sp}(\alpha_c) \subset \{z^{-1} w : z, w \in \mathrm{Sp}(c)\}$.

Proof. Let λ, ρ denote the left and right regular representations of A on A, i.e.

$$\lambda_a x = ax, \qquad \rho_a x = xa \qquad (a, x \in A).$$

Then $\lambda_a \rho_{a'} x = ax a' = \rho_{a'} \lambda_a x$, and so $\lambda_a \rho_{a'} = \rho_{a'} \lambda_a$ $(a, a' \in A)$. Also $\delta_b = \rho_b - \lambda_b$, $\alpha_c = \lambda_{c^{-1}} \rho_c$.
(i) Let B denote a maximal commutative subset of $BL(A)$ containing the commutative subset $\{\lambda_b, \rho_b\}$, and $\Phi(B)$ its carrier space. We have by Theorem 15.4 and Propositions 16.9, 5.4,

$$\begin{aligned}
\mathrm{Sp}(\delta_b) = \mathrm{Sp}(BL(A), \delta_b) &= \mathrm{Sp}(B, \delta_b) = \{\phi(\delta_b) : \phi \in \Phi(B)\} \\
&= \{\phi(\rho_b) - \phi(\lambda_b) : \phi \in \Phi(B)\} \\
&\subset \{z - w : z \in \mathrm{Sp}(B, \rho_b), \ w \in \mathrm{Sp}(B, \lambda_b)\} \\
&= \{z - w : z \in \mathrm{Sp}(BL(A), \rho_b), \ w \in \mathrm{Sp}(BL(A), \lambda_b)\} \\
&= \{z - w : z, w \in \mathrm{Sp}(b)\}.
\end{aligned}$$

(ii) Let B denote a maximal commutative subset of $BL(A)$ containing $\lambda_{c^{-1}}, \rho_c$, and argue as in (i). $\Box$

As immediate applications of the elementary properties of derivations and automorphisms we give some further properties of roots and logarithms (see §8 and Hille [196]), and prove a theorem on commutators.

Definition 10. Let K be a subset of $\mathbb{C}$, and let m be a positive integer. Then K is *irrotational* (mod $2\pi/m$) if

$$\alpha, \beta \in K , \qquad \alpha^m = \beta^m \Rightarrow \alpha = \beta .$$

K is *incongruent* (mod $2\pi i$) if

$$\alpha, \beta \in K , \qquad \exp\alpha = \exp\beta \Rightarrow \alpha = \beta .$$

Proposition 11. *Let* $a, b \in A$, *let* m *be a positive integer, let*

$$b^m = a^m \in \mathrm{Inv}(A) ,$$

and let $\mathrm{Sp}(a)$ *be irrotational* (mod $2\pi/m$). *Then*
   (i) $a \in \{b\}^{cc}$ *(and so* $ab = ba$*)*;
   (ii) *there exist complex numbers* $\lambda_1, \ldots, \lambda_r$ *with* $\lambda_j^m = 1$ $(j = 1, \ldots, r)$, *and non-zero idempotents* $e_1, \ldots, e_r$ *belonging to* $\{b\}^{cc}$ *such that* $e_j e_k = 0$ $(j \neq k)$, $1 = e_1 + \cdots + e_r$, *and*

$$a e_j = \lambda_j b e_j \qquad (j = 1, \ldots, r) .$$

Proof. (i) We have $a, b \in \mathrm{Inv}(A)$. Let $x \in \{b\}^c$. Then

$$(\alpha_a)^m x = a^{-m} x a^m = b^{-m} x b^m = x ,$$

and therefore, with $\omega_k = \exp(2\pi k i/m)$,

$$\prod_{k=1}^{m-1} (\omega_k I - \alpha_a)(I - \alpha_a) x = 0 .$$

If $\lambda \in \mathrm{Sp}(\alpha_a)$ with $\lambda^m = 1$, then, by Proposition 9, $\lambda = z^{-1} w$ with $z, w \in \mathrm{Sp}(a)$, and we have $z^m = w^m$, $z = w$, $\lambda = 1$. Therefore $\omega_k \notin \mathrm{Sp}(\alpha_a)$ $(k = 1, \ldots, m-1)$, $\omega_k I - \alpha_a \in \mathrm{Inv}(BL(A))$, and therefore $(I - \alpha_a) x = 0$, i.e. $ax = xa$. Since x is an arbitrary element of $\{b\}^c$, this proves that $a \in \{b\}^{cc}$.
   (ii) Let $q = ab^{-1}$. Then, since $ab = ba$, $q^m = a^m b^{-m} = 1$. By Proposition 8.11 applied to q with $f(z) = z^m - 1$, there exist $\lambda_1, \ldots, \lambda_r$, $e_1, \ldots, e_r$ with the stated properties, and with $e_1, \ldots, e_r$ belonging to the closed linear span of $\{(z - q)^{-1} : z \in \mathbb{C} \setminus \mathrm{Sp}(q)\}$. Given $z \in \mathbb{C} \setminus \mathrm{Sp}(q)$ and $y \in \{b\}^c$, we have $y \in \{a\}^c$, and therefore in turn $y \in \{q\}^c$, $(z - q)^{-1} \in \{y\}^c$. Therefore $e_j \in \{b\}^{cc}$ $(j = 1, \ldots, r)$. $\Box$

Proposition 12. *Let* $a, b \in A$, *let* $\mathrm{Sp}(a)$ *be incongruent* (mod $2\pi i$), *and let*

$$\exp(a) = \exp(b).$$

Then (i) $a \in \{b\}^{cc}$ *(and so* $ab = ba$*)*;

(ii) *there exist integers* $n_1, \ldots, n_r$, *and non-zero idempotents* $e_1, \ldots, e_r$ *belonging to* $\{b\}^{cc}$ *such that* $e_j e_k = 0$ $(j \neq k)$, $1 = e_1 + \cdots + e_r$, *and* $ae_j = (b + 2\pi i n_j)e_j$ $(j = 1, \ldots, r)$.

Proof. (i) Let $x \in \{b\}^c$. By Proposition 8,

$$\exp(\delta_a)x = \alpha_{\exp(a)}x = \alpha_{\exp(b)}x = \exp(-b)x\exp(b) = x. \tag{2}$$

If $z, w \in \mathrm{Sp}(a)$ and $z = w + 2k\pi i$ with $k \in \mathbb{Z}$, then, since $\mathrm{Sp}(a)$ is irrotational, $k = 0$. By Proposition 9 (i), $\mathrm{Sp}(\delta_a) \subset \{z - w : z, w \in \mathrm{Sp}(a)\}$, and therefore

$$\mathrm{Sp}(\delta_a) \cap 2\pi i(\mathbb{Z} \setminus \{0\}) = \emptyset. \tag{3}$$

Let g be defined on $\mathbb{C}$ by

$$g(z) = \begin{cases} z^{-1}(\exp(z) - 1) & (z \neq 0), \\ 1 & (z = 0). \end{cases}$$

Then g is an entire function with $2\pi i(\mathbb{Z} \setminus \{0\})$ as its set of zeros. Therefore, by (3)

$$g(z) \neq 0 \quad (z \in \mathrm{Sp}(\delta_a)),$$

from which, $0 \notin \mathrm{Sp}(g(\delta_a))$, $g(\delta_a) \in \mathrm{Inv}(BL(A))$. Since $g(z)z = \exp(z) - 1$ $(z \in \mathbb{C})$, the functional calculus gives

$$g(\delta_a)\delta_a = \exp(\delta_a) - I.$$

Therefore, by (2),

$$g(\delta_a)\delta_a x = \exp(\delta_a)x - x = 0.$$

Since $g(\delta_a)$ is invertible, this gives $\delta_a x = 0$, i.e. $xa - ax = 0$. Since x is an arbitrary element of $\{b\}^c$, we have proved that $a \in \{b\}^{cc}$.

(ii) Since $ab = ba$, we have $\exp(a - b) = \exp(a)(\exp(b))^{-1} = 1$. Application of Proposition 8.11 to the element $a - b$ with $f(z) = \exp(z) - 1$ completes the proof, the proof that the idempotents belong to $\{b\}^{cc}$ being just as in the proof of Proposition 11. $\square$

For more general results related to Propositions 11 and 12 see Paterson [303]. Note also that if A in Proposition 12 is a Banach star algebra then we have $\exp(a^*) = \exp(b^*)$ by the method of proof in Proposition 12.12.

The following proposition was conjectured by Kaplansky, and proved independently by Kleinecke [246] and Shirokov [351].

Proposition 13. *Let* $a, b \in A$ *and* $a(ab-ba)=(ab-ba)a$. *Then* $r(ab-ba)=0$.

Proof. Let $D=\delta_a$. Then, by hypothesis $D^2 b=0$. Therefore, by Proposition 4 (iii),

$$D^n(b^n)=n!(Db)^n \qquad (n=1,2,\ldots).$$

Also, $\|D\| \leqslant 2\|a\|$, and so

$$\|(Db)^n\| \leqslant \frac{1}{n!} 2^n \|a\|^n \|b^n\|,$$

$$\|(Db)^n\|^{\frac{1}{n}} \leqslant (n!)^{-\frac{1}{n}} 2\|a\| \|b\| \qquad (n=1,2,\ldots). \quad \square$$

We have seen (Proposition 7) that the exponential of a continuous derivation on A is a continuous automorphism. Conversely, it was proved by Zeller-Meier [434] that the logarithms of certain continuous automorphisms are continuous derivations. For the proof of this we need the following lemma.

Lemma 14. *Let* $a \in A$ *with* $\mathrm{Sp}(a) \subset \{z \in \mathbb{C} : \mathrm{Re}\, z > 0\}$, *and let* $b = \log a$. *Then*

$$\delta_b = \log \alpha_a,$$

and δ_b *is an operator norm limit of polynomials in* α_a.

Proof. $\mathrm{Sp}(a) \subset D = \{z : \mathrm{Re}\, z > 0\} \subset \mathbb{C} \setminus \mathbb{R}_-$, and so $b = \log a$ and $\exp b = a$ (Proposition 8.3). For each $z \in D$, we have $z = r e^{i\theta}$ with $r > 0$ and $-\pi/2 < \theta < \pi/2$. Thus $-\pi/2 < \mathrm{Im}(\log z) < \pi/2$. Therefore

$$\mathrm{Sp}(b) \subset \{z : -\pi/2 < \mathrm{Im}\, z < \pi/2\},$$

and by Proposition 9 (i), $\mathrm{Sp}(\delta_b) \subset \{z : -\pi < \mathrm{Im}\, z < \pi\}$. By Proposition 8.3 applied to the Banach algebra $BL(A)$, we now have $\log(\exp(\delta_b))=\delta_b$. But, by Proposition 8, $\exp(\delta_b)=\alpha_{\exp(b)}=\alpha_a$, and so $\delta_b=\log(\alpha_a)$.

By Proposition 9 (ii), $\mathrm{Sp}(\alpha_a) \subset \{z^{-1} w : z, w \in D\} \subset \mathbb{C} \setminus \mathbb{R}_-$. Therefore, by Proposition 8.3, there exists a sequence $\{p_n\}$ of complex polynomials such that $\lim_{n \to \infty} p_n(\alpha_a)=\log(\alpha_a)=\delta_b$. $\quad \square$

Theorem 15. *Let* ϕ *be a continuous automorphism on* A *such that* $\mathrm{Sp}(\phi) \subset \{z \in \mathbb{C} : \mathrm{Re}\, z > 0\}$. *Then* $\log \phi$ *is a continuous derivation on* A.

Proof. Let $D = \log \phi$, let $B = BL(A)$, let λ denote the left regular representation of A on A, and let $X = \{\lambda_a : a \in A\}$. Since

$$\|\lambda_a\| \leqslant \|a\| \leqslant \|\lambda_a\| \|1\| \qquad (a \in A),$$

X is a closed subalgebra of B. Since $\phi^{-1}(a)x = \phi^{-1}(a\phi(x))$, we have $\lambda_{\phi^{-1}(a)} = \phi^{-1}\lambda_a\phi$, and so the inner automorphism α_ϕ on B satisfies

$$\lambda_{\phi^{-1}(a)} = \alpha_\phi\lambda_a \quad (a \in A),$$

which shows that $\alpha_\phi X \subset X$. By Lemma 14, it follows that $\delta_D X \subset X$. Thus, for each $a \in A$, there exists $a' \in A$ such that $D\lambda_a - \lambda_a D = \lambda_{a'}$, i.e.

$$D(ab) - aDb = a'b \quad (b \in A).$$

When $b = 1$, this gives $a' = Da - aD1$, and it is therefore sufficient to prove that $D1 = 0$.

By Proposition 8.3, there exists a sequence $\{p_n\}$ of complex polynomials such that $\lim_{n\to\infty} p_n(z) = \log z$ uniformly on a neighbourhood of $\mathrm{Sp}(\phi)$ and $\lim_{n\to\infty} p_n(\phi) = \log\phi = D$. We have $1 \in \mathrm{Sp}(\phi)$ since $\phi(1) = 1$. Therefore $\lim_{n\to\infty} p_n(1) = \log 1 = 0$. Finally therefore,

$$D1 = \lim_{n\to\infty} p_n(\phi)(1) = \lim_{n\to\infty} p_n(1) = 0. \quad \square$$

Remarks. (1) Theorem 15 holds also for complex Banach algebras without unit.

(2) Theorem 15 is applicable in particular to any continuous automorphism ϕ on A such that $\|\phi - I\| < 1$. Moreover, if ϕ is an isometric automorphism on A, then $\mathrm{Sp}(\phi) \subset \{z : |z| = 1\}$, and in this case we have $\mathrm{Sp}(\phi) \subset \{z : \mathrm{Re}\,z > 0\}$ whenever $\|\phi - I\| < \sqrt{2}$.

(3) Kamowitz and Scheinberg [228] have proved that, if ϕ is an automorphism of a commutative semi-simple complex Banach algebra, then either $\phi^n = I$ for some positive integer n, in which case $\mathrm{Sp}(\phi)$ consists of a finite union of finite subgroups of the unit circle Γ, or else $\mathrm{Sp}(\phi) \supset \Gamma$. In particular, if ϕ is also isometric, then either $\phi^n = I$ for some n or else $\mathrm{Sp}(\phi) = \Gamma$. They also note from the theorem below and Theorem 15 that if the automorphism ϕ satisfies $r(I - \phi) < 1$, then $\phi = I$.

We have noted already the trivial fact that commutative algebras do not have non-zero inner derivations. The rest of this section is concerned with showing that derivations are rare on commutative Banach algebras.

Notation. For the rest of this section A will denote a complex commutative Banach algebra (with or without unit).

Theorem 16. (Singer and Wermer [360].) *Let* $\mathrm{rad}(A)$ *denote the radical of* A, *and let* D *be a continuous derivation on* A. *Then* $Da \in \mathrm{rad}(A)$ $(a \in A)$.

Proof. Let $\phi \in \Phi_A$ and $z \in \mathbb{C}$. Then zD is a continuous derivation on A, and so, by Proposition 7, $\exp(zD)$ is a continuous automorphism on A. Therefore the mapping $a \to \phi(\exp(zD)a)$ is a multiplicative linear functional on A. Therefore, given $a \in A$,

$$|\phi(\exp(zD)a)| \leqslant \|a\| \qquad (z \in \mathbb{C});$$

and so the mapping $z \to \phi(\exp(zD)a)$ is a bounded entire function. Since $\phi(Da)$ is the coefficient of z in the power series expansion of this function it follows that $\phi(Da) = 0$. Since ϕ is an arbitrary element of Φ_A, $Da \in \operatorname{rad}(A)$. $\quad \square$

Remark. A. M. Sinclair has shown us the following alternative proof of Theorem 16. Given $a \in A$, $\phi \in \Phi_A$, let $x = a - \phi(a)$. Then $Dx = Da$, and, by a combinatorial argument (see Sinclair [353]), $\phi(D^n x^n) = n! \, \phi((Dx)^n)$. Hence

$$|\phi(Dx)| = |\phi(D^n x^n)|^{\frac{1}{n}} (n!)^{-\frac{1}{n}} \leqslant \|D\| \, \|x\| \, (n!)^{-\frac{1}{n}},$$

and therefore $\phi(Da) = 0$, as required.

It follows at once from Theorem 16 that there are no non-zero continuous derivations on a semi-simple complex commutative Banach algebra. However by a theorem of Johnson [215], which we now prove, there are no non-zero derivations (continuous or otherwise) on such an algebra. We need two preliminary lemmas which have independent interest.

Lemma 17. *Let* $\phi_1, \ldots, \phi_n$ *be distinct elements of* Φ_A. *Then the mapping* $a \to (\phi_1(a), \ldots, \phi_n(a))$ *maps* A *onto* $\mathbb{C}^n$.

Proof. Given distinct elements ϕ, ψ of Φ_A, the kernel of ψ is not contained in the kernel of ϕ, and so there exists an element $x \in A$ with $\phi(x) = 1$ and $\psi(x) = 0$. In particular, for $i = 2, \ldots, n$, there exist elements $y_i \in A$ with $\phi_1(y_i) = 1$ and $\phi_i(y_i) = 0$. Let $a_1 = y_2 y_3 \ldots y_n$. Then $\phi_1(a_1) = 1$ and $\phi_i(a_1) = 0$ $(i = 2, \ldots, n)$. Thus the natural basis element $(1, 0, \ldots, 0)$ of $\mathbb{C}^n$ belongs to the image of A under the mapping $a \to (\phi_1(a), \ldots, \phi_n(a))$, and similarly the other natural basis elements of $\mathbb{C}^n$ belong to this image. $\quad \square$

Corollary 18. Φ_A *is a linearly independent subset of* A'.

Lemma 19. *Let* $\{\phi_n\}$ *be a sequence of distinct elements of* Φ_A, *and let* $k \in \mathbb{N}$. *Then there exists* $y_k \in A$ *such that*

$$\phi_i(y_k) = 0 \quad (1 \leqslant i \leqslant k - 1), \qquad \phi_i(y_k) \neq 0 \quad (i \geqslant k).$$

Proof. For $j = k, k+1, \ldots$, Lemma 17 allows the choice of $z_j \in A$ such that

$$\phi_i(z_j) = 0 \quad (1 \leqslant i \leqslant j-1), \qquad \phi_j(z_j) = 1 .$$

Define positive numbers c_j by

$$c_j^{-1} = 3^j (1 + \|z_k\|)(1 + \|z_{k+1}\|) \ldots (1 + \|z_j\|),$$

and let $y_k = \sum_{j=k}^{\infty} c_j z_j$. Then $\phi_i(y_k) = 0 \ (1 \leqslant i \leqslant k-1)$, and with $i \geqslant k$, we have $\phi_i(y_k) = c_i + \lambda$, where

$$|\lambda| = \left| \sum_{j=i+1}^{\infty} c_j \phi_i(z_j) \right| \leqslant \sum_{j=i+1}^{\infty} c_j \|z_j\| \leqslant c_i \sum_{n=1}^{\infty} 3^{-n} = \frac{1}{2} c_i ,$$

since $c_j \|z_j\| \leqslant 3^{i-j} c_i \ (j > i)$. □

Lemma 20. *Let D be a derivation on A, and for $\phi \in \Phi_A$ let $D'\phi$ be defined by $(D'\phi)(a) = \phi(Da) \ (a \in A)$. Then $\{\phi \in \Phi_A : D'\phi \notin A'\}$ is a finite set.*

Proof. We may assume that A has a unit element. Since $(D'\phi)(ab) = \phi(aDb + (Da)b)$, we have

$$(D'\phi)(ab) = \phi(a)(D'\phi)(b) \qquad (a \in A, b \in \ker \phi), \tag{4}$$

and

$$(D'\phi)(ab) = 0 \qquad (a, b \in \ker \phi). \tag{5}$$

Suppose that there exists a sequence $\{\phi_n\}$ of distinct elements of Φ_A such that $D'\phi_n \notin A' \ (n \in \mathbb{N})$. For $k \in \mathbb{N}$, let y_k be as in Lemma 19, and let $w_k = y_1^2 y_2^2 \ldots y_k^2$. Since $(D'\phi_n)(1) = 0$, $D'\phi_n$ is unbounded on $\ker \phi_n$, and therefore there exists $x \in \ker \phi_n$ with $\|x\|$ arbitrarily small and $|(D'\phi_n)(x)|$ arbitrarily large. Since $\phi_n(w_n) \neq 0$, we may therefore choose $\{x_n : n \in \mathbb{N}\}$ inductively so that

(i) $x_n \in \ker \phi_n$,
(ii) $\|x_n\| \leqslant 2^{-n} \|w_n\|^{-1}$,
(iii) $|\phi_n(w_n)| \, |(D'\phi_n)(x_n)| > n + |\sum_{r=1}^{n-1} (D'\phi_n)(w_r x_r)|$.

Let $z = \sum_{n=1}^{\infty} w_n x_n$. For each $p \in \mathbb{N}$, we may write z in the form

$$z = \sum_{k=1}^{p} w_k x_k + y_{p+1}^2 z_p .$$

Since y_{p+1} and $y_{p+1} z_p$ belong to $\ker \phi_p$, (5) gives $(D'\phi_p)(y_{p+1}^2 z_p) = 0$. Since also $x_p \in \ker \phi_p$, (4) gives $(D'\phi_p)(w_p x_p) = \phi_p(w_p)(D'\phi_p)(x_p)$, and therefore, by (iii), $|(D'\phi_p)(z)| \geqslant p$. Since $\|\phi_p\| \leqslant 1$, this gives

$$\|Dz\| \geqslant p \qquad (p \in \mathbb{N}). \quad □$$

Theorem 21. (Johnson [215]). *Let A be semi-simple. Then 0 is the only derivation on A.*

Proof. Let D be a derivation on A. By Theorem 16, it is enough to prove that D is continuous. Suppose that $x_n, y \in A$ with

$$\lim_{n \to \infty} x_n = 0, \quad \lim_{n \to \infty} D x_n = y.$$

By the closed graph theorem, it is enough to prove that $y = 0$.
If $\phi \in \Phi_A$ and $D' \phi \in A'$, then

$$\phi(y) = \lim_{n \to \infty} \phi(D x_n) = \lim_{n \to \infty} (D' \phi)(x_n) = 0.$$

By Lemma 20, we may suppose that $\{\phi \in \Phi_A : D' \phi \notin A'\} = \{\phi_1, \dots, \phi_n\}$ with $\phi_1, \dots, \phi_n$ distinct elements of Φ_A. Suppose that $\phi_1(y) \neq 0$. Then, by Lemma 17, there exists $z \in A$ with $\phi_1(z) = 1$, $\phi_j(z) = 0$ $(j = 2, \dots, n)$. Let $e = (\phi_1(y))^{-1} y z$. Then $\phi_1(e) = 1$, and $\phi(e) = 0$ $(\phi \in \Phi_A \setminus \{\phi_1\})$. Therefore $\phi(x e - \phi_1(x) e) = 0$ $(\phi \in \Phi_A, x \in A)$, and semi-simplicity of A gives

$$x e = \phi_1(x) e \quad (x \in A). \tag{6}$$

In particular $e^2 = e$, and so $De = 0$, by Proposition 5. It now follows from (6) that

$$(Dx)e = D(xe) = \phi_1(x) De = 0 \quad (x \in A),$$

and so $ye = \lim_{n \to \infty} (D x_n) e = 0$. But then

$$e = e^2 = (ye)(\phi_1(y))^{-1} z = 0,$$

contradicting $\phi_1(e) = 1$. Therefore $\phi_1(y) = 0$. Similarly $\phi_j(y) = 0$ for $j = 2, \dots, n$, and so $\phi(y) = 0$ $(\phi \in \Phi_A)$, $y = 0$. ☐

Corollary 22. *The algebra $C^\infty[0,1]$ of functions with continuous derivatives of all orders on $[0,1]$ cannot be normed so that it becomes a Banach algebra.*

Proof. The algebra $C^\infty[0,1]$ is semi-simple since the point evaluation functionals are multiplicative linear functionals; and the mapping $f \to f'$ is a non-zero derivation on the algebra. ☐

Corollary 23. *Let D be an open subset of $\mathbb{C}$. Then either $H^\infty(D)$ consists of constant functions only, or there exists $f \in H^\infty(D)$ with unbounded derivative on D.*

Proof. $H^\infty(D)$ is a semi-simple Banach algebra. If $f' \in H^\infty(D)$ for all $f \in H^\infty(D)$, then $f \to f'$ is a derivation on $H^\infty(D)$ which is non-zero if $H^\infty(D)$ contains a non-constant function. ☐

Johnson and Sinclair [219] show that all derivations on a semi-simple Banach algebra (not necessarily commutative) are continuous.

The situation is rather more complicated for commutative Banach algebras which are not semi-simple. Let $\{\alpha_n\}$ be a sequence of strictly positive real numbers such that $\alpha_{m+n} \leqslant \alpha_m \alpha_n$ for all $m,n \in \mathbb{N}$, and

$$\lim_{n \to \infty} \alpha_n^{\frac{1}{n}} = 0. \tag{7}$$

Let B denote the Banach algebra $l^1(\mathbb{N}, \alpha)$ as defined in Example 1.23 with S the additive semi-group $\mathbb{N}$. Let a be the element of B defined by $a(1)=1$, $a(n)=0$ $(n>1)$. The smallest closed subalgebra of B containing a is clearly B itself.

Proposition 24. *Let B be as above.*
  (i) $\mathrm{rad}(B) = B$.
  (ii) *Every derivation on B is continuous.*
  (iii) *A derivation D on B is uniquely determined by the value of Da.*

Proof. (i) We have $r(a)=0$ by (7). Since B is commutative it follows that $r(b)=0$ $(b \in B)$.

(ii) Let D be a derivation on B. Let ϕ_n be the continuous linear functional on B defined by

$$\phi_n(b) = b(n) \qquad (b \in B),$$

and let f_n be the linear functional on B defined by $f_n = \phi_n \circ D$. Let $k \in \mathbb{N}$. Clearly $a^k B \subset \ker(\phi_j)$ for $j=1, \ldots, k$, and since

$$\sum_{n=k+1}^{\infty} |b(n-k)| \alpha_n \leqslant \alpha_k \sum_{n=1}^{\infty} |b(n)| \alpha_n,$$

it follows that

$$a^k B = \bigcap_{j=1}^{k} \ker(\phi_j). \tag{8}$$

We have, for $b \in B$,

$$D(a^k b) = k a^{k-1}(Da)b + a^k Db = \lambda a^{k+1} + a^{k+1} c$$

for some $\lambda \in \mathbb{C}$, $c \in B$. Therefore $f_k(a^k b)=0$ $(b \in B)$, and (8) then gives

$$f_k = \sum_{j=1}^{k} f_k(a^j) \phi_j,$$

so that f_k is continuous.

Let $\{x_m\} \subset B$, and let $\lim_{m \to \infty} x_m = 0$, $\lim_{m \to \infty} Dx_m = y$. For each $k \in \mathbb{N}$ we have

$$\phi_k(y) = \lim_{m \to \infty} \phi_k(Dx_m) = \lim_{m \to \infty} f_k(x_m) = 0.$$

Therefore $y=0$, and the closed graph theorem applies.

(iii) This follows from (ii) and Proposition 4 (ii). □

Example 25. Let B be as in Proposition 24 with $\alpha_n = 1/n!$. Given $k \in \mathbb{N}$, $k \geqslant 2$, let D be defined on B by

$$Db = \sum_{n=1}^{\infty} b(n) n \, a^{n+k-1} \qquad (b \in B).$$

Then D is a derivation on B and $Da = a^k$. It is straightforward to verify in this case that there is a one-one correspondence between aB and the set of all derivations on B.

Example 26. (Newman [288]). Let B be as in Proposition 24 with $\alpha_n = \{\log(n+1)\}^{-n}$. Let D be a derivation on B and let $Da = \sum_{n=1}^{\infty} c_n a^n$. Then

$$Da^k = \sum_{n=1}^{\infty} k c_n a^{n+k-1} \qquad (k \in \mathbb{N}),$$

and since D is continuous,

$$k|c_n|\{\log(n+k)\}^{-n-k+1} \leqslant \|D\| \, \|a^k\| \, .$$

This gives, for $k \in \mathbb{N}$,

$$|c_n| \leqslant k^{-1} \|D\| \{\log(k+n)\}^{n-1} \left[\frac{\log(k+n)}{\log(k+1)} \right]^k$$
$$\leqslant k^{-1} \|D\| \{\log(k+n)\}^{n-1} e^{n-1} \, .$$

Therefore $c_n = 0$ $(n \in \mathbb{N})$, $D = 0$.

§ 19. Generators and Joint Spectra

Throughout this section A will denote a complex commutative Banach algebra with unit, Φ_A its carrier space.

Definition 1. Given a finite subset $\{a_1, \ldots, a_n\}$ of A, we denote by $A(a_1, \ldots, a_n)$ the least closed subalgebra of A containing the set $\{1, a_1, \ldots, a_n\}$, and we call $A(a_1, \ldots, a_n)$ the subalgebra *generated* by $\{a_1, \ldots, a_n\}$. The elements $\{a_1, \ldots, a_n\}$ are called a set of *joint generators* for A if $A(a_1, \ldots, a_n) = A$. In particular, an element a of A is a *generator* for A if $A(a) = A$. An algebra that possesses a generator is said to be *monothetic*. A is said to be *finitely generated* if there exists a finite set of joint generators for A.

Theorem 2. *Let u be a generator for A. Then the mapping*

$$\phi \to \phi(u) \qquad (\phi \in \Phi_A)$$

is a homeomorphism of the carrier space Φ_A onto $\mathrm{Sp}(u)$.

Proof. By Proposition 16.9 or Theorem 17.4, $\{\phi(u): \phi \in \Phi_A\} = \mathrm{Sp}(u)$, and the mapping $\phi \to \phi(u)$ is continuous by definition of the A-topology. Therefore it is enough to prove that the mapping is injective. Suppose that $\phi, \psi \in \Phi_A$ with $\phi(u) = \psi(u)$, and let $B = \{x \in A: \phi(x) = \psi(x)\}$. Since ϕ, ψ are multiplicative linear functionals and continuous, B is a closed subalgebra of A containing $1, u$. Therefore $B = A$, $\phi = \psi$. ☐

Example 3. Let A denote the disc algebra $\mathscr{A}(\varDelta)$ (Example 1.16), and u the element of A given by $u(z) = z$ ($z \in \varDelta$). Then u is a generator for A; for, given $a \in A$ and $0 < \rho < 1$, let a_ρ be defined by $a_\rho(z) = a(\rho z)$ ($z \in \varDelta$). Then $a_\rho \in A(u)$ and $\lim_{\rho \to 1-0} \|a - a_\rho\|_\infty = 0$, so that $a \in A(u)$. It follows from Theorem 2 that the mapping $\phi \to \phi(u)$ is a homeomorphism of Φ_A onto $\varDelta$. Also

$$\phi(a) = a(\phi(u)) \qquad (a \in A, \phi \in \Phi_A);$$

for given $a \in A$, we have $a(z) = \sum_{n=0}^\infty \alpha_n z^n$ ($|z| < 1$), and so $a_\rho = \sum_{n=0}^\infty \alpha_n \rho^n u^n$, $\phi(a_\rho) = a_\rho(\phi(u)) = a(\rho \phi(u))$ ($0 < \rho < 1$).

Example 4. Let A denote the Wiener algebra W (Example 1.19), and let u be the element of A given by

$$u(t) = \exp(it) \qquad (0 \leqslant t \leqslant 2\pi).$$

Then $u \in \mathrm{Inv}(A)$ and $u^{-1}(t) = \exp(-it)$ ($0 \leqslant t \leqslant 2\pi$). Given $a \in A$, we have

$$a(t) = \sum_{k \in \mathbb{Z}} \alpha_k \exp(ikt) \qquad (0 \leqslant t \leqslant 2\pi),$$

with $\sum_{k \in \mathbb{Z}} |\alpha_k| = \|a\|$. Thus

$$a = \sum_{k \in \mathbb{Z}} \alpha_k u^k, \tag{1}$$

with the series convergent in norm, and so $\{u, u^{-1}\}$ is a set of joint generators for A.

We observe next that $\mathrm{Sp}(u) = \Gamma = \{z \in \mathbb{C}: |z| = 1\}$. For, since W is an algebra of functions, we have $\Gamma \subset \mathrm{Sp}(u)$; and on the other hand the inequalities $r(u) \leqslant \|u\| = 1$, $r(u^{-1}) \leqslant \|u^{-1}\| = 1$, show that $|z| \leqslant 1$ and $|z^{-1}| \leqslant 1$ whenever $z \in \mathrm{Sp}(u)$.

By (1), given $\phi \in \Phi_A$ and $a \in A$, we have

$$\phi(a) = \sum_{k \in \mathbb{Z}} \alpha_k (\phi(u))^k = a(t),$$

where $t\in[0,2\pi]$ with $e^{it}=\phi(u)$. Thus $\mathrm{Sp}(a)=\{a(t):t\in[0,2\pi]\}$, and in particular $a\in\mathrm{Inv}(A)$ if and only if $a(t)\neq0$ $(t\in[0,2\pi])$, which is Wiener's theorem on absolutely convergent Fourier series.

Theorem 5. *Let u be a generator for A. Then $\mathbb{C}\backslash\mathrm{Sp}(u)$ is connected.*

Proof. Suppose that the open set $\mathbb{C}\backslash\mathrm{Sp}(u)$ has a non-void bounded component G, and let $\zeta_0\in G$. Let A_0 denote the least subalgebra of A containing 1 and u. Given a polynomial $p\in P(\mathbb{C})$, we have by the maximum modulus principle (using $\partial G\subset\mathrm{Sp}(u)$) and the spectral mapping theorem,

$$\begin{aligned}|p(\zeta_0)| &\leqslant \max\{|p(\zeta)|:\zeta\in\partial G\}\\ &\leqslant \max\{|p(\zeta)|:\zeta\in\mathrm{Sp}(u)\}\\ &= \max\{|\lambda|:\lambda\in\mathrm{Sp}(p(u))\}\\ &= r(p(u))\leqslant\|p(u)\|.\end{aligned}\tag{2}$$

Given $b\in A_0$, there exists $p\in P(\mathbb{C})$ with $p(u)=b$, and (2) shows that if $p_1(u)=p_2(u)$, then $p_1(\zeta_0)=p_2(\zeta_0)$. Therefore we can define a homomorphism ϕ_0 of A_0 into $\mathbb{C}$ by taking

$$\phi_0(b)=p(\zeta_0),$$

with p chosen so that $p(u)=b$. By (2), ϕ_0 is continuous on A_0, and therefore extends by continuity to a unique homomorphism ϕ of A into $\mathbb{C}$. We have $\phi_0(1)=1$, and so $\phi\in\Phi_A$. Since $u=p(u)$, where p is the polynomial $p(z)=z$ $(z\in\mathbb{C})$, we have

$$\phi(u)=\phi_0(u)=p(\zeta_0)=\zeta_0.$$

Therefore $\zeta_0\in\mathrm{Sp}(u)$, which is a contradiction. □

Theorem 5 characterizes the spectra of generators, as the following proposition shows.

Proposition 6. *Let K be a non-void compact set in $\mathbb{C}$ with $\mathbb{C}\backslash K$ connected. Then there exists a commutative complex Banach algebra A with unit, and with a generator u, such that*

$$\mathrm{Sp}(u)=K.$$

Proof. Let u denote the function defined on K by $u(z)=z$ $(z\in K)$, and let A be the least closed subalgebra of $C(K)$ containing 1 and u. It is clear that $K\subset\mathrm{Sp}(A,u)$. If $\lambda\notin K$, we have

$$\inf\{|\lambda-u(z)|:z\in K\}=\kappa>0,$$

and therefore

$$\|(\lambda-u)f\|_\infty\geqslant\kappa\|f\|_\infty\qquad(f\in A).$$

Therefore $\lambda - u$ is not a topological divisor of zero in A, and we have proved that $\partial \operatorname{Sp}(A, u) \subset K$. Let $G = \operatorname{int} \operatorname{Sp}(A, u)$, $H = \mathbb{C} \setminus \operatorname{Sp}(A, u)$. Then G, H are mutually disjoint open sets,

$$\mathbb{C} \setminus K \subset \mathbb{C} \setminus \partial \operatorname{Sp}(A, u) = G \cup H,$$

and $(\mathbb{C} \setminus K) \cap H$ is not void. Therefore the connectedness of $\mathbb{C} \setminus K$ gives $\mathbb{C} \setminus K \subset H$, $\operatorname{Sp}(A, u) \subset K$. □

Notation. For the rest of this section, n denotes a positive integer, $\mathbb{C}^n$ and A^n the Cartesian products of n copies of $\mathbb{C}$ and A respectively, and $P(\mathbb{C}^n)$ the set of complex polynomials in n complex variables. Given $p \in P(\mathbb{C}^n)$ and $\mathbf{a} \in A^n$, $p(\mathbf{a})$ is the element of A defined in the natural way by replacing the variables $z_1, \ldots, z_n$ by the elements $a_1, \ldots, a_n$.

Definition 7. Given $\mathbf{a} = (a_1, a_2, \ldots, a_n) \in A^n$, the *joint spectrum* of $\mathbf{a}$ is the subset $\operatorname{Sp}(A, \mathbf{a})$ of $\mathbb{C}^n$ given by

$$\operatorname{Sp}(A, \mathbf{a}) = \{\phi(\mathbf{a}) : \phi \in \Phi_A\},$$

where $\phi(\mathbf{a}) = (\phi(a_1), \ldots, \phi(a_n))$. As in the case $n = 1$, we write $\operatorname{Sp}(\mathbf{a})$ for $\operatorname{Sp}(A, \mathbf{a})$ when no confusion is possible.

Proposition 8. *Let* $\mathbf{a} \in A^n$, $\lambda \in \mathbb{C}^n$, *and let*

$$J = \left\{ \sum_{k=1}^{n} (\lambda_k - a_k) b_k : \mathbf{b} \in A^n \right\}.$$

Then the following statements are equivalent:
   (i) $\lambda \in \operatorname{Sp}(\mathbf{a})$;
   (ii) *J is a proper ideal of A;*
   (iii) *$J \subset \operatorname{Sing} A$;*
   (iv) *$1 \notin J$.*

Proof. (i) $\Rightarrow$ (ii). If $\lambda = \phi(\mathbf{a})$, then $\phi(\lambda_k - a_k) = 0$ $(k = 1, \ldots, n)$ and so J is contained in the kernel of ϕ.

(ii) $\Rightarrow$ (i). If J is a proper ideal it is contained in the kernel of some $\phi \in \Phi_A$. Then taking $b_j = 1$ $(j = k)$, $b_j = 0$ $(j \neq k)$, we have $\phi(\lambda_k - a_k) = 0$. Thus $\lambda = \phi(\mathbf{a}) \in \operatorname{Sp}(\mathbf{a})$.

(ii) $\Leftrightarrow$ (iii) $\Leftrightarrow$ (iv). Clear. □

Proposition 9. *Let* $\{a_1, \ldots, a_n\}$ *be a set of joint generators for A and let* $\lambda \in \mathbb{C}^n \setminus \operatorname{Sp}(\mathbf{a})$. *Then there exists a polynomial* $p \in P(\mathbb{C}^n)$ *such that*

$$|p(\lambda)| > \|p(\mathbf{a})\|.$$

Proof. By Proposition 8, there exists $\mathbf{b} = (b_1, \ldots, b_n) \in A^n$ with $\sum_{k=1}^{n} (\lambda_k - a_k) b_k = 1$. Since $\{a_1, \ldots, a_n\}$ is a set of joint generators for A, the elements b_k can be approximated arbitrarily closely by elements of the form $p(\mathbf{a})$ with $p \in P(\mathbb{C}^n)$. Thus there exist polynomials $p_k \in P(\mathbb{C}^n)$ such that

$$\left\| 1 - \sum_{k=1}^{n} (\lambda_k - a_k) p_k(\mathbf{a}) \right\| < 1.$$

Let $p(\mathbf{z}) = 1 - \sum_{k=1}^{n} (\lambda_k - z_k) p_k(\mathbf{z})$. Then $p \in P(\mathbb{C}^n)$, $p(\lambda) = 1$, and $\|p(\mathbf{a})\| < 1$. $\quad\square$

Notation. Given a compact subset K of a topological space E, we denote by $|\cdot|_K$ the semi-norm on $C(E)$ defined by

$$|f|_K = \sup \{|f(x)| : x \in K\}.$$

Definition 10. A compact subset K of $\mathbb{C}^n$ is *polynomially convex* if K contains all points $\mathbf{z} \in \mathbb{C}^n$ such that

$$|p(\mathbf{z})| \leqslant |p|_K \qquad (p \in P(\mathbb{C}^n)).$$

Theorem 11. *Let $\{a_1, \dots, a_n\}$ be a set of joint generators for A. Then $\operatorname{Sp}(\mathbf{a})$ is a non-void polynomially convex compact subset of $\mathbb{C}^n$.*

Proof. That $\operatorname{Sp}(\mathbf{a})$ is a non-void compact subset of $\mathbb{C}^n$ is immediate from the fact that Φ_A is a non-void and compact set in the A-topology together with the continuity of the mapping $\phi \to \phi(\mathbf{a})$. That $\operatorname{Sp}(\mathbf{a})$ is polynomially convex follows from Proposition 9, for, given $\mathbf{z} \in \operatorname{Sp}(\mathbf{a})$, we have $\mathbf{z} = \phi(\mathbf{a})$ for some $\phi \in \Phi_A$, and so

$$|p(\mathbf{z})| = |p(\phi(\mathbf{a}))| = |\phi(p(\mathbf{a}))| \leqslant \|p(\mathbf{a})\|,$$

from which $|p|_{\operatorname{Sp}(\mathbf{a})} \leqslant \|p(\mathbf{a})\|$ $(p \in P(\mathbb{C}^n))$. $\quad\square$

The properties in Theorem 11 characterize the joint spectra of sets of joint generators, as the following proposition shows.

Proposition 12. *Let K be a non-void polynomially convex compact subset of $\mathbb{C}^n$. Then there exists a complex commutative Banach algebra A with unit and with a set $\{u_1, \dots, u_n\}$ of joint generators such that $\operatorname{Sp}(\mathbf{u}) = K$.*

Proof. Given $j \in \{1, \dots, n\}$, let u_j denote the function on K defined by $u_j(\mathbf{z}) = z_j$ $(\mathbf{z} \in K)$. Let A be the least closed subalgebra of $C(K)$ containing $\{1, u_1, \dots, u_n\}$. Given $\zeta \in K$, the evaluation functional $f \to f(\zeta)$ is a non-zero homomorphism of A into $\mathbb{C}$, ϕ_ζ say, and we have $\phi_\zeta(u_j) = \zeta_j$. Therefore $\zeta = \phi_\zeta(\mathbf{u}) \in \operatorname{Sp}(\mathbf{u})$ $(\zeta \in K)$; $K \subset \operatorname{Sp}(\mathbf{u})$.

On the other hand, if $\zeta \in \mathbb{C}^n \setminus K$, then, by polynomial convexity of K, there exists $p \in P(\mathbb{C}^n)$ such that

$$|p(\zeta)| > |p|_K.$$

Let $a=p(\mathbf{u})$. Then $a\in A$, and

$$a(\mathbf{z}) = p(\mathbf{z}) \qquad (\mathbf{z}\in K).$$

Therefore $|p(\zeta)|>\|a\|_\infty$. It follows that $\zeta\notin\mathrm{Sp}(\mathbf{u})$, for otherwise there exists $\phi\in\Phi_A$ with $\zeta=\phi(\mathbf{u})$, and then

$$|p(\zeta)| = |p(\phi(\mathbf{u}))| = |\phi(p(\mathbf{u}))| = |\phi(a)| \leqslant \|a\|_\infty. \qquad \Box$$

Example 13. With the notation of Example 4, we have seen that $\{u,u^{-1}\}$ is a set of joint generators for W. Then $\mathrm{Sp}(u,u^{-1})=\{(z,z^{-1}):|z|=1\}$. By Theorem 11, $K=\mathrm{Sp}(u,u^{-1})$ is polynomially convex. To see this directly, suppose that $\zeta\in\mathbb{C}^2\setminus K$. If $|\zeta_1|>1$, take $p(\mathbf{z})=z_1$. Then $|p(\zeta)|>1=|p|_K$. Similarly if $|\zeta_2|>1$, take $p(\mathbf{z})=z_2$. If $|\zeta_1|\leqslant 1$ and $|\zeta_2|\leqslant 1$, take $p(\mathbf{z})=1-z_1 z_2$. Then $|p|_K=0$ and $|p(\zeta)|>0$ unless $\zeta_1\zeta_2=1$, in which case we should have $|\zeta_1|=1$ and $\zeta_2=1/\zeta_1$.

Remark. Theorem 2 is due to Gelfand, and the remainder of the section to Shilov [344], [346].

§ 20. A Functional Calculus for Several Banach Algebra Elements

Let A be a complex commutative Banach with unit, let n be a positive integer, let A^n denote the Cartesian product of n copies of A, and let $H(D)$ denote the algebra of complex holomorphic functions on an open subset D of $\mathbb{C}^n$.

We construct a functional calculus for holomorphic functions of n variables analogous to the functional calculus in § 7, but weaker and less explicit. Given $\mathbf{a}=(a_1,\ldots,a_n)\in A^n$, an open neighbourhood D of the joint spectrum $\mathrm{Sp}(\mathbf{a})$, and $f\in H(D)$, we prove the existence of $y\in A$ such that

$$\phi(y)=f(\phi(a_1),\ldots,\phi(a_n)) \qquad (\phi\in\Phi_A).$$

When A is semi-simple, there exists exactly one such element y and we obtain a continuous homomorphism of $H(D)$ into A extending the natural homomorphism of $P(\mathbb{C}^n)$ into A.

Let Δ^n denote the closed polydisc in $\mathbb{C}^n$,

$$\Delta^n=\{\mathbf{z}\in\mathbb{C}^n:|z_k|\leqslant 1 \ (k=1,\ldots,n)\}.$$

We follow several other authors in basing the functional calculus on the following fundamental theorem from the theory of holomorphic functions of several complex variables.

Theorem 1 (The Oka extension theorem [289]). *Let* $n, m \in \mathbb{N}$, $p_1, \ldots, p_m \in P(\mathbb{C}^n)$, *and let* π *denote the mapping of* $\mathbb{C}^n$ *into* $\mathbb{C}^{n+m}$ *given by*

$$\pi(\mathbf{z}) = (z_1, \ldots, z_n, p_1(\mathbf{z}), \ldots, p_m(\mathbf{z})) \qquad (\mathbf{z} = (z_1, \ldots, z_n) \in \mathbb{C}^n).$$

Given f holomorphic on an open neighbourhood of $\pi^{-1}(\Delta^{n+m})$, there exists F holomorphic on an open neighbourhood of Δ^{n+m} such that

$$F(\pi(\mathbf{z})) = f(\mathbf{z}) \qquad (\mathbf{z} \in \pi^{-1}(\Delta^{n+m})).$$

Remark. For the proof of Theorem 1, we refer the reader to Wermer [410, pp. 38—42] and to Gunning and Rossi [182, pp. 39—41].

Corollary 2. *Let* $n, m \in \mathbb{N}$, $p_1, \ldots, p_m \in P(\mathbb{C}^n)$, $\varepsilon_k > 0$ $(1 \leqslant k \leqslant m+n)$,

$$\Gamma = \{\mathbf{z} \in \mathbb{C}^{n+m} : |z_k| \leqslant \varepsilon_k \, (1 \leqslant k \leqslant m+n)\},$$

and let

$$\pi(\mathbf{z}) = (z_1, \ldots, z_n, p_1(\mathbf{z}), \ldots, p_m(\mathbf{z})) \qquad (\mathbf{z} = (z_1, \ldots, z_n) \in \mathbb{C}^n).$$

Then, given f holomorphic on an open neighbourhood of $\pi^{-1}(\Gamma)$, there exists F holomorphic on an open neighbourhood of Γ such that

$$F(\pi(\mathbf{z})) = f(\mathbf{z}) \qquad (\mathbf{z} \in \pi^{-1}(\Gamma)).$$

Proof. Let

$$\psi(w_1, \ldots, w_n) = (\varepsilon_1 w_1, \ldots, \varepsilon_n w_n), \quad \phi(z_1, \ldots, z_{n+m}) = (\varepsilon_1^{-1} z_1, \ldots, \varepsilon_{n+m}^{-1} z_{n+m}),$$

$$q_j(\mathbf{w}) = \varepsilon_{n+j}^{-1} p_j(\psi(\mathbf{w})) \qquad (j = 1, \ldots, m),$$

and

$$\sigma(\mathbf{w}) = (w_1, \ldots, w_n, q_1(\mathbf{w}), \ldots, q_m(\mathbf{w})) \qquad (\mathbf{w} \in \mathbb{C}^n).$$

Then $\sigma(\mathbf{w}) = \phi(\pi(\psi(\mathbf{w})))$, $\Delta^{n+m} = \phi(\Gamma)$,

$$\sigma^{-1}(\Delta^{n+m}) = \psi^{-1} \pi^{-1}(\Gamma).$$

Let U be an open neighbourhood of $\pi^{-1}(\Gamma)$ in $\mathbb{C}^n$, and let $f \in H(U)$. Then $\psi^{-1}(U)$ is an open neighbourhood of $\sigma^{-1}(\Delta^{n+m})$, and $f \circ \psi \in H(\psi^{-1}(U))$. By Theorem 1, there exists G holomorphic on an open neighbourhood V of Δ^{n+m} such that $G(\sigma(\mathbf{w})) = (f \circ \psi)(\mathbf{w})$ $(\mathbf{w} \in \sigma^{-1}(\Delta^{n+m}))$. Let $F = G \circ \phi$. Then F is holomorphic on $\phi^{-1}(V)$ which is an open neighbourhood of Γ, and for $\mathbf{z} \in \pi^{-1}(\Gamma)$, we have $\mathbf{z} = \psi(\mathbf{w})$ with $\mathbf{w} \in \sigma^{-1}(\Delta^{n+m})$, and so

$$f(\mathbf{z}) = f(\psi(\mathbf{w})) = G(\sigma(\mathbf{w})) = G(\phi(\pi(\psi(\mathbf{w})))) = F(\pi(\mathbf{z})). \quad \Box$$

Lemma 3. *Let* $\mathbf{a} = (a_1, \ldots, a_n) \in A^n$, *and let U be an open neighbourhood of $\mathrm{Sp}(\mathbf{a})$ in $\mathbb{C}^n$. Then there exists a finitely generated closed subalgebra B of A containing $1, a_1, \ldots, a_n$ such that*

$$\mathrm{Sp}(B, \mathbf{a}) \subset U.$$

Proof. Let $\Gamma=\{\mathbf{z}\in\mathbf{C}^n: |z_k|\leqslant\|a_k\| \ (1\leqslant k\leqslant n)\}$. By Proposition 19.8, given $\mathbf{z}\in\mathbf{C}^n\backslash\mathrm{Sp}(\mathbf{a})$, there exists $\mathbf{b}\in A^n$ with

$$\sum_{j=1}^{n}(z_j-a_j)b_j=1 . \tag{1}$$

Let $B(\mathbf{z})$ be the least closed subalgebra of A containing $1, a_1,\dots,a_n, b_1,\dots,b_n$. Then $\mathbf{b}\in B(\mathbf{z})^n$, and so by (1) and Proposition 19.8 again, $\mathbf{z}\notin\mathrm{Sp}(B(\mathbf{z}),\mathbf{a})$. Therefore there exists an open neighbourhood $G(\mathbf{z})$ of $\mathbf{z}$ that has void intersection with $\mathrm{Sp}(B(\mathbf{z}),\mathbf{a})$. Let $\mathbf{w}^1,\dots,\mathbf{w}^m$ be chosen so that the open sets $G(\mathbf{w}^1),\dots,G(\mathbf{w}^m)$ cover the compact set $\Gamma\backslash U$. Each of the algebras $B(\mathbf{w}^k)$ is finitely generated, and so there exists a finitely generated closed subalgebra B of A such that $B(\mathbf{w}^k)\subset B \ (k=1,\dots,m)$. Since $\mathrm{Sp}(B,\mathbf{a})$ $\subset\mathrm{Sp}(B(\mathbf{w}^k),\mathbf{a})$, we have

$$\mathrm{Sp}(B,\mathbf{a})\cap G(\mathbf{w}^k)=\emptyset \quad (k=1,\dots,m),$$

and therefore $\mathrm{Sp}(B,\mathbf{a})\cap(\Gamma\backslash U)=\emptyset$. But $\mathrm{Sp}(B,\mathbf{a})\subset\Gamma$, and so $\mathrm{Sp}(B,\mathbf{a})\subset U$. ☐

Theorem 4. *Let* $\mathbf{a}=(a_1,\dots,a_n)\in A^n$, *and let D be an open neighbourhood of* $\mathrm{Sp}(\mathbf{a})$ *in* $\mathbf{C}^n$. *Then there exist* $a_{n+1},\dots,a_N\in A$ *such that, given* $f\in H(D)$, *there exists F holomorphic on an open neighbourhood of the polydisc*

$$\{\mathbf{z}\in\mathbf{C}^N: |z_k|\leqslant 1+\|a_k\| \ (k=1,\dots,N)\}$$

with

$$f(\phi(a_1),\dots,\phi(a_n))=F(\phi(a_1),\dots,\phi(a_N)) \quad (\phi\in\Phi_A).$$

Proof. By Lemma 3, there exists a finitely generated closed subalgebra B of A containing $1,a_1,\dots,a_n$ such that $\mathrm{Sp}(B,\mathbf{a})\subset D$. Choose elements $a_{n+1},\dots,a_r$ of B so that $\{a_1,\dots,a_r\}$ is a set of joint generators for B, let $\mathbf{b}=(a_1,\dots,a_r)$, and let σ be the natural projection $\sigma(z_1,\dots,z_r)$ $=(z_1,\dots,z_n)$ of $\mathbf{C}^r$ onto $\mathbf{C}^n$. By Proposition 19.9, given $\mathbf{z}\in\mathbf{C}^r\backslash\mathrm{Sp}(\mathbf{b})$, there exists $p\in P(\mathbf{C}^r)$ with $|p(\mathbf{z})|>1+\|p(\mathbf{b})\|$, and so there exists an open neighbourhood $V(p,\mathbf{z})$ of $\mathbf{z}$ such that $|p(\mathbf{w})|>1+\|p(\mathbf{b})\| \ (\mathbf{w}\in V(p,\mathbf{z}))$. Let

$$\Omega=\{\mathbf{z}\in\mathbf{C}^r: |z_k|\leqslant 1+\|a_k\| \ (k=1,\dots,r)\} .$$

Then $\Omega\backslash\sigma^{-1}(D)$ is compact, and, since $\sigma(\mathrm{Sp}(\mathbf{b}))=\mathrm{Sp}(\mathbf{a})$, $\sigma^{-1}(D)$ contains $\mathrm{Sp}(\mathbf{b})$. Therefore $\Omega\backslash\sigma^{-1}(D)$ can be finitely covered by open sets of the form $V(p,\mathbf{z})$. Thus there exists a finite subset $\{p_1,\dots,p_m\}$ of $P(\mathbf{C}^r)$ such that, for each $\mathbf{z}\in\Omega\backslash\sigma^{-1}(D)$, we have $|p_k(\mathbf{z})|>1+\|p_k(\mathbf{b})\|$ for at least one $k\in\{1,\dots,m\}$. Let $N=r+m$, $a_{r+k}=p_k(\mathbf{b}) \ (k=1,\dots,m)$,

$$\Gamma=\{\mathbf{z}\in\mathbf{C}^N: |z_k|\leqslant 1+\|a_k\| \ (k=1,\dots,N)\} ,$$

and let π be the mapping of $\mathbb{C}^r$ into $\mathbb{C}^N$ given by

$$\pi(\mathbf{z}) = (z_1, \ldots, z_r, p_1(\mathbf{z}), \ldots, p_m(\mathbf{z})) \qquad (\mathbf{z} \in \mathbb{C}^r) \,.$$

If $\pi(\mathbf{z}) \in \Gamma$, we have $\mathbf{z} \in \Omega$ and $|p_k(\mathbf{z})| \leqslant 1 + \|p_k(\mathbf{b})\|$ $(k = 1, \ldots, m)$. Thus

$$\pi^{-1}(\Gamma) \subset \sigma^{-1}(D) \,.$$

Since $f \circ \sigma$ is holomorphic on the open neighbourhood $\sigma^{-1}(D)$ of $\pi^{-1}(\Gamma)$, Corollary 2 gives the existence of F holomorphic on an open neighbourhood of Γ such that

$$F(\pi(\mathbf{z})) = (f \circ \sigma)(\mathbf{z}) \qquad (\mathbf{z} \in \pi^{-1}(\Gamma)) \,.$$

Let $\phi \in \Phi_A$, and $z_k = \phi(a_k)$ $(k = 1, \ldots, r)$. Then

$$p_k(\mathbf{z}) = \phi(p_k(\mathbf{b})) = \phi(a_{r+k}) \qquad (k = 1, \ldots, m) \,,$$

and so $\pi(\mathbf{z}) = (\phi(a_1), \ldots, \phi(a_N))$. We have $|z_k| \leqslant \|a_k\|$ $(k = 1, \ldots, r)$, and $|p_k(\mathbf{z})| \leqslant \|a_{r+k}\|$ $(k = 1, \ldots, m)$, and so $\pi(\mathbf{z}) \in \Gamma$, i.e. $\mathbf{z} \in \pi^{-1}(\Gamma)$. Therefore

$$f(\phi(a_1), \ldots, \phi(a_n)) = (f \circ \sigma)(\mathbf{z}) = F(\pi(\mathbf{z}))$$
$$= F(\phi(a_1), \ldots, \phi(a_N)) \,. \quad \square$$

Theorem 5. *Let* $\mathbf{a} = (a_1, \ldots, a_n) \in A^n$, *let* D *be an open neighbourhood of* $\mathrm{Sp}(\mathbf{a})$ *in* $\mathbb{C}^n$, *and let* $f \in H(D)$. *Then there exists* $y \in A$ *such that*

$$\phi(y) = f(\phi(a_1), \ldots, \phi(a_n)) \qquad (\phi \in \Phi_A) \,.$$

Proof. By Theorem 4, there exist $a_{n+1}, \ldots, a_N \in A$ and F holomorphic on an open neighbourhood of the polydisc

$$\Gamma = \{\mathbf{z} \in \mathbb{C}^N : |z_k| \leqslant 1 + \|a_k\| \ (k = 1, \ldots, N)\} \,,$$

such that

$$f(\phi(a_1), \ldots, \phi(a_n)) = F(\phi(a_1), \ldots, \phi(a_N)) \qquad (\phi \in \Phi_A) \,.$$

The function F has a power series expansion

$$F(\mathbf{z}) = \sum_{\mathbf{k} \in \mathbb{N}^N} \alpha_{\mathbf{k}} z_1^{k_1} \ldots z_N^{k_N}$$

(with $\mathbf{k} = (k_1, \ldots, k_N)$), convergent on a neighbourhood of Γ and therefore absolutely convergent on Γ. Therefore the series

$$\sum_{\mathbf{k}} |\alpha_{\mathbf{k}}| \, \|a_1\|^{k_1} \ldots \|a_N\|^{k_N}$$

converges, and so the series $\sum_{\mathbf{k}} \alpha_{\mathbf{k}} a_1^{k_1} \ldots a_N^{k_N}$ converges in norm to an element y of A. Then, for $\phi \in \Phi_A$,

$$\phi(y) = \sum_{\mathbf{k}} \alpha_{\mathbf{k}} (\phi(a_1))^{k_1} \ldots (\phi(a_N))^{k_N} = F(\phi(a_1), \ldots, \phi(a_N)) \,. \quad \square$$

Corollary 6. *Let A be semi-simple, let $\mathbf{a} = (a_1, \ldots, a_n) \in A^n$, and let D be an open neighbourhood of $\mathrm{Sp}(\mathbf{a})$ in $\mathbb{C}^n$. Then there exists a homomorphism $f \to \mathbf{f}$ of $H(D)$ into A which extends the natural homomorphism of $P(\mathbb{C}^n)$ into A and is continuous with respect to the topology of uniform convergence on compact subsets of D.*

Proof. Given $f \in H(D)$, there exists $y \in A$ with

$$\phi(y) = f(\phi(a_1), \ldots, \phi(a_n)) \qquad (\phi \in \Phi_A); \tag{2}$$

and, since Φ_A separates the points of A, there is exactly one element y of A satisfying (2), which we denote by $\mathbf{f}$. Given $p \in P(\mathbb{C}^n)$, we have

$$\phi(p(a_1, \ldots, a_n)) = p(\phi(a_1), \ldots, \phi(a_n)) \qquad (\phi \in \Phi_A),$$

and so $\mathbf{p} = p(a_1, \ldots, a_n)$. That the mapping $f \to \mathbf{f}$ is a homomorphism also follows at once from the fact that each $\phi \in \Phi_A$ is a homomorphism together with the uniqueness of y satisfying (2).

The linear space $H(D)$ with the topology of uniform convergence on compact subsets of D is an F-space, and so the closed graph theorem (Dunford and Schwartz [130] Part I, p. 57) can be applied to prove the continuity of the mapping $f \to \mathbf{f}$. Suppose that $f_n \in H(D)$, $\lim\limits_{n \to \infty} f_n = 0$, $\lim\limits_{n \to \infty} \mathbf{f}_n = x$. Then $\phi(x) = \lim\limits_{n \to \infty} \phi(\mathbf{f}_n) = \lim\limits_{n \to \infty} f_n(\phi(a_1), \ldots, \phi(a_n)) = 0$ $(\phi \in \Phi_A)$. Therefore $x = 0$, by the semi-simplicity of A. ∎

Theorem 5 was proved by Shilov [346] under the hypothesis that A be finitely generated, and was extended to the general case by Arens and Calderon [28]. Stronger forms were proved by Waelbroeck [402] and Guichardet [181], for which see also Bourbaki [84].

§ 21. Functions Analytic on a Neighbourhood of the Carrier Space

Let A be a complex commutative Banach algebra with unit, A' its dual space, and Φ_A its carrier space. We shall be concerned with complex valued functions defined on an open neighbourhood D of Φ_A in A' that are analytic on D in the sense of the following definition.

Definition 1. Let D be an open subset of a linear topological space X over $\mathbb{C}$. Given $n \in \mathbb{N}$, and $\mathbf{x} = (x_0, x_1, \ldots, x_n) \in X^{n+1}$, let $\pi_{\mathbf{x}}$ be the mapping of $\mathbb{C}^n$ into X defined by

$$\pi_{\mathbf{x}}(\mathbf{z}) = x_0 + z_1 x_1 + \cdots + z_n x_n \qquad (\mathbf{z} = (z_1, \ldots, z_n) \in \mathbb{C}^n).$$

A complex valued function f defined on D is said to be *analytic* on D if the following conditions are satisfied:

(i) f is locally bounded on D, i.e. bounded on some neighbourhood of each point of D;

(ii) given $n \in \mathbb{N}$ and $\mathbf{x} = (x_0, \ldots, x_n) \in X^{n+1}$, the function $f \circ \pi_{\mathbf{x}}$ is holomorphic on $\pi_{\mathbf{x}}^{-1}(D)$.

Example 2. Let D_1, D_2 be disjoint open subsets of a linear topological space X over $\mathbb{C}$, let $D = D_1 \cup D_2$ and let f be defined on D by $f(x) = 1$ $(x \in D_1)$, $f(x) = 0$ $(x \in D_2)$. Obviously, f is locally bounded on D; and, given $\mathbf{x} \in X^{n+1}$, $\pi_{\mathbf{x}}^{-1}(D) = \pi_{\mathbf{x}}^{-1}(D_1) \cup \pi_{\mathbf{x}}^{-1}(D_2)$ expresses $\pi_{\mathbf{x}}^{-1}(D)$ as a union of disjoint open sets on each of which $f \circ \pi_{\mathbf{x}}$ is constant valued and therefore holomorphic. Thus f is analytic on D.

Lemma 3. *Let X, Y be complex linear spaces, let $\langle , \rangle$ be a bilinear form on $X \times Y$, let D be a subset of X that is open in the weak topology $w(X, Y)$, and let f be a complex valued function defined on D that is analytic with respect to the topology $w(X, Y)$. Given $x_0 \in D$, there exists a finite subset E of Y such that the set U, defined by*

$$U = \{ x \in X : |\langle x - x_0, y \rangle| < 1 \ (y \in E) \}, \tag{1}$$

has the following properties.

(i) *$U \subset D$ and f is bounded on D.*

(ii) *If $x_1 \in U$, $x_2 \in X$, and $\langle x_1, y \rangle = \langle x_2, y \rangle$ $(y \in E)$, then $x_2 \in U$ and $f(x_1) = f(x_2)$.*

Proof. Let $x_0 \in D$. Since f is locally bounded on the $w(X, Y)$ open set D, there exists a finite subset E of Y such that the set U defined by (1) satisfies (i). We prove that it also satisfies (ii). Let $x_1 \in U$, $x_2 \in X$, and suppose that $\langle x_1, y \rangle = \langle x_2, y \rangle$ $(y \in E)$. Then, for all $z \in \mathbb{C}$,

$$|\langle x_1 + z(x_2 - x_1) - x_0, y \rangle| = |\langle x_1 - x_0, y \rangle| < 1 \qquad (y \in E).$$

Thus $x_1 + z(x_2 - x_1) \in U$ for all $z \in \mathbb{C}$, and, in particular, $x_2 \in U$. Let

$$\pi(z) = x_1 + z(x_2 - x_1) \qquad (z \in \mathbb{C}).$$

Then $f \circ \pi$ is holomorphic on $\pi^{-1}(U) = \mathbb{C}$, and is bounded there. Therefore, by Liouville's theorem $f \circ \pi$ is constant valued on $\mathbb{C}$; and so

$$f(x_1) = (f \circ \pi)(0) = (f \circ \pi)(1) = f(x_2). \quad \square$$

Theorem 4. *With respect to the weak* topology on A', let D be an open neighbourhood of Φ_A in A' and let f be a complex valued function analytic on D. Then there exists $u \in A$ such that*

$$f(\phi) = \phi(u) \qquad (\phi \in \Phi_A).$$

Proof. The weak* topology on A' is the $\sigma(A', A)$ topology with respect to the natural bilinear form on $A' \times A$. Therefore, by Lemma 3, to each point ϕ_0 of D corresponds a finite subset $E(\phi_0)$ of A such that the set $U(\phi_0)$, defined by

$$U(\phi_0) = \{\phi \in A' : |\phi(y) - \phi_0(y)| < 1 \ (y \in E(\phi_0))\},$$

has the following properties.

(i) $U(\phi_0) \subset D$ and f is bounded on $U(\phi_0)$.

(ii) If $\phi \in U(\phi_0)$, $\psi \in A'$, and $\phi(y) = \psi(y)$ $(y \in E(\phi_0))$, then $\psi \in U(\phi_0)$ and $f(\phi) = f(\psi)$.

Choose $\phi_1, \dots, \phi_m \in D$ such that the open sets $U(\phi_k)$ $(k = 1, \dots, m)$ cover the compact set Φ_A, and take $E = \bigcup_{k=1}^m E(\phi_k)$, $U = \bigcup_{k=1}^m U(\phi_k)$. Then U is a weak* open set, $\Phi_A \subset U \subset D$, f is bounded on U, and the following statement holds.

(iii) If $\phi \in U$, $\psi \in A'$, and $\phi(y) = \psi(y)$ $(y \in E)$, then $\psi \in U$ and $f(\phi) = f(\psi)$.

Clearly we can replace E by a linearly independent subset without disturbing (iii). Assume then that $E = \{a_1, \dots, a_n\}$ with $a_1, \dots, a_n$ linearly independent, let $\mathbf{a} = (a_1, \dots, a_n)$, and define a mapping T of A' into $\mathbb{C}^n$ by $T\phi = (\phi(a_1), \dots, \phi(a_n))$ $(\phi \in A')$. Choose $0_1, \dots, 0_n \in A'$ such that

$$0_i(a_j) = \begin{cases} 1 & (i = j), \\ 0 & (i \neq j) \end{cases}$$

and let π be the mapping of $\mathbb{C}^n$ into A' given by

$$\pi(\mathbf{z}) = z_1 0_1 + \cdots + z_n 0_n \qquad (\mathbf{z} = (z_1, \dots, z_n) \in \mathbb{C}^n).$$

With this notation, (iii) becomes

$$\phi \in U, \ \psi \in A', \ T\phi = T\psi \ \Rightarrow \ \psi \in U \text{ and } f(\phi) = f(\psi), \tag{2}$$

and we have

$$(T \circ \pi)(\mathbf{z}) = \mathbf{z} \qquad (\mathbf{z} \in \mathbb{C}^n). \tag{3}$$

If $\psi = (\pi \circ T)(\phi)$ with $\phi \in U$, then, by (3),

$$T\psi = ((T \circ \pi) \circ T)(\phi) = T\phi,$$

and so, by (2), $\psi \in U$ and $f(\phi) = f(\psi)$. Therefore we have

$$(\pi \circ T)(U) \subset U \quad \text{and} \quad (f \circ \pi \circ T)(\phi) = f(\phi) \quad (\phi \in U). \tag{4}$$

Since f is analytic on U, $f \circ \pi$ is holomorphic on $\pi^{-1}(U)$, which is an open set in $\mathbb{C}^n$. Also, by (4),

$$\mathrm{Sp}(\mathbf{a}) = T\Phi_A \subset TU \subset \pi^{-1}(U).$$

Therefore, by the Shilov-Arens-Calderon theorem (Theorem 20.5), there exists an element u of A such that

$$\phi(u) = (f \circ \pi)(\phi(a_1), \dots, \phi(a_n)) \qquad (\phi \in \Phi_A).$$

But $\Phi_A \subset U$, and, by (4),

$$(f \circ \pi)(\phi(a_1), \ldots, \phi(a_n)) = (f \circ \pi \circ T)(\phi) = f(\phi) \qquad (\phi \in U). \quad \square$$

Remarks. (1) It is easy to verify that, given $a \in A$, the function f, defined on A' by $f(\phi) = \phi(a)$ $(\phi \in A')$, is analytic on A' with respect to the weak* topology. Thus, by Theorem 4, any complex valued function on Φ_A that has an analytic extension to some weak* neighbourhood of Φ_A in A' also has an analytic extension to the whole of A'.

(2) Theorem 4 is due to Allan [12].

Theorem 5 (Shilov [346]). *Let* $\Phi_A = \Omega_1 \cup \Omega_2$, *where* Ω_1, Ω_2 *are disjoint non-void compact sets. Then* A *contains an idempotent* e *with*

$$\phi(e) = 1 \ (\phi \in \Omega_1), \qquad \phi(e) = 0 \ (\phi \in \Omega_2).$$

Proof. There exist disjoint weak* open neighbourhoods U_1, U_2 of Ω_1, Ω_2 respectively in A'. Let $D = U_1 \cup U_2$ and define f on D by $f(\phi) = 1 \ (\phi \in U_1)$, $f(\phi) = 0 \ (\phi \in U_2)$. As in Example 2, f is analytic on D, which is an open neighbourhood in A' of Φ_A. Therefore, by Theorem 4, there exists $u \in A$ with

$$\phi(u) = f(\phi) \qquad (\phi \in \Phi_A).$$

Therefore $\mathrm{Sp}(u) = \{0,1\}$, and by Proposition 7.9 there exists an idempotent e in A with $ue - e$ and $u(1-e)$ quasi-nilpotent. Therefore

$$\phi(e) = \phi(ue) = \phi(u) \qquad (\phi \in \Phi_A). \quad \square$$

Alternatively we may observe that $r(u^2 - u) = 0$ and apply Theorem 8.14 to finish the proof.

Corollary 6. *Let* B *be a semi-simple complex commutative Banach algebra without unit. Then* Φ_B *is not compact.*

Proof. Let A denote the unitization $B + \mathbb{C}$ of B, and suppose that Φ_B is compact. Let ψ be defined on A by $\psi((x, \alpha)) = \alpha \ (x \in B, \alpha \in \mathbb{C})$, and let $\Omega_2 = \{\psi\}$, $\Omega_1 = \Phi_A \setminus \Omega_2$. Given $\phi \in \Phi_A$, let $(T\phi)(x) = \phi((x, 0)) \ (x \in B)$. Then T maps Ω_1 homeomorphically onto Φ_B, and so Ω_1 and Ω_2 are disjoint non-void compact subsets with their union Φ_A. By Theorem 5 there exists an idempotent $f = (e, \alpha)$ in A with $\phi(f) = 1 \ (\phi \in \Omega_1)$ and $\phi(f) = 0 \ (\phi \in \Omega_2)$. We have $\alpha = \psi(f) = 0$, $f = (e, 0)$, $e^2 = e \neq 0$. Moreover $\phi(e) = 1 \ (\phi \in \Phi_B)$, and so $\phi(x - xe) = 0 \ (\phi \in \Phi_B, x \in B)$. Since B is semi-simple, this gives $x - xe = 0 \ (x \in B)$. $\quad \square$

As another application of Theorem 4, we return to the existence of square roots (see § 8). We recall that $a^\wedge$ denotes the Gelfand representation of a on Φ_A.

Theorem 7. *Let $a \in A$ be invertible, and suppose that there exists $h \in C(\Phi_A)$ such that $h^2 = a^\wedge$. Then there exists $b \in A$ with $b^2 = a$.*

Proof. We may assume that $\inf\{|h(\phi)|: \phi \in \Phi_A\} = 1$, and therefore that $\inf\{|a^\wedge(\phi)|: \phi \in \Phi_A\} = 1$. Let

$$S = \{(\phi, \phi') \in \Phi_A \times \Phi_A : |h(\phi) + h(\phi')| \leqslant 1\} \,.$$

If $(\phi, \phi') \in S$, we have $\phi \neq \phi'$, and so there exists $x \in A$ with $\phi(x) \neq \phi'(x)$. Therefore $\psi(x) \neq \psi'(x)$ for all (ψ, ψ') in some open neighbourhood of (ϕ, ϕ') in the product space $\Phi_A \times \Phi_A$. Since S is a compact subset of $\Phi_A \times \Phi_A$, it can be covered by a finite union of such neighbourhoods. Therefore there exist $x_1, \ldots, x_n \in A$ such that, for all $(\phi, \phi') \in S$, we have $\phi(x_k) \neq \phi'(x_k)$ for some $k \in \{1, \ldots, n\}$. Let p be defined on A' by

$$p(\phi) = \sum_{k=1}^{n} |\phi(x_k)| \,.$$

Then p is a weak* continuous semi-norm on A', and

$$p(\phi - \phi') > 0 \qquad ((\phi, \phi') \in S) \,.$$

Let $2\delta = \inf\{p(\phi - \phi'): (\phi, \phi') \in S\}$. Then $\delta > 0$, and

$$(\phi, \phi') \in \Phi_A \times \Phi_A, \ p(\phi - \phi') < 2\delta \ \Rightarrow \ |h(\phi) + h(\phi')| > 1 \,. \tag{5}$$

Given $\phi_0 \in \Phi_A$, let

$$U(\phi_0) = \{\phi \in A' : p(\phi - \phi_0) < \delta \ \text{and} \ |\phi(a) - \phi_0(a)| < \tfrac{1}{4}\} \,,$$

and for $\phi \in U(\phi_0)$, let

$$g(\phi_0, \phi) = h(\phi_0)[1 + \lambda(\phi_0, \phi)] \,,$$

where

$$\lambda(\phi_0, \phi) = \sum_{k=1}^{\infty} \frac{1}{k!} \cdot \frac{1}{2} \cdot \left(\frac{1}{2} - 1\right) \cdots \left(\frac{1}{2} - k + 1\right) (\phi(a) - \phi_0(a))^k (\phi_0(a))^{-k} \,.$$

For all $\phi \in U(\phi_0)$, we have

$$|\lambda(\phi_0, \phi)| < \tfrac{1}{3} \,, \tag{6}$$

$$(g(\phi_0, \phi))^2 = \phi(a) \,. \tag{7}$$

We prove next that, given $\phi_1, \phi_2 \in \Phi_A$,

$$\phi \in U(\phi_1) \cap U(\phi_2) \Rightarrow g(\phi_1, \phi) = g(\phi_2, \phi) \,. \tag{8}$$

Let $\phi \in U(\phi_1) \cap U(\phi_2)$. Then $p(\phi_1 - \phi_2) < 2\delta$ and $|\phi_1(a) - \phi_2(a)| < \tfrac{1}{2}$. By (5), we have $|h(\phi_1) + h(\phi_2)| > 1$, and so

$$|h(\phi_1) - h(\phi_2)| = |\phi_1(a) - \phi_2(a)| \cdot |h(\phi_1) + h(\phi_2)|^{-1} < \tfrac{1}{2} \,.$$

With (6), this gives

$$\left|\frac{h(\phi_1)}{h(\phi_2)} - 1\right| < \frac{1}{2}, \quad \left|\frac{1+\lambda(\phi_2,\phi)}{1+\lambda(\phi_1,\phi)} - 1\right| < 1. \tag{9}$$

By (7) and (9) we now have

$$\frac{1+\lambda(\phi_2,\phi)}{1+\lambda(\phi_1,\phi)} = \frac{h(\phi_1)}{h(\phi_2)},$$

and (8) is proved.

Now choose $\phi_1,\dots,\phi_m \in \Phi_A$ so that $\Phi_A \subset D = \bigcup_{k=1}^m U(\phi_k)$, and define f on D by taking $f(\phi) = g(\phi_k,\phi)$ on $U(\phi_k)$. Then D is a weak* open neighbourhood of Φ_A in A', and f is analytic on D. Therefore, by Theorem 4, there exists $u \in A$ with

$$\phi(u) = f(\phi) \quad (\phi \in \Phi_A).$$

By (7), this gives

$$\phi(u^2) = (\phi(u))^2 = \phi(a) \quad (\phi \in \Phi_A). \tag{10}$$

If A is semi-simple, we take $b = u$. In the general case, let $v = a - u^2$. By (10), $\phi(v) = 0$ ($\phi \in \Phi_A$), and $\phi(u) \neq 0$ ($\phi \in \Phi_A$). Thus $u \in \mathrm{Inv}(A)$, and we have $a = u^2(1+x)$ with $x = u^{-2}v$. We have $\phi(x) = 0$ ($\phi \in \Phi_A$), so $r(x) = 0$, and, by Proposition 8.13, there exists $y \in A$ with $1+x = (1+y)^2$. Take $b = u(1+y)$. ∎

Theorem 7 is a special case of a theorem of Arens and Calderon [28] on the existence of a solution x in A of an algebraic equation of the form

$$\sum_{k=0}^n a_k x^k = 0$$

with $a_0, \dots, a_n$ given elements of A. Such a solution exists if there exists a function $h \in C(\Phi_A)$ such that $\sum_{k=0}^n a_k^\wedge h^k = 0$, and

$$\sum_{k=1}^n k a_k^\wedge(\phi) h^{k-1}(\phi) \neq 0 \quad (\phi \in \Phi_A).$$

For a proof of this theorem see Hörmander [201, Theorem 3.2.5]. Our proof of Theorem 7 is a variant of the proof in Wermer [410, Theorem 8.4].

Theorem 7 is also related to results of Katznelson [238—240]. A complex function F on $\mathbb{C}$ is said to *operate* on A if for each $a \in A$ there exists $b \in A$ such that $F \circ a^\wedge = b^\wedge$. Katznelson shows that A is isomorphic to $C(\Phi_A)$ under the Gelfand representation if either of the following conditions holds:

(i) all continuous functions operate on A,

(ii) A is semi-simple, complex conjugation operates on A, and elements a such that $a^\wedge \geqslant 0$ have square roots in A. See also DeLeeuw and Katznelson [113].

§ 22. The Shilov Boundary

X will denote a compact Hausdorff space. Given a subset E of X and $f \in C(X)$, we denote $\sup\{|f(x)|: x \in E\}$ by $|f|_E$. Throughout this section A will denote a complex commutative Banach algebra with unit element. For a more general theory concerning locally compact spaces and Banach algebras without unit element the reader is referred to Rickart [321, pp. 132—148].

Definition 1. Let F be a subset of $C(X)$. A *maximizing set* for F is a closed subset E of X such that $|f|_E = |f|_X$ ($f \in F$).

Lemma 2. *Let F be a subalgebra of $C(X)$ that contains the constant functions and separates the points of X, let S be the intersection of all maximizing sets for F, and let $x_0 \in X \setminus S$. Then there exists an open neighbourhood U of x_0 such that $E \setminus U$ is a maximizing set for F, whenever E is a maximizing set for F.*

Proof. Since $x_0 \notin S$, there exists a maximizing set E_0 for F such that $x_0 \notin E_0$. Since F contains the constant functions and separates the points of X, for each $y \in E_0$ there exists $f \in F$ with $f(x_0) = 0$ and $f(y) = 2$. Then $\{x \in X: |f(x)| > 1\}$ is an open neighbourhood of y, and so the compact set E_0 can be covered by a finite union of open sets of this form. Thus there exist $f_1, \ldots, f_r \in F$ such that $f_k(x_0) = 0$ ($k = 1, \ldots, r$) and for each $y \in E_0$ we have $|f_k(y)| > 1$ for some $k \in \{1, \ldots, r\}$. Let

$$U = \{x \in X: |f_k(x)| < 1 \ (k = 1, \ldots, r)\}.$$

Then U is an open neighbourhood of x_0 and $U \cap E_0 = \emptyset$.

Let E be a maximizing set for F, and suppose that $E \setminus U$ is not a maximizing set for F. Then there exists $f \in F$ with

$$|f|_X = 1 > |f|_{E \setminus U}.$$

Let $M = \max\{|f_k|_X: 1 \leqslant k \leqslant r\}$, and choose n such that

$$(|f|_{E \setminus U})^n M < 1.$$

Then $|f^n f_k|_{E \setminus U} < 1$ ($1 \leqslant k \leqslant r$), and also $|f^n(x) f_k(x)| < 1$ ($x \in U, 1 \leqslant k \leqslant r$). Therefore, E being a maximizing set, we have

$$|f^n f_k|_X = |f^n f_k|_E < 1 \qquad (1 \leqslant k \leqslant r).$$

Since E_0 is a maximizing set, there exists $y \in E_0$ with $|f(y)| = 1$, and so

$$|f_k(y)| = |f^n(y) f_k(y)| \leqslant |f^n f_k|_X < 1 \qquad (1 \leqslant k \leqslant r).$$

This gives $y \in E_0 \cap U$, which is contradictory. $\square$

Theorem 3. *Let F be a subalgebra of $C(X)$ that contains the constant functions and separates the points of X. Then the intersection of all maximizing sets for F is a maximizing set for F.*

Proof. Let S denote the intersection of all maximizing sets for F, let $f \in F$, and let $K = \{x \in X : |f(x)| = |f|_X\}$. It is enough to prove that $S \cap K \neq \emptyset$. Suppose that $S \cap K = \emptyset$. Then each $x_0 \in K$ has an open neighbourhood U as in Lemma 2. Since K is compact it is covered by a finite union of such open sets $U_1, \ldots, U_n$. X is a maximizing set, and therefore so are in turn $X \setminus U_1, X \setminus (U_1 \cup U_2), \ldots, X \setminus (U_1 \cup \cdots \cup U_n) = E$, say. But $K \cap E = \emptyset$, and therefore $|f|_E < |f|_X$, a contradiction. $\square$

Let $a^\wedge$ denote the Gelfand representation of $a \in A$ on Φ_A, and let $A^\wedge = \{a^\wedge : a \in A\}$. Then $A^\wedge$ is a subalgebra of $C(\Phi_A)$ containing the constant functions and separating the points of Φ_A. Thus Theorem 3 is applicable to $A^\wedge$.

Definition 4. The *Shilov boundary* for A is the intersection of all maximizing sets for the algebra $A^\wedge$; it is denoted by $\check{S}(A)$.

By Theorem 3, $\check{S}(A)$ is itself a maximizing set for $A^\wedge$.

Example 5. Let $\Delta = \{z \in \mathbb{C} : |z| \leqslant 1\}$, $\Gamma = \{z \in \mathbb{C} : |z| = 1\}$, let $n \in \mathbb{N}$, and let A denote the algebra of functions belonging to $C(\Delta^n)$ that are holomorphic on the interior of Δ^n. Let $u_k(\mathbf{z}) = z_k$ $(\mathbf{z} = (z_1, \ldots, z_n) \in \Delta^n)$. Then $\{u_1, \ldots, u_n\}$ is a set of joint generators for A, and so Φ_A can be identified with $\mathrm{Sp}(\mathbf{u})$, i.e. with Δ^n. By the maximum principle Γ^n is a maximizing set for A. Let $\lambda \in \Gamma^n$, and let $f(\mathbf{z}) = \prod_{k=1}^n \exp(z_k / \lambda_k)$. Then $|f(\mathbf{z})| = |f|_{\Delta^n}$ if and only if $\mathbf{z} = \lambda$, and so λ belongs to every maximizing set for A. Thus $\check{S}(A)$ can be identified with Γ^n.

Note that since A' is an infinite dimensional Banach space, Φ_A has void interior as a subset of A' and is therefore its own topological boundary in A'. Thus $\check{S}(A)$ is a proper subset of this topological boundary. When $n > 1$, $\check{S}(A)$ is identified with Γ^n which is also a proper subset of the topological boundary of Δ^n relative to $\mathbb{C}^n$.

Example 6. Let X be a compact metric space with distance function d, and let $A = C(X)$. Then $\Phi(A)$ can be identified with X, and under this identification $\check{S}(A) = X$. For given $x_0 \in X$, let $f(x) = [1 + d(x, x_0)]^{-1}$, then $|f(x)| = |f|_X$ if and only if $x = x_0$. Thus x_0 belongs to every maximizing set for A.

Let $\partial\mathrm{Sp}(a)$ denote the topological boundary of $\mathrm{Sp}(a)$ as a subset of $\mathbb{C}$.

Proposition 7. *Let* $a\in A$. *Then* $\partial\mathrm{Sp}(a)\subset a^{\wedge}(\check{S}(A))$.

Proof. It is sufficient to prove that if $0\in\partial\mathrm{Sp}(a)$, then $0\in a^{\wedge}(\check{S}(a))$. Suppose then that $0\in\partial\mathrm{Sp}(a)\backslash a^{\wedge}(\check{S}(A))$. There exist $\phi_0\in\Phi_A$ and $\delta>0$ such that $a^{\wedge}(\phi_0)=0$ and

$$|a^{\wedge}(\phi)|\geqslant 2\delta \quad (\phi\in\check{S}(A)).$$

Choose $\lambda\in\mathbb{C}\backslash\mathrm{Sp}(a)$ with $|\lambda|<\delta$. Then $\lambda-a\in\mathrm{Inv}(A)$, and we take $u=(\lambda-a)^{-1}$. Then $|a^{\wedge}(\phi)-\lambda|\geqslant 2\delta-|\lambda|>\delta$ $(\phi\in\check{S}(A))$, and so

$$r(u)=|u^{\wedge}|_{\Phi_A}=|u^{\wedge}|_{\check{S}(A)}$$
$$=\left[\inf\{|\lambda-a^{\wedge}(\phi)|:\phi\in\check{S}(A)\}\right]^{-1}<\delta^{-1}.$$

Whereas also

$$r(u)\geqslant|u^{\wedge}(\phi_0)|=|\lambda^{-1}|>\delta^{-1}. \quad \square$$

Corollary 8. *Let* W *be a component of the open set* $\mathbb{C}\backslash a^{\wedge}(\check{S}(A))$ *that has non-void intersection with* $\mathrm{Sp}(a)$. *Then* $W\subset\mathrm{Sp}(a)$.

Proof. If $W\cap(\mathbb{C}\backslash\mathrm{Sp}(a))$ is non-void, $W\cap\partial\mathrm{Sp}(a)$ is non-void, contradicting Proposition 7. $\quad\square$

Example 9. Let A be the algebra considered in Example 5 of functions continuous on the closed polydisc Δ^n and holomorphic in its interior. Let Φ_A be identified with Δ^n and $\check{S}(A)$ with Γ^n. Let $f\in A$ and let W be a component of the open set $\mathbb{C}\backslash f(\Gamma^n)$. If W has non-void intersection with $f(\Delta^n)$, then $W\subset f(\Delta^n)$. Also $\partial f(\Delta^n)\subset f(\Gamma^n)$.

Corollary 10. *Suppose that for some* $a\in A$, $\mathrm{Sp}(a)\not\subset a^{\wedge}(\check{S}(A))$. *Then* $\check{S}(A)$ *and* $\Phi_A\backslash\check{S}(A)$ *are both uncountable.*

Proof. Let $\lambda\in\mathrm{Sp}(a)\backslash a^{\wedge}(\check{S}(A))$, and let W be the component of the open set $\mathbb{C}\backslash a^{\wedge}(\check{S}(A))$ containing λ. Then, by Corollary 8, $W\subset\mathrm{Sp}(a)\backslash a^{\wedge}(\check{S}(A))$, i. e. $W\subset\{a^{\wedge}(\phi):\phi\in\Phi_A\backslash\check{S}(A)\}$. Therefore $\Phi_A\backslash\check{S}(A)$ is uncountable. Since $W\subset\mathrm{Sp}(a)$, $\partial\mathrm{Sp}(a)$ is uncountable, and therefore $a^{\wedge}(\check{S}(A))$ is uncountable, by Proposition 7. $\quad\square$

We end this section by stating without proof the local maximum modulus principle due to Rossi [322]. For the proof see Gunning and Rossi [182, pp. 62—63] and Wermer [410, pp. 52—55].

Theorem 11. *Let* U *be an open subset of* $\Phi_A\backslash\check{S}(A)$. *Then for all* $a\in A$ *and* $\phi\in U$,

$$|a^{\wedge}(\phi)|\leqslant|a^{\wedge}|_{\partial U}.$$

The concept of Shilov boundary and the basic Theorem 3 are due to Shilov [343]. The proof of Theorem 3 given here is that given in Wermer [410, pp. 50—51], where it is attributed to Hörmander [201, Theorem 3.1.18].

§ 23. The Hull-Kernel Topology

Let A denote a complex commutative Banach algebra with or without unit element. We recall that the carrier space Φ_A is the set of all (non-zero) multiplicative linear functionals on A, and is a locally compact Hausdorff space with respect to the A-topology. We now consider a second topology on Φ_A which in general differs from the A-topology but coincides with it for certain important algebras A.

Definition 1. Given a subset E of Φ_A, the *kernel* of E, $\ker(E)$, is defined by

$$\ker(E) = \cap \{\ker(\phi): \phi \in E\}.$$

Given a subset J of A, the *hull* of J, $\mathrm{hul}(J)$, is defined by

$$\mathrm{hul}(J) = \{\phi \in \Phi_A: \ker \phi \supset J\}.$$

We denote $\mathrm{hul}(\ker(E))$ by $E^\sim$.

Lemma 2. *For subsets E, E_1, E_2 of Φ_A,*
(i) $\ker(E) = \ker(E^\sim)$, (ii) $E \subset E^\sim = (E^\sim)^\sim$, (iii) $(E_1 \cup E_2)^\sim = E_1^\sim \cup E_2^\sim$.

Proof. (i) and (ii). We have $\ker(E) \subset \ker(\psi)$ $(\psi \in E^\sim)$, and so $\ker(E) \subset \ker(E^\sim)$. It is obvious that $E \subset E^\sim$ and therefore $\ker(E) \supset \ker(E^\sim)$. (i) and (ii) are now clear.
(iii). It is obvious that $E_1^\sim \cup E_2^\sim \subset (E_1 \cup E_2)^\sim$. On the other hand, let $\phi \in \Phi_A \backslash (E_1^\sim \cup E_2^\sim)$. Then there exist $x_i \in \ker(E_i)$ with $\phi(x_i) \neq 0$ $(i=1,2)$, and we have $x_1 x_2 \in \ker(E_1 \cup E_2)$, but $\phi(x_1 x_2) \neq 0$. Therefore $\phi \notin (E_1 \cup E_2)^\sim$. □

Definition 3. Lemma 2 shows that the mapping $E \to E^\sim$ is a Kuratowski closure operation on the subsets of Φ_A. The corresponding topology on Φ_A is called the *hull-kernel topology*.

Proposition 4. (i) $\phi_0 \in \Phi_A \backslash E^\sim$ *if and only if there exists $a \in A$ such that $a^\wedge(\phi) = 0$ $(\phi \in E)$ and $a^\wedge(\phi_0) \neq 0$.*
(ii) *The A-topology contains the hull-kernel topology.*
(iii) *The hull of a modular ideal is compact in the A-topology (and so also in the hull-kernel topology).*

Proof. (i) Clear.

(ii) If E is closed in the hull-kernel topology, then, by (i), it is closed in the A-topology.

(iii) Let J be a modular ideal of A with modular unit e. Being the intersection for $j \in J$ of the sets $\{\phi \in \Phi_A : \phi(j) = 0\}$, $\mathrm{hul}(J)$ is closed in the A-topology. Also,

$$\phi(x) - \phi(x)\phi(e) = \phi(x - xe) = 0 \qquad (x \in A, \phi \in \mathrm{hul}(J)),$$

and $\phi(x) \neq 0$ for some $x \in A$. Therefore $\phi(e) = 1$ $(\phi \in \mathrm{hul}(J))$, and so $\mathrm{hul}(J)$ is contained in the set $\{\phi \in \Phi_A : |\phi(e)| \geq 1\}$, which is compact in the A-topology (see Proposition 17.2). □

Proposition 5. *Let E be a non-void subset of Φ_A that is closed in the hull-kernel topology. Let $B = A/\mathrm{ker}(E)$, let π denote the canonical mapping of A onto B, and let π^* denote the dual mapping of Φ_B into Φ_A given by*

$$(\pi^* \psi)(a) = \psi(\pi a) \qquad (\psi \in \Phi_B, a \in A).$$

Then (i) *π^* is a homeomorphism of Φ_B with the B-topology onto E with the A-topology,*

(ii) *B is semi-simple.*

Proof. (i) Since E is non-void, $\mathrm{ker}(E)$ is a proper closed ideal of A, B is a (non-zero) complex commutative Banach algebra. For $\psi \in \Phi_B$, we have $(\pi^* \psi)(a) = \psi(\pi a) = 0$ $(a \in \mathrm{ker}(E))$, and so $\mathrm{ker}(\pi^* \psi) \supset \mathrm{ker}(E)$, $\pi^* \psi \in \mathrm{hul}(\mathrm{ker}(E)) = E$. Moreover, if $\phi \in E$, we have $\mathrm{ker}(E) \subset \mathrm{ker}(\phi)$, and a well defined element ψ of Φ_B is obtained by taking $\psi(b) = \phi(a)$ $(a \in b \in B)$. Thus π^* is a bijection of Φ_B onto E. Given $a_1, \ldots, a_n \in A$ and $b_k = \pi a_k$ $(k = 1, \ldots, n)$, we have

$$|\psi(b_k) - \psi_0(b_k)| < \varepsilon \, (k = 1, \ldots, n) \Leftrightarrow |(\pi^* \psi)(a_k) - (\pi^* \psi_0)(a_k)| < \varepsilon \, (k = 1, \ldots, n).$$

Therefore π^* is a homeomorphism with respect to the B-topology on Φ_B and the A-topology on E.

(ii) If $\psi(b) = 0$ $(\psi \in \Phi_B)$, then $b = \pi a$ for some $a \in A$, and $(\pi^* \psi)(a) = 0$ $(\psi \in \Phi_B)$. Thus $a \in \mathrm{ker}(E)$, $b = 0$. □

Corollary 6. *Let E be a subset of Φ_A that is closed in the hull-kernel topology and compact in the A-topology. Then $\mathrm{ker}(E)$ is a modular ideal of A.*

Proof. If E is void, $\mathrm{ker}(E) = A$, which is a modular ideal of A. Otherwise, let B, π be as in Proposition 5. Then Φ_B is compact and B is semi-simple. Therefore B has a unit element, by Corollary 21.6. Choose $e \in A$ with $\pi e = 1$. Then e is a modular unit for $\mathrm{ker}(E)$, for

$$\pi(x - xe) = \pi x - \pi x \cdot \pi e = 0 \qquad (x \in A). \quad □$$

Remark. Given that E is compact in the A-topology, we know that E is compact in the hull-kernel topology, but we cannot conclude that E is closed in the hull-kernel topology, since that topology need not be Hausdorff.

Definition 7. A is said to be *completely regular* if the hull-kernel topology on Φ_A satisfies the following two conditions:

(i) Φ_A is a Hausdorff space;

(ii) each element of Φ_A has a neighbourhood V such that $\ker(V)$ is a modular ideal.

If $\ker(V)$ is a modular ideal, it follows at once from Proposition 4(iii) that $V^\sim = \mathrm{hul}(\ker(V))$ is compact in the hull-kernel topology. Thus the carrier space of a completely regular algebra is a locally compact Hausdorff space in the hull-kernel topology as well as in the A-topology.

Theorem 8. *A is completely regular if and only if the hull-kernel topology coincides with the A-topology.*

Proof. In this proof, τ_A and τ_h will denote the A-topology and the hull-kernel topology respectively. Suppose first that A is completely regular, let F be a τ_A-closed subset of Φ_A, and let $\phi_0 \in \Phi_A \setminus F$. Then ϕ_0 has a τ_h-neighbourhood V such that $\ker(V)$ is a modular ideal, and, since $\ker(\mathrm{int}\, V) \supset \ker(V)$ we may suppose that V is τ_h-open. Let e be a modular unit for $\ker(V)$. Then $\phi(x) = \phi(xe)$ $(x \in A, \phi \in V)$, and so

$$\phi(e) = 1 \qquad (\phi \in V). \tag{1}$$

Let $F_0 = \{\phi \in F : \phi(e) = 1\}$. Since $\{\phi \in \Phi_A : |e^{\wedge}(\phi)| \geqslant 1\}$ is τ_A-compact and F is τ_A-closed, F_0 is τ_A-compact, and so is also τ_h-compact. Since the topology τ_h is Hausdorff, each point of F_0 has a τ_h-open neighbourhood U with $\phi_0 \notin U^\sim = \mathrm{hul}(\ker(U))$. We cover F_0 with a finite union of such neighbourhoods $U_1, \ldots, U_n$ say. Since $\phi_0 \notin \mathrm{hul}(\ker(U_i))$, there exists $v_i \in \ker(U_i)$ with $\phi_0(v_i) \neq 0$. Also since $\Phi_A \setminus V$ is τ_h-closed and $\phi_0 \notin \Phi_A \setminus V$, there exists $v_0 \in \ker(\Phi_A \setminus V)$ with $\phi_0(v_0) \neq 0$. Let $v = v_0 v_1 \ldots v_n$. Then $\phi_0(v) \neq 0$, and $v \in \ker(U_1 \cup \cdots \cup U_n \cup (\Phi_A \setminus V)) \subset \ker(F_0 \cup (\Phi_A \setminus V))$. By (1), we have $V \cap F \subset F_0$, and therefore $F \subset F_0 \cup (\Phi_A \setminus V)$. We have now proved that $v \in \ker(F)$ and $\phi_0(v) \neq 0$. Therefore $\phi_0 \notin F^\sim = \mathrm{hul}(\ker(F))$; F is τ_h-closed, $\tau_A \subset \tau_h$.

Conversely, suppose that $\tau_A = \tau_h$. Then, obviously, τ_h is a Hausdorff topology. Let $\phi_0 \in \Phi_A$. Since $\phi_0 \neq 0$, there exists $u \in A$ with $\phi_0(u) = 2$. Let $V = \{\phi \in \Phi_A : |\phi(u)| > 1\}$. Then V is an open neighbourhood of ϕ_0 and $V^- \subset \{\phi \in \Phi_A : |\phi(u)| \geqslant 1\}$ which is closed and compact for both topologies. By Corollary 6, $\ker(V^-)$ is a modular ideal of A. $\quad\Box$

Remark. Since Corollary 6 depends on the results of § 21, it is of interest to complete the proof of Theorem 8 without using Corollary 6. Let $B = A/\ker(V^-)$, let π be the canonical mapping of A onto B, and let π^* be the dual mapping. By Proposition 5, π^* maps Φ_B onto V^-, and B is semi-simple. Let $b = \pi u$. Then

$$|\psi(b)| = |(\pi^* \psi)(u)| \geqslant 1 \qquad (\psi \in \Phi_B) .$$

Therefore, by Proposition 17.9, B has a unit element πe, say, with $e \in A$. Then e is a modular unit for $\ker(V^-)$. □

Corollary 9. *The following statements are equivalent:*
(i) *A is completely regular.*
(ii) *For each subset F of Φ_A closed in the A-topology and each $\phi_0 \in \Phi_A \backslash F$ there exists $a \in A$ with*

$$a^\wedge(\phi) = 0 \quad (\phi \in F) , \qquad a^\wedge(\phi_0) \neq 0 .$$

Proof. Theorem 8 and Proposition 4(i). □

An obvious example of a completely regular algebra is the algebra $C(X)$ for a compact Hausdorff space X. A much less obvious example of a completely regular algebra is the algebra $L^1(G)$ for a locally compact Abelian group G. For a proof that $L^1(G)$ is completely regular, a fact of great significance for Harmonic Analysis, see Godement [167].

It is obvious from Corollary 9 that the Shilov boundary $\check{S}(A)$ is the whole of Φ_A whenever A is completely regular. Therefore, in particular, the disc algebra $\mathscr{A}(\varDelta)$ is not completely regular.

Chapter III. Representation Theory

§ 24. Algebraic Preliminaries

In this section A denotes an algebra over $\mathbb{F}$ and we consider the purely algebraic theory of irreducible left A-modules.

Definition 1. Let X be a linear space over $\mathbb{F}$. A *representation* of A on X is a homomorphism of A into $L(X)$. Given a representation π of A on X, the *corresponding left A-module* is the linear space X with the module multiplication given by

$$a x = \pi(a) x \quad (a \in A, x \in X). \tag{1}$$

Conversely, given a left A-module X, the *corresponding representation* of A on X is the homomorphism π of A into $L(X)$ defined by (1).

We shall make frequent use of the simple observation that the kernel of a representation π is given in terms of the corresponding left A-module X by

$$\ker(\pi) = \{a \in A : a X = \{0\}\} .$$

Example 2. Let L be a left ideal of A, and let $a \to a'$ denote the canonical mapping of A onto $A - L$. Then $A - L$ is a left A-module under the module multiplication

$$a x = (a b)' \quad (a \in A, b \in x \in A - L).$$

This is called the *regular left A-module $A - L$*. The corresponding representation is the *left regular representation on $A - L$*. It is clear that the kernel of this representation is the set

$$\{a : (a A)' = \{0\}\} = \{a : a A \subset L\} .$$

Multiplicative linear functionals have limited significance in the study of non-commutative Banach algebras since they necessarily vanish on all commutators $a b - b a$. The class of homomorphisms replacing them in the general theory is the class of irreducible representations, which we now define.

Definition 3. A left A-module X is *non-trivial* if $A X \neq \{0\}$. An *irreducible left A-module* is a non-trivial left A-module X such that X

and $\{0\}$ are the only A-submodules of X. A representation of A is *irreducible* if the corresponding left A-module is irreducible.

Let x_0 be an element of a left A-module X. We denote by $\ker(x_0)$ the left ideal of A given by

$$\ker(x_0) = \{a \in A : a x_0 = 0\} \,.$$

We say that x_0 is a *cyclic vector* if $A x_0 = X$. We denote by $\mathrm{id}(x_0)$ the subset of A given by

$$\mathrm{id}(x_0) = \{e \in A : e x_0 = x_0\} \,.$$

Proposition 4. *Let X be a left A-module and let $x_0 \in X \setminus \{0\}$.*

(i) *Each element of $\mathrm{id}(x_0)$ is a right modular unit for the left ideal $\ker(x_0)$.*

(ii) *If x_0 is a cyclic vector, then $\mathrm{id}(x_0)$ is non-void and $\ker(x_0)$ is a modular left ideal.*

(iii) *If X is irreducible, then x_0 is a cyclic vector and $\ker(x_0)$ is a maximal modular left ideal.*

Proof. (i) Let $e \in \mathrm{id}(x_0)$, $a \in A$. Then, for all $a \in A$,

$$(a - ae) x_0 = a x_0 - a x_0 = 0 \,,$$

and so $a - ae \in \ker(x_0)$.

(ii) If x_0 is cyclic, $A x_0 = X$ and so there exists $e \in A$ with $e x_0 = x_0$, i.e. $e \in \mathrm{id}(x_0)$. That $\ker(x_0)$ is modular now follows at once from (i).

(iii) Let X be irreducible. Since $A x_0$ is an A-submodule, either $A x_0 = \{0\}$ or $A x_0 = X$. If $A x_0 = \{0\}$, $N = \{x \in X : A x = \{0\}\}$ is a non-zero A-submodule, and so $N = X$. But this is impossible, since X is non-trivial. Therefore $A x_0 = X$, x_0 is cyclic. By (ii), $\ker(x_0)$ is a modular left ideal. Let J be a left ideal of A with $\ker(x_0) \subsetneqq J$. Then $J x_0$ is an A-submodule and $J x_0 \neq \{0\}$. Therefore $J x_0 = X$. Therefore there exists $e \in J \cap \mathrm{id}(x_0)$. By (i), e is a right modular unit for $\ker(x_0)$ and therefore also for J. But then since $e \in J$, we have $J = A$.   □

Proposition 5. *Let J be a modular left ideal of A with $J \neq A$, let e be a right modular unit for J, let x_0 denote the canonical image of e in $A - J$, and let X denote the regular left A-module $A - J$. Then*

(i) *x_0 is a cyclic vector and $\ker(x_0) = J$;*

(ii) *if J is a maximal modular left ideal, then X is irreducible.*

Proof. (i) Let $a \to a'$ denote the canonical mapping of A onto $A - J$. For all $a \in A$, we have $a - ae \in J$ and so $a' - a x_0 = (a - ae)' = 0$, $X = A x_0$. If $a \in \ker(x_0)$, then $(ae)' = a x_0 = 0$, and so $ae \in J$. But then $a = (a - ae) + ae \in J$.

(ii) Let Y be an A-submodule of X with $Y \neq \{0\}$, and let $K = \{a \in A : a' \in Y\}$. Then K is a left ideal with $J \subsetneq K$. If J is maximal, we have $K = A$. But then $x_0 = e' \in Y$, $X = A x_0 \subset Y$. □

Notation. For the rest of this section X will denote an irreducible left A-module, and $\mathfrak{D}$ will denote the subset of $L(X)$ given by

$$\mathfrak{D} = \{T \in L(X) : a T x = T(a x) \ (a \in A, x \in X)\} .$$

Proposition 6. (Schur's lemma). $\mathfrak{D}$ *contains the identity operator and is a division subalgebra of* $L(X)$ (i.e. $\mathfrak{D}$ *is a subalgebra of* $L(X)$ *and each non-zero element is invertible in* $\mathfrak{D}$).

Proof. That $\mathfrak{D}$ is a subalgebra of $L(X)$ is clear. Let $T \in \mathfrak{D} \setminus \{0\}$. We have, for all $a \in A$,

$$a T X = T a X \subset T X ;$$

and so TX is a non-zero A-submodule of X. Therefore $TX = X$. Let $N = \{x \in X : Tx = 0\}$. Then, for all $a \in A$,

$$T a x = a T x = 0 \quad (x \in N).$$

Therefore N is an A-submodule of X. Since $T \neq \{0\}$, $N \neq X$, and therefore $N = \{0\}$. We have now proved that T is a bijective mapping of X onto X. Therefore T has an inverse $T^{-1} \in L(X)$, and $T^{-1} \in \mathfrak{D}$, since

$$a T^{-1} x = T^{-1} T(a T^{-1} x) = T^{-1} a(T T^{-1} x) = T^{-1}(a x) . \quad □$$

Our next objective is the Jacobson density theorem, Theorem 10. To see the significance of this theorem, it should be observed that if the division algebra $\mathfrak{D}$ coincides with the scalar field $\mathbb{F}$, then for every finite dimensional linear subspace X_0 of X and every linear mapping $T \in L(X_0, X)$ the theorem gives the existence of $a \in A$ such that $T = \pi(a)|_{X_0}$.

Definition 7. Vectors $x_1, \ldots, x_n \in X$ are $\mathfrak{D}$-*independent* if $D_1, \ldots, D_n \in \mathfrak{D}$, $D_1 x_1 + \cdots + D_n x_n = 0 \Rightarrow D_1 = D_2 = \cdots = D_n = 0$.

Lemma 8. *Let* x_1, x_2 *be* $\mathfrak{D}$-*independent vectors in* X. *Then there exists* $a \in A$ *with* $a x_1 = 0$ *and* $a x_2 \neq 0$.

Proof. We have $x_k \neq 0$, and since X is irreducible it follows from Proposition 4(iii) that $\ker(x_k)$ is a maximal modular left ideal M_k say. Suppose there does not exist any element $a \in A$ with $a x_1 = 0$ but $a x_2 \neq 0$. Then $M_1 \subset M_2$, and so $M_1 = M_2$.

Let Y denote the regular left A-module $A - M_1$. By Proposition 5(ii), Y is irreducible. Let P, Q be the mappings of Y into X defined by

$$Py = a x_1, \quad Qy = a x_2 \quad (a \in y \in Y).$$

If $a, b \in y \in Y$, we have $a - b \in M_1 = M_2$, $a - b \in \ker(x_1) = \ker(x_2)$, and so $a x_1 = b x_1$, $a x_2 = b x_2$. Thus P, Q are well defined elements of $L(Y, X)$. Moreover, since x_1, x_2 are cyclic vectors, $P Y = Q Y = X$.

Let $a \to a'$ denote the canonical mapping of A onto $A - M_1 = Y$. Then for all $a, b \in A$,

$$b P a' = b(a x_1) = (b a) x_1 = P(b a)' = P(b a').$$

Then $b P y = P b y$ ($y \in Y, b \in A$), and similarly $b Q y = Q b y$. Let $D = Q P^{-1}$. Then for all $a \in A$, $x \in X$,

$$D(a x) = Q P^{-1}(a P P^{-1} x) = Q P^{-1}(P a P^{-1} x) = Q a P^{-1} x = a D x.$$

Therefore $D \in \mathfrak{D}$. Also, for all $a \in A$,

$$D(a x_1) = Q P^{-1} P a' = Q a' = a x_2.$$

Then $a(D x_1 - x_2) = D a x_1 - a x_2 = 0$, and so $D x_1 - x_2 = 0$, since otherwise $D x_1 - x_2$ is cyclic and we have $X = A(D x_1 - x_2) = \{0\}$. We have now contradicted the hypothesis that x_1, x_2 are $\mathfrak{D}$-independent.   □

Lemma 9. Let $x_1, \ldots, x_n$ be $\mathfrak{D}$-independent vectors in X. Then there exists $a \in A$ with

$$a x_k = 0 \quad (1 \leqslant k \leqslant n - 1), \qquad a x_n \neq 0.$$

Proof. The proof is by induction on n. Lemma 8 is the case $n = 2$. Let $n > 2$ and assume that the Lemma holds with n replaced by $n - 1$. There exists $a \in A$ with

$$a x_k = 0 \quad (k = 1, \ldots, n - 2), \qquad a x_n \neq 0.$$

If $a x_{n-1} = 0$, there is nothing to prove. Suppose that $a x_{n-1} \neq 0$. If $a x_{n-1}$ and $a x_n$ are $\mathfrak{D}$-independent, then, by Lemma 8, there exists $b \in A$ with

$$b(a x_{n-1}) = 0, \qquad b(a x_n) \neq 0.$$

But then $(b a) x_k = 0$ ($1 \leqslant k \leqslant n - 1$) and $(b a) x_n \neq 0$. Therefore, we may assume that $a x_{n-1}$ and $a x_n$ are $\mathfrak{D}$-dependent, and so there exists $S \in \mathfrak{D}$ with $S a x_{n-1} = a x_n$. The vectors $x_1, \ldots, x_{n-2}$, $S x_{n-1} - x_n$ are $\mathfrak{D}$-independent, and so there exists $c \in A$ with

$$c x_k = 0 \quad (1 \leqslant k \leqslant n - 2), \qquad c(S x_{n-1} - x_n) \neq 0.$$

If $c x_{n-1} = 0$, we have $c S x_{n-1} = 0$ and so $c x_n \neq 0$, and there is nothing more to prove. Assume that $c x_{n-1} \neq 0$. If $c x_{n-1}$ and $c x_n$ are $\mathfrak{D}$-independent, there exists $d \in A$ with $d c x_{n-1} = 0$ and $d c x_n \neq 0$. But then $d c x_k = 0$ ($1 \leqslant k \leqslant n - 1$), $d c x_n \neq 0$, and there is nothing more to prove. We may therefore assume that there exists $T \in \mathfrak{D}$ with

$$T c x_{n-1} = c x_n.$$

Since $Scx_{n-1} \neq cx_n$, we have $(S-T)cx_{n-1} \neq 0$, $S-T \neq 0$. Since $cx_{n-1} \neq 0$, there exists $e \in A$ with $ecx_{n-1} = ax_{n-1}$. Let $f = a - ec$. Then $fx_k = 0$ $(1 \leq k \leq n-1)$, and

$$fx_n = ax_n - ecx_n = Sax_{n-1} - eTcx_{n-1}$$
$$= Sax_{n-1} - Tecx_{n-1}$$
$$= (S-T)ax_{n-1} \neq 0,$$

since $S-T$ is invertible. □

Theorem 10. *Let* $x_1, \ldots, x_n$ *be* $\mathfrak{D}$*-independent vectors in* X, *and let* $y_1, \ldots, y_n \in X$. *Then there exists* $a \in A$ *with*

$$ax_k = y_k \quad (1 \leq k \leq n).$$

Proof. By Lemma 9, there exists $b \in A$ with

$$bx_k = 0 \quad (1 \leq k \leq n-1), \quad bx_n \neq 0.$$

Then, since bx_n is cyclic, there exists $c \in A$ with $cbx_n = y_n$. Taking $a_n = cb$, we have

$$a_n x_k = 0 \quad (1 \leq k \leq n-1), \quad a_n x_n = y_n.$$

Similarly, for each $j \in \{1, \ldots, n\}$ there exists $a_j \in A$ such that

$$a_j x_k = 0 \quad (k \neq j), \quad a_j x_j = y_j.$$

Finally, take $a = a_1 + \cdots + a_n$. □

Definition 11. Given a left ideal L of A the *quotient* of L is the bi-ideal $L : A$ defined by

$$L : A = \{a \in A : aA \subset L\}.$$

The quotient of a maximal modular left ideal is called a *primitive ideal*.

We list some elementary properties of these concepts in the next proposition.

Proposition 12. (i) *If* L *is a modular left ideal, then* $L : A \subset L$, *and* $L : A$ *is the kernel of the left regular representation on* $A - L$.

(ii) *The primitive ideals of* A *are the kernels of the irreducible representations of* A.

(iii) *Let* P *be a primitive ideal and let* L_1, L_2 *be left ideals with* $L_1 L_2 \subset P$. *Then either* $L_1 \subset P$ *or* $L_2 \subset P$.

(iv) *A primitive ideal is the intersection of the maximal modular left ideals containing it.*

Proof. (i) Let e be a right modular unit for L, and let $a \in L : A$. Then

$$a = (a - ae) + ae \in L.$$

That $L:A$ is the kernel of the left regular representation on $A-L$ has been proved in Example 2.

(ii) Given a primitive ideal P, we have $P=L:A$ with L a maximal modular left ideal. Then P is the kernel of the left regular representation on $A-L$, which is irreducible. Conversely, let X be an irreducible left A-module, and let $x_0 \in X\setminus\{0\}$. Then $\ker(x_0)$ is a maximal modular left ideal L, and the kernel of the corresponding representation is the set $\{a \in A : aX = \{0\}\}$, which is $L:A$, since

$$aX = \{0\} \Leftrightarrow a\,A\,x_0 = \{0\} \Leftrightarrow a\,A \subset \ker(x_0) \Leftrightarrow a \in L:A\,.$$

(iii) Suppose that $L_2 \not\subset P$ and that $P=M:A$ with M a maximal modular left ideal. Then $L_2\,A \not\subset M$, and so[1]

$$L_2\,A + M = A\,.$$

By (i), $P \subset M$, and so

$$L_1\,A = L_1(L_2\,A + M) \subset L_1 L_2\,A + L_1\,M \subset P + M = M\,,$$

from which $L_1 \subset M:A = P$.

(iv) Let P be a primitive ideal. By (ii), there exists an irreducible left A-module X such that $P = \{a \in A : aX = \{0\}\}$. This shows that $P = \bigcap\{\ker(x) : x \in X\setminus\{0\}\}$, and each $\ker(x)$ with $x \in X\setminus\{0\}$ is a maximal modular left ideal.

Definition 13. The (Jacobson) *radical* of A is the intersection of the kernels of all irreducible representations of A. It is denoted by $\mathrm{rad}(A)$, and, by the usual convention, $\mathrm{rad}(A)=A$ if there are no irreducible representations of A. A is said to be *semi-simple* if $\mathrm{rad}(A)=\{0\}$, and to be a *radical algebra* if $\mathrm{rad}(A)=A$.

Proposition 14. (i) $\mathrm{rad}(A)$ *is the intersection of the primitive ideals of A.*
(ii) $\mathrm{rad}(A)$ *is the intersection of the maximal modular left ideals of A.*

Proof. (i) Proposition 12(ii).
(ii) Proposition 12(iv).   ☐

It follows at once from Proposition 14(ii), that for commutative Banach algebras Definition 13 and Definition 17.6 give the same radical. We show now that $\mathrm{rad}(A)$ can also be characterized in terms of quasi-inverses.

Lemma 15. *Let every element of a left ideal J of A have a left quasi-inverse. Then $J \subset \mathrm{q-Inv}(A)$.*

Proof. Let $a \in J$. Then there exists $b \in A$ with $b \circ a = 0$. Therefore $b = ba - a \in J$, and so b has a left quasi-inverse as well as its right quasi-

---

[1] In this proof, for left ideals B, C of A, BC denotes the linear span of $\{bc : b \in B, c \in C\}$ (cf. page 45).

inverse a. Then b has the quasi-inverse a, and consequently a has the quasi-inverse b. ☐

Proposition 16. (i) $\mathrm{rad}(A) \subset q - \mathrm{Inv}(A)$.
 (ii) $\mathrm{rad}(A) = \{q \in A : Aq \subset q - \mathrm{Inv}(A)\}$.
 (iii) *Let J be a left ideal every element of which has a left quasi-inverse. Then* $J \subset \mathrm{rad}(A)$.

Proof. (i) Let s have no left quasi-inverse. Then $s \notin J = \{as - a : a \in A\}$. J is a modular left ideal with right modular unit s. Therefore there exists a maximal modular left ideal M with $J \subset M$, and $s \notin M$. Therefore, by Proposition 14(ii), $s \notin \mathrm{rad}(A)$. Therefore each element of $\mathrm{rad}(A)$ has a left quasi-inverse. Since $\mathrm{rad}(A)$ is a left ideal, Lemma 15 now gives $\mathrm{rad}(A) \subset q - \mathrm{Inv}(A)$.
 (ii) If $q \in \mathrm{rad}(A)$, then $Aq \subset \mathrm{rad}(A)$, and so $Aq \subset q - \mathrm{Inv}(A)$, by (i).
 Suppose on the other hand that $q \notin \mathrm{rad}(A)$. Then there exists an irreducible left A-module X such that $qX \neq \{0\}$. Choose $x_0 \in X$ with $qx_0 \neq 0$. Then qx_0 is a cyclic vector, and so there exists $a \in A$ with $aqx_0 = x_0$. Let $e = aq$. Then, by Proposition 4(i), e is a right modular unit for the ideal $\ker(x_0)$. Therefore e does not have a left quasi-inverse, $aq \notin q - \mathrm{Inv}(A)$.
 (iii) By Lemma 15, for all $j \in J$, $Aj \subset J \subset q - \mathrm{Inv}(A)$; and so, by (ii), $j \in \mathrm{rad}(A)$. ☐

Corollary 17. $\mathrm{rad}(A) = \{q \in A : qA \subset q - \mathrm{Inv}(A)\}$.

Proof. By Proposition 3.6, qa has a left quasi-inverse if and only if aq has a left quasi-inverse, and similarly for right quasi-inverses. ☐

There is a lack of symmetry in the definition of $\mathrm{rad}(A)$ in that it is defined in terms of left A-modules or, equivalently, in terms of maximal modular left ideals. If $B = \mathrm{rev}(A)$, i.e. A with the reversed product $x \cdot y = yx$, it is not evident that $\mathrm{rad}(B) = \mathrm{rad}(A)$. However Proposition 16(iii) and Corollary 17 show that these radicals coincide. There is a similar lack of symmetry in the definition of the primitive ideals of A, and it is an open question whether the primitive ideals of A coincide with the primitive ideals of B. For rings it is known that they need not coincide, see Bergman [47].

Corollary 18. $\mathrm{rad}(A)$ *is the union of all left (right) ideals all of whose elements are quasi-invertible.*

Proof. Proposition 16 and Corollary 17. ☐

Corollary 19. $\mathrm{rad}(A) : A = \mathrm{rad}(A)$.

Proof. If $x \in \mathrm{rad}(A) : A$, we have in turn $xA \subset \mathrm{rad}(A) \subset \mathrm{q} - \mathrm{Inv}(A)$, $x \in \mathrm{rad}(A)$. The opposite inclusion is obvious since $\mathrm{rad}(A)$ is a bi-ideal. ☐

Corollary 20. *Let B be a bi-ideal of A. Then* $\mathrm{rad}(B) = B \cap \mathrm{rad}(A)$.

Proof. By Corollary 19, $\mathrm{rad}(B) = \{x \in B : xB \subset \mathrm{rad}(B)\}$. If $x \in \mathrm{rad}(B)$ and $a \in A$, we have $xa \in B$ and $xaB \subset xB \subset \mathrm{rad}(B)$. Then $xa \in \mathrm{rad}(B)$, and $\mathrm{rad}(B)$ is a right ideal of A. Since $\mathrm{rad}(B) \subset \mathrm{q} - \mathrm{Inv}(B) \subset \mathrm{q} - \mathrm{Inv}(A)$, it follows, by Corollary 18, that $\mathrm{rad}(B) \subset \mathrm{rad}(A)$. Thus $\mathrm{rad}(B) \subset B \cap \mathrm{rad}(A)$.

Suppose on the other hand that $x \in B \cap \mathrm{rad}(A)$, and let $b \in B$. Then bx has a quasi-inverse y in A, and since $y = ybx - bx$, $y \in B$. Thus $x \in \mathrm{rad}(B)$. ☐

Proposition 21. $A/\mathrm{rad}(A)$ *is semi-simple.*

Proof. Let $B = A/\mathrm{rad}(A)$, let X be an irreducible left A-module, and let π denote the corresponding irreducible representation of A on X. If $a_1, a_2 \in b \in B$, then $a_1 - a_2 \in \mathrm{rad}(A) \subset \ker(\pi)$, and so $(a_1 - a_2)X = \{0\}$, $a_1 x = a_2 x$ $(x \in X)$. Therefore a well defined left module multiplication is given by

$$bx = ax \quad (a \in b \in B, x \in X).$$

It is clear that the left B-module X so obtained is irreducible. Therefore, if $b \in \mathrm{rad}(B)$, we have in turn $bX = \{0\}$, $aX = \{0\}$ $(a \in b)$. Since this holds for every irreducible left A-module X, we have $a \in \mathrm{rad}(A)$ whenever $a \in b$, and so $b = 0$, $\mathrm{rad}(B) = \{0\}$. ☐

§ 25. Irreducible Representations of Banach Algebras

Let A be a Banach algebra over $\mathbb{F}$. Since maximal modular left ideals of A are closed, it is clear that primitive ideals are closed, $\mathrm{rad}(A)$ is closed, and $A/\mathrm{rad}(A)$ is a semi-simple Banach algebra. The radical can also be characterized in terms of the spectral radius, as in the following proposition.

Proposition 1. (i) *If* $q \in \mathrm{rad}(A)$, *then* $r(q) = r(aq) = r(qa) = 0$ $(a \in A)$.
(ii) *If* $r(aq) = 0$ $(a \in A)$, *or if* $r(qa) = 0$ $(a \in A)$, *then* $q \in \mathrm{rad}(A)$.

Proof. (i) Let $q \in \mathrm{rad}(A)$. Since $\mathrm{rad}(A)$ is a bi-ideal, it is enough to prove that $r(q) = 0$. Suppose first that $\mathbb{F} = \mathbb{C}$. Then by Proposition 24.16(i), $\zeta^{-1}q \in \mathrm{q} - \mathrm{Inv}(A)$ for all non-zero complex numbers ζ, so $\mathrm{Sp}(q) = \{0\}$, $r(q) = 0$. Suppose now that $\mathbb{F} = \mathbb{R}$, and let $\alpha, \beta \in \mathbb{R}$ with $\alpha + i\beta = \zeta \neq 0$. Then $|\zeta|^{-2}(2\alpha q - q^2) \in \mathrm{rad}(A)$, and therefore, by Proposition 24.16(i) again, $|\zeta|^{-2}(2\alpha q - q^2) \in \mathrm{q} - \mathrm{Inv}(A)$. Therefore in this case also $\mathrm{Sp}(q) = \{0\}$, $r(q) = 0$.

(ii) $r(aq) = 0$ implies $aq \in \mathrm{q} - \mathrm{Inv}(A)$. Apply Proposition 24.16(ii). ☐

Remarks. (1) Proposition 1(i) does not need A to be complete.

(2) Proposition 1 shows that any closed subalgebra of a radical Banach algebra is a radical algebra.

Notation. Let X be an irreducible left A-module, let $\mathfrak{D}$ be the division subalgebra of $L(X)$ given by

$$\mathfrak{D} = \{T \in L(X) : aTx = T(ax) \ (a \in A, \ x \in X)\},$$

(see Proposition 24.6), and let π be the representation corresponding to the left A-module X.

Lemma 2. *Let* $x_0 \in X \setminus \{0\}$, $M = \ker(x_0)$, *and, for each* $y \in A - M$, *let*

$$Uy = ax_0 \qquad (a \in y).$$

Then the regular left A-module $A - M$ is a Banach left A-module, and U is a module isomorphism of $A - M$ onto X. If X is a Banach left A-module, then U is also a homeomorphism.

Proof. Being a maximal modular left ideal, M is closed, and therefore $A - M$ with the canonical norm is a Banach left A-module. If $a_1, a_2 \in y \in A - M$, then $a_1 - a_2 \in M = \ker(x_0)$, and so $a_1 x_0 = a_2 x_0$. Thus U is a well defined linear mapping of $A - M$ into X which is surjective since X is irreducible. If $Uy = 0$ and $a \in y$, then $ax_0 = 0$, $a \in \ker(x_0) = M$, $y = 0$. Therefore U is bijective. For all $b \in A$ and $a \in y \in A - M$, we have $bUy = bax_0 = U(by)$, and so U is a module isomorphism.

Suppose now that X is a Banach left A-module. Then there exists a positive constant κ with $\|ax\| \leqslant \kappa \|a\| \|x\|$ $(a \in A, \ x \in X)$. Therefore

$$\|Uy\| \leqslant \kappa \|a\| \|x_0\| \qquad (a \in y \in A - M).$$

By definition of the canonical norm on $A - M$ this gives

$$\|Uy\| \leqslant \kappa \|y\| \|x_0\| \qquad (y \in A - M).$$

Thus U is a continuous linear isomorphism of the Banach space $A - M$ onto the Banach space X, and therefore, by Banach's isomorphism theorem, U is a homeomorphism. $\square$

Proposition 3. (i) $\mathfrak{D}$ *is a normed division algebra over* $\mathbb{F}$.

(ii) *If X is a Banach left A-module, then* $\mathfrak{D} \subset BL(X)$.

Proof. (i) Let $x_0 \in X \setminus \{0\}$, M, U be as in Lemma 2, let $T \in \mathfrak{D}$, and let $z = U^{-1} Tx_0$. We prove that, for all $y \in A - M$, we have

$$\|U^{-1} TU y\| \leqslant \|y\| \|z\|. \tag{1}$$

For arbitrary $a \in y$, we have

$$U^{-1} TU y = U^{-1} Tax_0 = U^{-1} a Tx_0 = a U^{-1} Tx_0 = az.$$

Therefore $\|U^{-1}TUy\| = \|az\| \leqslant \|a\|\,\|z\|$. Since this inequality holds for all $a \in y$, (1) is proved. It follows that $U^{-1}TU \in BL(A-M)$, and we may therefore define a norm $|\cdot|$ on $\mathfrak{D}$ by taking

$$|T| = \|U^{-1}TU\|.$$

It is easy to see that this defines an algebra-norm on $\mathfrak{D}$.

(ii) Let X be a Banach left A-module. Then for $T \in \mathfrak{D}$, we have $U^{-1}TU \in BL(A-M)$ and $U \in BL(A-M, X)$, $U^{-1} \in BL(X, A-M)$. Therefore $T = U(U^{-1}TU)U^{-1} \in BL(X)$. □

Corollary 4. $\mathfrak{D}$ is $\mathbb{R}$, $\mathbb{C}$, or $\mathbb{H}$.

Proof. Theorem 14.7. □

Corollary 5. *Let A be a Banach algebra over $\mathbb{C}$. Then $\mathfrak{D} = \mathbb{C}$, and A is strictly dense on X, i.e. given $x_1, \ldots, x_n, y_1, \ldots, y_n \in X$ with $x_1, \ldots, x_n$ linearly independent, there exists $a \in A$ such that*

$$a x_k = y_k \qquad (1 \leqslant k \leqslant n).$$

Proof. Theorem 24.10. □

We come now to an important theorem of Johnson [211] on the continuity of irreducible representations, the proof of which uses the following lemma.

Lemma 6. *Let A_1 denote the closed unit ball in A, let $x_0 \in X \setminus \{0\}$, and let L be a closed left ideal of A with $L \not\subset \ker(x_0)$. Then there exists $\kappa > 0$ such that*

$$\kappa A_1 x_0 \subset (L \cap A_1) x_0.$$

Proof. Let $M = \ker(x_0)$, and let a' denote the M-coset of a. Since M is a maximal left ideal and $L \not\subset M$,

$$L + M = A.$$

Therefore the canonical mapping $a \rightarrow a'$ maps the Banach space L onto the Banach space $A - M$. Therefore, by the open mapping theorem, there exists $\kappa > 0$ such that for every $y \in A - M$ with $\|y\| \leqslant \kappa$ there exists $b \in L \cap A_1$ with $b' = y$. Given $a \in \kappa A_1$, we have $\|a'\| \leqslant \kappa$ and so there exists $b \in L \cap A_1$ with $b' = a'$. Then $b - a \in \ker(x_0)$, and so $a x_0 = b x_0$. □

Theorem 7. *Let π be an irreducible representation of a Banach algebra A on a normed linear space X such that $\pi(a) \in BL(X)$ $(a \in A)$. Then π is continuous.*

Proof. We first reduce the proof to the case when $\ker(\pi) = \{0\}$. Let $K = \ker(\pi)$. Then

$$K = \{a \in A : a X = \{0\}\} = \bigcap \{\ker(x) : x \in X \setminus \{0\}\}.$$

Since $\ker(x)$ is a maximal modular ideal for each $x \in X \setminus \{0\}$ and since $\ker(\pi)$ is a primitive ideal (Proposition 24.12(ii)), K is a closed bi-ideal, and so $B = A/K$ is a Banach algebra. Let τ be defined on B by

$$\tau(b)x = \pi(a)x \qquad (a \in b \in B, \ x \in X).$$

Then $\tau(b)$ is a well defined linear operator on X, and

$$\|\tau(b)x\| = \|\pi(a)x\| \leqslant \|\pi(a)\| \ \|x\|,$$

so that $\tau(b) \in BL(X)$. It is straightforward to check that τ is an irreducible representation of B on X, that $\ker(\tau) = \{0\}$, and that the continuity of τ would imply the continuity of π.

We assume that $\ker(\pi) = \{0\}$. We assume further that X has infinite dimension. For if X has finite dimension, then $L(X)$ has finite dimension and so has A, since $\ker(\pi) = \{0\}$. But then the continuity of π is elementary.

Given $x \in X$, let $\sigma(x)$ be the linear mapping of A into X defined by

$$\sigma(x)a = ax \qquad (a \in A).$$

Let $Y = \{x \in X : \sigma(x) \in BL(A, X)\}$. Then Y is an A-submodule of X, for, if $y \in Y$ and $b \in A$, we have

$$\sigma(by)a = aby = \sigma(y)(ab),$$
$$\|\sigma(by)a\| \leqslant \|\sigma(y)\| \ \|ab\| \leqslant \|\sigma(y)\| \ \|a\| \ \|b\|.$$

Since X is irreducible, we therefore have $Y = X$ or $Y = \{0\}$. If $Y = X$ and X_1 denotes the closed unit ball of X, we have

$$\|\sigma(x)a\| = \|ax\| = \|\pi(a)x\| \leqslant \|\pi(a)\| \qquad (x \in X_1, \ a \in A).$$

Therefore, by the principle of uniform boundedness, there exists a positive constant M with $\|\sigma(x)\| \leqslant M$ $(x \in X_1)$. But this gives $\|ax\| \leqslant M \|a\|$ $(a \in A, \ x \in X_1)$, i.e. $\|\pi(a)\| \leqslant M \|a\|$ $(a \in A)$, π is continuous.

We assume now that $Y = \{0\}$ and take A_1 to denote the closed unit ball in A. Then $A_1 x$ is unbounded whenever $x \in X \setminus \{0\}$. Since X has infinite dimension and $\mathfrak{D}$ has dimension at most 4, X contains an infinite sequence $\{x_n\}$ of $\mathfrak{D}$-independent vectors, and we may take $\|x_n\| = 1$ $(n \in \mathbb{N})$. Let $M_n = \ker(x_n)$ and let $L_n = M_1 \cap \cdots \cap M_{n-1}$. By Lemma 24.9, there exists $a \in A$ with $ax_k = 0$ $(1 \leqslant k \leqslant n-1)$ and $ax_n \neq 0$; i.e. $a \in L_n \setminus M_n$. Therefore $L_n \not\subset M_n$ and, since $A_1 x_n$ is unbounded, Lemma 6 shows that $(L_n \cap A_1)x_n$ is unbounded.

We choose $a_1, a_2, \ldots$ in turn with $a_n \in L_n$, $\|a_n\| < 2^{-n}$, and

$$\|a_n x_n\| > n + \|(a_1 + \cdots + a_{n-1})x_n\|.$$

Let $b=\sum_{k=1}^{\infty} a_k$ and $b_n=\sum_{k=n+1}^{\infty} a_k$. We have $a_k \in M_n$ $(k>n)$, and, since M_n is closed, it follows that $b_n \in M_n$. Therefore $b_n x_n = 0$. But $b = a_1 + \cdots + a_n + b_n$, and so $b x_n = a_1 x_n + \cdots + a_n x_n$,

$$\|b x_n\| \geqslant \|a_n x_n\| - \|(a_1 + \cdots + a_{n-1}) x_n\| > n \qquad (n \in \mathbb{N}).$$

Since $\pi(b) \in BL(X)$, this is impossible. □

Corollary 8. *Let π be an irreducible representation of a Banach algebra A on a normed linear space X such that $\pi(a) \in BL(X)$ $(a \in A)$. Then X is a normed left A-module.*

Proof. Since π is continuous, we have

$$\|a x\| = \|\pi(a) x\| \leqslant \|\pi(a)\| \, \|x\| \leqslant \|\pi\| \, \|a\| \, \|x\| \qquad (a \in A, \, x \in X). \quad □$$

Theorem 9 (Johnson [211]). *Let $(A, \|\cdot\|)$ be a semi-simple Banach algebra, and let $\|\cdot\|'$ be a second algebra-norm on A with respect to which A is complete. Then $\|\cdot\|'$ is equivalent to $\|\cdot\|$.*

Proof. Let M be a maximal modular left ideal of A, let $X = A - M$, and let $\|\cdot\|$, $\|\cdot\|'$ denote the canonical norms on X derived from $\|\cdot\|$, $\|\cdot\|'$ respectively on A. Then X is a Banach space with respect to each of these canonical norms. Let π denote the left regular representation of A on X. Then π is an irreducible representation of the Banach algebra $(A, \|\cdot\|)$ on the normed linear space $(X, \|\cdot\|')$, and since

$$\|\pi(a) x\|' = \|a x\|' \leqslant \|a\|' \, \|x\|' \qquad (a \in A, \, x \in X),$$

$\pi(a) \in BL(X, \|\cdot\|')$ for each $a \in A$. Therefore, by Theorem 7, π is continuous, i.e. there exists a positive constant κ such that

$$\|\pi(a) x\|' \leqslant \kappa \|a\| \, \|x\|' \qquad (a \in A, \, x \in X). \tag{2}$$

Let $a \to a'$ denote the canonical mapping of A onto $A - M$, and let u be a right modular unit for M. Then, for all $a \in A$, $au - a \in M$, and so

$$\pi(a) u = (au)' = a'.$$

Therefore, by (2),

$$\|x\|' = \|a'\|' = \|\pi(a) u\|' \leqslant \kappa \|a\| \, \|u\|' \qquad (a \in x \in X).$$

Since this holds for all $a \in x$, we have,

$$\|x\|' \leqslant \kappa \|x\| \, \|u\|' \qquad (x \in X),$$

and it follows that the two canonical norms $\|\cdot\|$, $\|\cdot\|'$ are equivalent on $A - M$.

Suppose now that $a, a_n \in A$ $(n \in \mathbb{N})$ and that $\lim_{n \to \infty} \|a_n\| = 0$ and $\lim_{n \to \infty} \|a_n - a\|' = 0$. Then $\lim_{n \to \infty} \|a_n'\| = 0$ and $\lim_{n \to \infty} \|a_n' - a'\|' = 0$. Therefore

$\lim\limits_{n\to\infty} \|a'_n - a'\| = 0$, $a' = 0$, $a \in M$. Since this holds for every maximal modular left ideal M, we have $a \in \mathrm{rad}(A) = \{0\}$. Therefore $a = 0$, and the closed graph theorem completes the proof. ☐

The following proposition is a useful complement for Theorem 9, though it lies much less deep.

Proposition 10. *Let π be an epimorphism of a Banach algebra A onto a semi-simple normed algebra. Then $\mathrm{ker}(\pi)$ is closed, and $\mathrm{rad}(A) \subset \mathrm{ker}(\pi)$.*

Proof. Let B be a semi-simple normed algebra with $\pi(A) = B$, let $K = \mathrm{ker}(\pi)$, and let $a \in K^-$. Given $x \in A$, we have $xa \in K^-$, and so there exists $k \in K$ with $\|xa - k\| < 1$. Therefore $xa - k \in \mathrm{q} - \mathrm{Inv}(A)$, i.e. there exists $y \in A$ with

$$(xa - k)y = y(xa - k) = (xa - k) + y.$$

Since π is a homomorphism and $\pi(k) = 0$, this gives

$$\pi(xa)\pi(y) = \pi(y)\pi(xa) = \pi(xa) + \pi(y),$$

which shows that $\pi(xa) \in \mathrm{q} - \mathrm{Inv}(B)$. Since $\pi(xa) = \pi(x)\pi(a)$ and $\pi(A) = B$, we have proved that $B\pi(a) \subset \mathrm{q} - \mathrm{Inv}(B)$. Therefore $\pi(a) \in \mathrm{rad}(B) = \{0\}$, by Proposition 24.16; $a \in K$.

Given $j \in \mathrm{rad}(A)$, we have $Aj \subset \mathrm{q} - \mathrm{Inv}(A)$ (by Proposition 24.16), and so

$$B\pi(j) = \pi(A)\pi(j) = \pi(Aj) \subset \mathrm{q} - \mathrm{Inv}(B).$$

Therefore $\pi(j) \in \mathrm{rad}(B) = \{0\}$, $j \in \mathrm{ker}(\pi)$. ☐

The equivalence of all complete algebra-norms on a semi-simple Banach algebra was an open question for many years. Theorem 9 for commutative Banach algebras (see Theorem 17.8) was proved by Gelfand [154] in 1948, and substantial results were obtained by Rickart [317], mainly using the concept of the *separating function* for two norms. It remains an open question whether every algebra-norm (not necessarily complete) on $C(E)$, for a compact Hausdorff space E, is equivalent to the uniform norm. See Bade and Curtis [32], Johnson [212].

There is another concept of irreducibility that is available for Banach algebras. Let A be a Banach algebra over $\mathbb{F}$ and let X be a Banach left A-module. We say that X is *topologically irreducible* if $A \neq \{0\}$ and $\{0\}, X$ are the only closed submodules of X. It is an open question whether

$$\mathfrak{D} = \{T \in BL(X) : aTx = T(ax) \ (a \in A, \ x \in X)\}$$

is then a division algebra. We say that $x_0 \in X$ is *topologically cyclic* if $(A x_0)^- = X$. Thus, for $A \neq \{0\}$, X is topologically irreducible if and only if each non-zero $x_0 \in X$ is topologically cyclic. We say that A is *topologically dense* if, given $\varepsilon > 0$, linearly independent $x_1, \ldots, x_n \in X$, and any $y_1, \ldots, y_n \in X$, there exists $a \in A$ such that

$$\| a x_i - y_i \| < \varepsilon \qquad (i = 1, \ldots, n).$$

Let $\mathbb{F} = \mathbb{C}$. It is an open question whether a topologically irreducible Banach left A-module X is topologically dense, even under the additional hypothesis that the algebra $\mathfrak{D}$, defined above, be a division algebra. Some positive results about topologically irreducible modules are given in Chapter V.

§ 26. The Structure Space of an Algebra

Let Π_A denote the set of all primitive ideals of an algebra A. Given $E \subset \Pi_A$ and $J \subset A$, let

$$\ker(E) = \bigcap \{P : P \in E\}, \qquad \operatorname{hul}(J) = \{P \in \Pi_A : P \supset J\},$$

and let

$$E^\sim = \operatorname{hul}(\ker(E)). \tag{1}$$

It is understood that $\ker(\emptyset) = A$.

Lemma 1. *Let* $J \subset A$, *and* $E = \operatorname{hul}(J)$. *Then* $E^\sim = E$.

Proof. We have $P \in E$ if and only if $P \supset J$. Therefore $\ker(E) \supset J$, and, if $P \in E^\sim$, we have $P \supset \ker(E) \supset J$, $P \in E$. Thus $E^\sim \subset E$, and the opposite inclusion obviously holds for all subsets of Π_A. □

Lemma 2. *Let* $E_1, E_2 \subset \Pi_A$. *Then* $(E_1 \cup E_2)^\sim = E_1^\sim \cup E_2^\sim$.

Proof. $\ker(E_1 \cup E_2) = \ker(E_1) \cap \ker(E_2)$, and so

$$(E_1 \cup E_2)^\sim = \operatorname{hul}(\ker(E_1 \cup E_2)) = \operatorname{hul}(\ker(E_1) \cap \ker(E_2)).$$

If $P \in E_1^\sim \cup E_2^\sim$, we have $P \supset \ker(E_1)$ or $P \supset \ker(E_2)$, and so $P \supset \ker(E_1) \cap \ker(E_2)$, $P \in (E_1 \cup E_2)^\sim$. Thus $E_1^\sim \cup E_2^\sim \subset (E_1 \cup E_2)^\sim$.

Suppose on the other hand that $P \in (E_1 \cup E_2)^\sim$. Then since $\ker(E)$ is a bi-ideal,

$$\ker(E_1)\ker(E_2) \subset \ker(E_1) \cap \ker(E_2) = \ker(E_1 \cup E_2) \subset P.$$

By Proposition 24.12 (iii), either $\ker(E_1) \subset P$ or $\ker(E_2) \subset P$, i.e. $P \in E_1^\sim \cup E_2^\sim$. □

Since $\emptyset^{\sim}=\emptyset$ and $\Pi_A^{\sim}=\Pi_A$, it follows from Lemmas 1 and 2 that the mapping $E \to E^{\sim}$ is a Kuratowski closure operation on the subsets of Π_A.

Definition 3. The *hull-kernel topology* on Π_A is the topology corresponding to the closure operation $E \to E^{\sim}$ defined by (1). The *structure space* of A is Π_A with the hull-kernel topology.

Proposition 4. *Let* $e \in A$, $A(1-e) \subset J \subset A$. *Then* $\mathrm{hul}(J)$ *is a compact subset of the structure space* Π_A.

Proof. By Lemma 1, $E = \mathrm{hul}(J)$ is closed. Let $\{F_\lambda : \lambda \in \Lambda\}$ be a set of closed subsets of E, let Φ denote the set of finite subsets of Λ, and suppose that

$$\bigcap \{F_\lambda : \lambda \in \Gamma\} \neq \emptyset \quad (\Gamma \in \Phi). \tag{2}$$

For each $\lambda \in \Lambda$, we have $F_\lambda \subset E$, and so

$$A(1-e) \subset J \subset \ker(E) \subset \ker(F_\lambda). \tag{3}$$

Given a subset Γ of Λ, let $K(\Gamma)$ denote the least bi-ideal of A containing $\ker(F_\lambda)$ for all $\lambda \in \Gamma$. If $\Gamma \in \Phi$, there exists $P \in \bigcap \{F_\lambda : \lambda \in \Gamma\}$, by (2), and then $\ker(F_\lambda) \subset P$ $(\lambda \in \Gamma)$, $K(\Gamma) \subset P$. By (3), e is a right modular unit for $K(\Gamma)$. Since $K(\Gamma) \neq A$, we therefore have $e \notin K(\Gamma)$. Since this holds for all $\Gamma \in \Phi$ and since $K(\Lambda) = \bigcup \{K(\Gamma) : \Gamma \in \Phi\}$, it follows that $e \notin K(\Lambda)$. Thus $K(\Lambda)$ is a proper modular left ideal, and is therefore contained in a maximal modular left ideal M. Let $Q = M : A$. Since $K(\Lambda)$ is a bi-ideal, $K(\Lambda) \subset Q$. Therefore $\ker(F_\lambda) \subset Q$, $Q \in \mathrm{hul}(\ker(F_\lambda)) = F_\lambda$ $(\lambda \in \Lambda)$, and so

$$\bigcap \{F_\lambda : \lambda \in \Lambda\} \neq \emptyset. \quad \square$$

Corollary 5. *If* A *has a unit element, the structure space of* A *is compact.*

Proof. Take $e=1$ and $J=\{0\}$ in Proposition 4. $\square$

Proposition 6. *The canonical mapping of* A *onto* $A/\mathrm{rad}(A)$ *induces a homeomorphism of their structure spaces.*

Proof. Let π denote the canonical mapping of A onto $B = A/\mathrm{rad}(A)$. Given $P \in \Pi_A$, we have $\mathrm{rad}(A) \subset P$, and so $\pi^{-1}(\pi(P)) = P$. Moreover, if X is an irreducible left A-module with P the kernel of the corresponding representation, then X becomes an irreducible left B-module under the multiplication given by

$$bx = ax \quad (a \in b \in B, \, x \in X),$$

and $\pi(P)$ is the kernel of the representation corresponding to this B-module. Routine checking now shows that the mapping $P \to \pi(P)$ is a bijection of Π_A onto Π_B. Given a subset E of Π_A, let $\pi(E)$ denote

$\{\pi(P):P\in E\}$. Then $\ker(\pi(E))=\pi(\ker(E))$, and so E is closed if and only if $\pi(E)$ is closed. $\quad\square$

In the next proposition, we determine the structure space of a finite dimensional complex algebra. This is intended mainly as an illustration of the structure space concept, but in the course of the proof we also establish the Wedderburn structure theorem for a semi-simple finite dimensional complex algebra.

Proposition 7. *Let A be a finite dimensional algebra over $\mathbb{C}$. The primitive ideals of A are the maximal bi-ideals, and the structure space is a finite set with the discrete topology.*

Proof. By Proposition 6, we may assume that A is semi-simple. Given $P\in\Pi_A$, the algebra A/P has a faithful representation on an irreducible left A-module X. It is clear that X has finite dimension and so, by norming X and transferring the operator norm to A/P, we may regard A/P as a Banach algebra over $\mathbb{C}$. By Corollary 25.5, A/P is isomorphic to $L(X)$. In particular A/P has a unit element and has no bi-ideals other than $\{0\}$ and A/P. Therefore P is a maximal bi-ideal of A.

Since A has finite dimension and is semi-simple, there exists a finite subset $\{P_1,...,P_m\}$ of Π_A with zero intersection, and we may suppose that this set is chosen so that m is as small as possible. Let $J_i=\bigcap\{P_k:k\neq i\}$. Then J_i is a non-zero bi-ideal and $P_i\cap J_i=\{0\}$. By maximality of P_i, we have $A=P_i\oplus J_i$; and so J_i is isomorphic to A/P_i. Therefore J_i is a minimal bi-ideal of A and has a unit element e_i. We prove that

$$A=J_1\oplus\cdots\oplus J_m.\qquad(4)$$

Given $a\in A$, there exist $j_1\in J_1$, $p_1\in P_1$ with $a=j_1+p_1$. Then there exist $j_2\in J_2$, $p_2\in P_2$ with $p_1=j_2+p_2$. Since $J_2\subset P_1$, we have $p_2=p_1-j_2\in P_1$, $p_2\in P_1\cap P_2$. Next, we have $p_2=j_3+p_3$ with $j_3\in J_3$ and $p_3\in P_1\cap P_2\cap P_3$, and so on. Eventually, we have $p_{n-1}=j_n+p_n$ with $j_n\in J_n$ and $p_n\in P_1\cap\cdots\cap P_n=\{0\}$. Thus $a=j_1+j_2+\cdots+j_n$, $A=J_1+\cdots+J_n$. Since the directness of the sum is easy to verify, this proves (4).

We prove next that

$$\Pi_A=\{P_1,...,P_m\}.\qquad(5)$$

Let $P\in\Pi_A$, with $J_1\not\subset P$. By minimality of J_1, we have $J_1\cap P=\{0\}$, and so $PJ_1=\{0\}=P_1J_1$. Given $p\in P$, we have $p=j+q$ with $j\in J_1$ and $q\in P_1$, and $j=je_1=(p-q)e_1=0$. Therefore $P\subset P_1$, and, by maximality, $P=P_1$. It follows that any primitive ideal not belonging to $\{P_1,...,P_m\}$ contains all the ideals J_i, which would contradict (4). Thus (5) is proved; Π_A is a finite set. To prove that the hull-kernel topology is discrete, it is now enough to prove that each singleton is closed. But this is clear, since the primitive ideals are maximal.

By (4), A has the unit element $e_1 + \cdots + e_m$. Therefore each maximal
bi-ideal is modular, is contained in a maximal modular left ideal, and
is therefore primitive. This completes the proof, and we have also proved
that A is the direct sum of its minimal bi-ideals.   □

If $E = \{P_0\}$ for some $P_0 \in \Pi_A$, then $E^\sim = \{P \in \Pi_A : P \supset P_0\}$. If there
exists $P \in \Pi_A$ with $P \supset P_0$ but $P \neq P_0$, then the singleton $\{P_0\}$ is not
a closed set, and so Π_A is not a T_1-space, still less a Hausdorff space.
For an example of this, take $A = BL(X)$ with X an infinite dimensional
Banach space, and let $K = KL(X)$, the set of compact linear operators
on X. Then K is a non-zero proper modular bi-ideal and is therefore
contained in a maximal modular left ideal M of A, and $K \subset P = M:A$.
Then P is a primitive ideal, as is also the zero ideal, P_0 say. In this connection, it is also of interest that if X is a separable Hilbert space, then
$\Pi_A = \{\{0\}, K\}$ (see Calkin [93]).

We remark also at this point that the quotient of a semi-simple
Banach algebra by a closed bi-ideal need not be semi-simple. For example,
let $X = L^1[0,1]$ and let W denote the set of weakly compact linear
operators on X. Then $K \subset W \subset A$, W is a closed bi-ideal of A, $W \neq K$,
$W^2 \subset K$ (see Dunford and Schwartz [130] pp. 507—510). Let J be
the image of W in A/K. Then J is a non-zero closed bi-ideal and $J^2 = \{0\}$;
in particular $J \subset \mathrm{rad}(A/K)$.

The topology of the structure space has not so far been very effectively
related to algebraic properties of general Banach algebras, though it
has proved highly significant for certain special classes of Banach algebras.

Let J be a maximal modular bi-ideal of an algebra A. Then J is
contained in a maximal modular left ideal M, and since $JA \subset J \subset M$,
we have $J \subset M:A$, and therefore, by maximality, $J = M:A$. Thus
maximal modular bi-ideals are primitive. The set Ξ_A of all maximal
modular bi-ideals is therefore a subset of Π_A, and with the (relative)
hull-kernel topology it is called the *strong structure space* of A. It is clear
that singletons are closed subsets of Ξ_A, but in general Ξ_A is not Hausdorff. It is easy to see that Proposition 4 holds with Ξ_A in place of Π_A,
only slight changes in the proof being necessary. Thus Corollary 5 also
holds for the strong structure space.

For a much fuller account of the structure space, the strong structure
space, and completely regular algebras we refer to Rickart [321], Willcox
[413]. The concept of the structure space of a general algebra (or of a
ring) is due to Jacobson [207].

We shall mainly be concerned with the structure space Π_A as the
index set for expressing a semi-simple algebra as a 'subdirect sum' of
'primitive algebras'.

Definition 8. An algebra A is said to be *primitive* if $\{0\}$ is a primitive ideal of A.

Proposition 9. *Let P be a primitive ideal of a Banach algebra A. Then A/P is a primitive Banach algebra.*

Proof. There exists a maximal modular left ideal M of A with $P = M : A$. We have remarked already that P is closed, since M is closed, and so $B = A/P$ is a Banach algebra. $X = A - M$ is an irreducible left A-module, and $P = \{a \in A : a X = \{0\}\}$. Given $a_1, a_2 \in b \in B$, we have $a_1 - a_2 \in P$, and so $a_1 x = a_2 x$ $(x \in X)$. Thus X is also a left B-module with the module multiplication given by $bx = ax$ $(a \in b \in B, x \in X)$. Clearly X is an irreducible left B-module, and the kernel of the corresponding representation of B is $\{0\}$. $\quad\square$

Definition 10. Let $\{A_\lambda : \lambda \in \Lambda\}$ denote a set of Banach algebras over the same scalar field. We denote by $\sum \{A_\lambda : \lambda \in \Lambda\}$ the set of functions f on Λ such that $f(\lambda) \in A_\lambda$ $(\lambda \in \Lambda)$ and

$$\|f\|_\infty = \sup \{\|f(\lambda)\| : \lambda \in \Lambda\} < \infty .$$

With pointwise addition, scalar multiplication, and product, and with the norm $\|\cdot\|_\infty$, $\sum \{A_\lambda : \lambda \in \Lambda\}$ is a Banach algebra which is called the *normed full direct sum* of $\{A_\lambda : \lambda \in \Lambda\}$. Given a subalgebra B of $\sum \{A_\lambda : \lambda \in \Lambda\}$, and given $\mu \in \Lambda$, the *coordinate homomorphism* of B into A_μ is the mapping π_μ defined by

$$\pi_\mu(f) = f(\mu) \quad (f \in B) .$$

A *normed subdirect sum* of $\{A_\lambda : \lambda \in \Lambda\}$ is a subalgebra B of the normed full direct sum $\sum \{A_\lambda : \lambda \in \Lambda\}$ such that for each $\mu \in \Lambda$ the coordinate homomorphism π_μ satisfies

$$\pi_\mu(B) = A_\mu .$$

Proposition 11. *Let A be a semi-simple Banach algebra, and, for each $a \in A$ and $P \in \Pi_A$, let $a^\wedge(P)$ denote the coset of a in A/P. Then the mapping $a \to a^\wedge$ is a norm reducing isomorphism of A onto a normed subdirect sum of the primitive Banach algebras $\{A/P : P \in \Pi_A\}$.*

Proof. We have

$$\|a^\wedge(P)\| \leqslant \|a\| \quad (a \in A, P \in \Pi_A),$$

and so $a^\wedge \in \sum \{A/P : P \in \Pi_A\}$, and the mapping $a \to a^\wedge$ is a norm reducing homomorphism into this normed full direct sum. The mapping is a monomorphism; for if $a^\wedge = 0$, then $a^\wedge(P) = 0$ $(P \in \Pi_A)$, and so

$$a \in \bigcap \{P : P \in \Pi_A\} = \mathrm{rad}(A) = \{0\} .$$

Let $B = A^\wedge$, the image of A under this monomorphism. For each $P \in \Pi_A$ the coordinate homomorphism π_P of B into A/P is given by

$$\pi_P(a^\wedge) = a^\wedge(P).$$

Thus $\pi_P(B) = A/P$, and B is a normed subdirect sum. ☐

We next consider conditions on the algebra-norm under which A is isometrically isomorphic to a normed subdirect sum of the primitive Banach algebras A/P with $P \in \Pi_A$. We recall that $S(A)$ denotes the unit sphere of A.

Definition 12. Let A be a Banach algebra. A has the $B^\#$-*property* if, for every $a \in S(A)$, there exists $a^\# \in S(A)$ with $r(a^\# a) = 1$. A has the *approximate $B^\#$-property* if, for every $a \in S(A)$ and $\varepsilon > 0$, there exists $a^\# \in S(A)$ with $r(a^\# a) > 1 - \varepsilon$.

It is easy to see that A has the $B^\#$-property if and only if, for every $a \in A \setminus \{0\}$ there exists $a^\# \in A \setminus \{0\}$ such that $\|(a^\# a)^n\|^{\frac{1}{n}} = \|a^\#\| \, \|a\| \ (n \in \mathbb{N})$. There is of course no assumption of uniqueness of the element $a^\#$.

Proposition 13. *Let A be a complex Banach algebra with the approximate $B^\#$-property. Then*

(i) *A is semi-simple;*

(ii) *A is isometrically isomorphic to a normed subdirect sum of the primitive Banach algebras $\{A/P : P \in \Pi_A\}$.*

Proof. (i) Let $a \in S(A)$. Then there exists $a^\# \in A$ with $r(a^\# a) > \frac{1}{2}$. Therefore $a \notin \mathrm{rad}(A)$, $\mathrm{rad}(A) = \{0\}$.

(ii) Let $a \in S(A)$, and $\varepsilon > 0$. Then there exists $a^\# \in S(A)$ with $r(a^\# a) > 1 - \varepsilon$, and there exists $\lambda \in \mathbb{C}$ with $|\lambda| = r(a^\# a)$ such that $\lambda^{-1} a^\# a$ is quasi-singular. Let $u = \lambda^{-1} a^\# a$. We have $1 \in \partial \mathrm{Sp}(u)$, and therefore by Corollary 3.10 and Proposition 9.2(iii) $A(1 - u) \neq A$, and so u is a right modular unit for a maximal modular left ideal L. Let $P = L : A$. By Proposition 24.12(i), we have $P \subset L$. Therefore

$$\|u\| \geqslant d(u, P) \geqslant d(u, L) \geqslant 1.$$

Let $x^\wedge(P)$ denote the canonical image of x in A/P. Then

$$\|u^\wedge(P)\| = d(u, P) \geqslant 1,$$

and so $\|a^{\#\wedge}(P)\| \, \|a^\wedge(P)\| \geqslant \|\lambda u^\wedge(P)\| \geqslant |\lambda| > 1 - \varepsilon$. Since $\|x^\wedge(P)\| \leqslant \|x\|$, it follows that $\|a^\wedge(P)\| > 1 - \varepsilon$, $\|a^\wedge\|_\infty = 1$. Thus the isomorphism $a \to a^\wedge$ is an isometry. ☐

In § 34 we shall need the following theorem of Kaplansky [234].

Theorem 14. *Let e be a non-zero idempotent in a Banach algebra A, and let $B = eAe$. Then the mapping $P \to P \cap B$ is a homeomorphism of $\Pi_A \setminus \mathrm{hul}(B)$ onto Π_B.*

Proof. Let $P \in \Pi_A \setminus \mathrm{hul}(B)$. Then there exists an irreducible left A-module X with

$$B \not\subseteq P = \{a \in A : aX = \{0\}\}.$$

Therefore $eX \neq \{0\}$. For each $x_0 \in eX \setminus \{0\}$, $Bx_0 = eAex_0 = eAx_0 = eX$. Therefore eX is an irreducible left B-module. The kernel of the corresponding representation is $\{b \in B : beX = \{0\}\} = \{b \in B : bX = \{0\}\} = B \cap P$, and so $B \cap P \in \Pi_B$.

Let $Q \in \Pi_B$. Then there exists a maximal modular left ideal M of B with $Q = M : B$. [1]Let $J = AM + A(1-e)$. Then J is a modular left ideal of A, and $J \neq A$, since $eJe \subset M \neq B$. Therefore J is contained in a maximal modular left ideal L of A, and $L : A = P \in \Pi_A$. We prove that $P \cap B = Q$.

We have $QAe = QB \subset M$, and so $AQAe \subset AM \subset J \subset L$. Since also $AQA(1-e) \subset A(1-e) \subset J \subset L$, this gives $AQA \subset L$, and so $Q = eQe \subset AQA \subset L : A = P$. Thus $Q \subset P \cap B$.

Since $e \notin L$, $M \subset L$ and $P \subset L$, we have $e \notin M + (P \cap B)$. But M being a maximal left ideal, this shows that $P \cap B \subset M$, and therefore $P \cap B \subset M : A = Q$. Thus $P \cap B = Q$, $P \not\supset B$, $P \in \Pi_A \setminus \mathrm{hul}(B)$.

The mapping $P \to P \cap B$ is injective. For, let $P_1, P_2 \in \Pi_A \setminus \mathrm{hul}(B)$ with $P_1 \cap B = P_2 \cap B$. Since $eP_2 e \subset P_2 \cap B$, we have

$$AeP_2 Ae \subset AeP_2 e \subset A(P_2 \cap B) = A(P_1 \cap B) \subset P_1.$$

Since $P_1 \notin \mathrm{hul}(B)$, $e \notin P_1$, $Ae \not\subset P_1$. Therefore, by Proposition 24.12(iii), $AeP_2 \subset P_1$. But then, for the same reason, $P_2 \subset P_1$.

It remains to prove the bicontinuity of the mapping $P \to P \cap B$. A subset E of $\Pi_A \setminus \mathrm{hul}(B)$ is closed in the relative hull-kernel topology if and only if $P_0 \in E$ whenever $P_0 \in \Pi_A \setminus \mathrm{hul}(B)$ and $P_0 \supset \bigcap \{P : P \in E\}$. Let F be a closed subset of Π_B, and let $E = \{P \in \Pi_A \setminus \mathrm{hul}(B) : P \cap B \in F\}$. If $P_0 \in \Pi_A \setminus \mathrm{hul}(B)$ and $P_0 \supset \bigcap \{P : P \in E\}$, then

$$P_0 \cap B \supset \bigcap \{P \cap B : P \in E\} = \bigcap \{Q : Q \in F\}.$$

Since F is closed, this gives $P_0 \cap B \in F$, $P_0 \in E$, E is relatively closed.

Suppose on the other hand that E is a relatively closed subset of $\Pi_A \setminus \mathrm{hul}(B)$, let $F = \{P \cap B : P \in E\}$, and let $Q_0 \in \Pi_B$ with $Q_0 \supset \ker(F)$. Then $Q_0 = P_0 \cap B$ for some $P_0 \in \Pi_A \setminus \mathrm{hul}(B)$, and

$$P_0 \supset Q_0 \supset \ker(F) = \bigcap \{P \cap B : P \in E\} = \ker(E) \cap B.$$

---

[1] See footnote on page 124.

Therefore $P_0 \supset e(\ker(E))e$, and, since P_0 and $\ker(E)$ are bi-ideals, this gives $Ae(\ker(E))Ae \subset P_0$. Therefore, by Proposition 24.12(iii), either $Ae \subset P_0$ or $(\ker(E))Ae \subset P_0$. The first alternative would give $B \subset P_0$, and so $P_0 \in \text{hul}(B)$. Therefore $(\ker(E))Ae \subset P_0$ and a repetition of the argument gives $\ker(E) \subset P_0$, and so $P_0 \in \dot{F}$. Thus $Q_0 \in F$, F is closed. $\square$

§ 27. *A*-Module Pairings

Much of the strength of the theories of commutative Banach algebras and of Banach star algebras derives from properties of certain linear functionals on the algebras. The rest of this chapter is mainly concerned with importing this source of strength into the general theory of Banach algebras.

Definition 1. A *pairing* is a triple $(X, Y, \langle , \rangle)$ where X, Y are linear spaces over $\mathbb{F}$ and $\langle , \rangle$ is a *non-degenerate* bilinear form on $X \times Y$; i.e. for $x \in X$ and $y \in Y$,

$$\langle x, Y \rangle = \{0\} \Rightarrow x = 0, \quad \langle X, y \rangle = \{0\} \Rightarrow y = 0.$$

A *normed pairing* is a pairing $(X, Y, \langle , \rangle)$ in which X, Y are normed linear spaces and $\langle , \rangle$ is continuous, i.e. there exists a positive constant κ such that

$$|\langle x, y \rangle| \leqslant \kappa \|x\| \|y\| \quad (x \in X, y \in Y).$$

A *Banach pairing* is a normed pairing $(X, Y, \langle , \rangle)$ in which X, Y are Banach spaces.

Remark. Our 'pairing' is 'separated pairing' in the terminology of Kelley and Namioka [241].

Definition 2. Given an algebra A, an *A-module pairing* is a pairing $(X, Y, \langle , \rangle)$ such that X is a left A-module, Y is a right A-module, and

$$\langle ax, y \rangle = \langle x, ya \rangle \quad (a \in A, x \in X, y \in Y).$$

Given a normed algebra A, a *Banach A-module pairing* is a Banach pairing $(X, Y, \langle , \rangle)$ which is also an A-module pairing and satisfies

$$\|ax\| \leqslant \kappa \|a\| \|x\|, \quad \|ya\| \leqslant \kappa \|y\| \|a\| \quad (a \in A, x \in X, y \in Y),$$

for some positive constant κ. Thus X is a Banach left A-module and Y is a Banach right A-module.

Example 3. Let A be a normed algebra and X a Banach left A-module. Let $Y = X'$, the dual space of X, and let $\langle , \rangle$ denote the natural bilinear form, i.e. $\langle x, y \rangle = y(x)$ $(x \in X, y \in Y)$. Then $(X, Y, \langle , \rangle)$ is a Banach pairing, and Y is a Banach right A-module (the dual module of X) with the multiplication defined by

$$(y\,a)(x)=y(a\,x)\quad(a\in A,\,x\in X,\,y\in Y)\,.$$

Clearly $(X,Y,\langle,\rangle)$ is a Banach A-module pairing.

Definition 4. Let $(X,Y,\langle,\rangle)$ be a normed pairing. Operators $S\in BL(X)$, $T\in BL(Y)$ are said to be *adjoint* with respect to $\langle,\rangle$ if

$$\langle S\,x,y\rangle=\langle x,T\,y\rangle\quad(x\in X,\,y\in Y)\,.$$

To each $S\in BL(X)$ corresponds at most one $T\in BL(Y)$ such that S,T are adjoint with respect to $\langle,\rangle$; this operator T, if it exists, is called the *adjoint of S with respect to* $\langle,\rangle$, and is denoted by S^*. Similarly each $T\in BL(Y)$ has at most one adjoint $T^*\in BL(X)$ with respect to $\langle,\rangle$. The set of operators $S\in BL(X)$ that have adjoints with respect to $\langle,\rangle$ is denoted by $BL(X,Y,\langle,\rangle)$, and the norm $\|\cdot\|_d$ is defined on $BL(X,Y,\langle,\rangle)$ by

$$\|S\|_d=\max\,(\|S\|,\,\|S^*\|)\,.$$

This will be the 'specified' norm on $BL(X,Y,\langle,\rangle)$.

Lemma 5. *Let* $(X,Y,\langle,\rangle)$ *be a Banach pairing. Then* $BL(X,Y,\langle,\rangle)$ *(with the norm* $\|\cdot\|_d$) *is a Banach algebra.*

Proof. It is clear that $BL(X,Y,\langle,\rangle)$ is a subalgebra of $BL(X)$ and that $\|\cdot\|_d$ is an algebra-norm. Let $\{S_n\}$ be a Cauchy sequence in $BL(X,Y,\langle,\rangle)$ with respect to $\|\cdot\|_d$. Then $\{S_n\}$ and $\{S_n^*\}$ are Cauchy sequences in $BL(X)$ and $BL(Y)$ respectively with respect to the operator norm $\|\cdot\|$. Therefore there exist $S\in BL(X)$, $T\in BL(Y)$ with $\lim_{n\to\infty}\|S_n-S\|=0$, $\lim_{n\to\infty}\|S_n^*-T\|=0$. Then, for all $x\in X$, $y\in Y$,

$$\langle S\,x,y\rangle=\lim_{n\to\infty}\,\langle S_n\,x,y\rangle=\lim_{n\to\infty}\,\langle x,S_n^*\,y\rangle=\langle x,T\,y\rangle\,.$$

Thus $S\in BL(X,Y,\langle,\rangle)$, $S^*=T$, and $\lim_{n\to\infty}\|S_n-S\|_d=0$. $\quad\square$

Notation. Given an algebra A and an A-module pairing $(X,Y,\langle,\rangle)$, let $\lambda(a)$, $\rho(a)$ be the linear operators on X,Y respectively defined (for each $a\in A$) by

$$\lambda(a)\,x=a\,x\quad(x\in X),\qquad\rho(a)\,y=y\,a\quad(y\in Y)\,.$$

Thus λ is the representation of A on X corresponding to the left A-module X and ρ is the anti-representation (representation of $\mathrm{rev}(A)$) of A on Y corresponding to the right A-module Y.

Proposition 6. *Let A be an algebra, and let $(X, Y, \langle, \rangle)$ be a Banach pairing and an A-module pairing. Then, for each $a \in A$, $\lambda(a) \in BL(X, Y, \langle, \rangle)$ and $(\lambda(a))^* = \rho(a)$.*

If A is a normed algebra and $(X, Y, \langle, \rangle)$ is a Banach A-module pairing, then λ is a continuous homomorphism of A into the Banach algebra $BL(X, Y, \langle, \rangle)$ (with the norm $\|\cdot\|_d$).

Proof. Let $a \in A$, $x_n, x, z \in X$ with $\lim_{n \to \infty} x_n = x$, $\lim_{n \to \infty} a x_n = z$. Then, for all $y \in Y$, we have

$$\langle z, y \rangle = \lim_{n \to \infty} \langle a x_n, y \rangle = \lim_{n \to \infty} \langle x_n, y a \rangle = \langle x, y a \rangle = \langle a x, y \rangle.$$

Therefore $a x = z$, and, by the closed graph theorem, $\lambda(a) \in BL(X)$. Similarly $\rho(a) \in BL(Y)$, and it is clear that $\lambda(a)$, $\rho(a)$ are adjoint with respect to $\langle, \rangle$. The rest is entirely straightforward. $\square$

Proposition 7. *Let A be a Banach algebra, let $(X, Y, \langle, \rangle)$ be a Banach pairing and an A-module pairing, and let X be irreducible. Then $(X, Y, \langle, \rangle)$ is a Banach A-module pairing.*

Proof. By Proposition 6, $\lambda(a) \in BL(X)$ ($a \in A$); and, since X is irreducible, it follows by Theorem 25.7 that λ is a continuous linear mapping of A into $BL(X)$; i.e. there exists a positive constant κ with

$$\|a x\| \leqslant \kappa \|a\| \|x\| \qquad (a \in A, x \in X). \tag{1}$$

This argument does not apply to Y, since we do not know that Y is irreducible. However, by Proposition 6, $\rho(a) \in BL(Y)$. Let $a_n, a \in A$ and $T \in BL(Y)$ with

$$\lim_{n \to \infty} a_n = a, \qquad \lim_{n \to \infty} \rho(a_n) = T.$$

Then, by (1), $\lim_{n \to \infty} a_n x = a x$ ($x \in X$), and so, for all $x \in X$, $y \in Y$,

$$\langle x, T y \rangle = \lim_{n \to \infty} \langle x, y a_n \rangle = \lim_{n \to \infty} \langle a_n x, y \rangle = \langle a x, y \rangle = \langle x, y a \rangle.$$

Therefore $T = \rho(a)$, and the closed graph theorem proves the continuity of ρ. $\square$

Definition 8. A *dual representation* of a normed algebra A on a Banach pairing $(X, Y, \langle, \rangle)$ is a continuous homomorphism of A into the Banach algebra $BL(X, Y, \langle, \rangle)$.

Given a Banach A-module pairing $(X, Y, \langle, \rangle)$, Proposition 6 shows that the mapping λ is a dual representation of A on $(X, Y, \langle, \rangle)$; this is called the *dual representation corresponding to the given Banach A-module pairing*. Conversely it is plain that if π is a dual representation of A on a Banach pairing $(X, Y, \langle, \rangle)$, then $(X, Y, \langle, \rangle)$ becomes a Banach A-module pairing with the module multiplication given by

$$a x = \pi(a) x, \qquad y a = (\pi(a))^* y \quad (a \in A, \, x \in X, \, y \in Y).$$

This is called the *Banach A-module pairing corresponding to the given dual representation.*

A Banach A-module pairing $(X, Y, \langle, \rangle)$ (or the corresponding dual representation) is said to be *dually irreducible* if X is an irreducible left A-module and Y is an irreducible right A-module.

Proposition 9. *Let A be a normed algebra, let $(X, Y, \langle, \rangle)$ be a Banach A-module pairing, and let X be irreducible. Then Y is weakly irreducible in the sense that $\{0\}$ and Y are the only A-submodules of Y that are closed in the weak topology $w(Y, X)$.*

Proof. Let M be a submodule of Y closed in the topology $w(Y, X)$. If $M \neq Y$, there exists $x \in X \setminus \{0\}$ with $\langle x, M \rangle = \{0\}$. Then $A x = X$, and so

$$\langle X, M \rangle = \langle A x, M \rangle = \langle x, M A \rangle \subset \langle x, M \rangle = \{0\},$$

from which $M = \{0\}$. □

Definition 10. Let $(X, Y, \langle, \rangle)$ be a normed pairing. The set of operators in $BL(X, Y, \langle, \rangle)$ that have finite rank (i.e. finite dimensional range) is denoted by $FBL(X, Y, \langle, \rangle)$. This is abbreviated to $FBL(X)$ when $Y = X'$ and $\langle, \rangle$ is the natural form. Given $u \in X$, $v \in Y$, let $u \otimes v$ denote the operator on X given by $(u \otimes v)(x) = \langle x, v \rangle u$ $(x \in X)$. Then $u \otimes v \in FBL(X, Y, \langle, \rangle)$, since it has the adjoint $(u \otimes v)^*$ given by

$$(u \otimes v)^*(y) = \langle u, y \rangle v \quad (y \in Y).$$

Proposition 11. *Let $(X, Y, \langle, \rangle)$ be a Banach pairing with $X \neq \{0\}$, let $A = BL(X, Y, \langle, \rangle)$, and let $B = FBL(X, Y, \langle, \rangle)$. Then $(X, Y, \langle, \rangle)$ is a dually irreducible Banach A-module pairing and a dually irreducible Banach B-module pairing.*

Proof. We regard X as a left A-module and Y as a right A-module with the module multiplications defined by

$$S x = S(x) \quad (S \in A, \, x \in X), \qquad y T = T^*(y) \quad (T \in A, \, y \in Y).$$

Clearly $(X, Y, \langle, \rangle)$ is then a Banach A-module pairing. It is now sufficient to prove that X and Y are irreducible B-modules.

Given $x_0 \in X \setminus \{0\}$, there exists $v \in Y$ with $\langle x_0, v \rangle = 1$, and then, for arbitrary $u \in X$, we have $(u \otimes v) x_0 = u$. Thus each $x_0 \in X \setminus \{0\}$ is cyclic for the algebra B, and X is an irreducible left B-module. Similarly Y is an irreducible right B-module. □

Suppose now that A is a Banach algebra over $\mathbb{C}$ and that $(X, Y, \langle, \rangle)$ is a dually irreducible Banach A-module pairing. By Corollary 25.5, we know that A is strictly dense on X and on Y. Strict density of A on X is equivalent to the following statement. Given a finite dimensional linear subspace X_0 of X and given $T_0 \in L(X_0, X)$, there exists $a \in A$ such that $\lambda(a)|_{X_0} = T_0$.

By Proposition 11, given such X_0, T_0 there exists $S \in BL(X, Y, \langle, \rangle)$ with $S|_{X_0} = T_0$. Thus strict density of A on X is also equivalent to the statement: given a finite dimensional linear space X_0 of X and given $S \in BL(X, Y, \langle, \rangle)$, there exists $a \in A$ such that

$$\lambda(a)|_{X_0} = S|_{X_0} .$$

The next proposition similarly expresses the dual irreducibility of $(X, Y, \langle, \rangle)$ in terms of the action of A on subspaces of X of finite dimension and of finite codimension.

Proposition 12. *Let A be a Banach algebra over $\mathbb{C}$, let $(X, Y, \langle, \rangle)$ be a dually irreducible Banach A-module pairing, let X_0 be a finite dimensional linear subspace of X, and let X_1 be a linear subspace of X that is closed in the weak topology $w(X, Y)$ and has finite codimension. Then, for every $S \in BL(X, Y, \langle, \rangle)$ there exist $a, b \in A$ such that $\lambda(a)|_{X_0} = S|_{X_0}$ and $(\lambda(b) - S) X \subset X_1$.*

Proof. The existence of the required $a \in A$ has been observed already. Since X_1 is $w(X, Y)$ closed and of finite codimension, there exists a finite dimensional linear subspace Y_1 of Y such that

$$X_1 = \{x \in X : \langle x, Y_1 \rangle = \{0\}\} .$$

Since A is strictly dense on the right A-module Y, there exists $b \in A$ with $\rho(b)|_{Y_1} = S^*|_{Y_1}$. Then, for all $x \in X$,

$$\langle (\lambda(b) - S) x, Y_1 \rangle = \langle x, (\rho(b) - S^*) Y_1 \rangle = \{0\} ;$$

and so $(\lambda(b) - S) X \subset X_1$. ☐

§ 28. The Dual Module of a Banach Algebra

In this section we continue the programme started in § 27 of bringing the dual space of A to bear on the algebraic structure of A, by developing some elementary properties of the dual module of A.

Let A be a Banach algebra over $\mathbb{F}$, and A' its dual Banach A bimodule, i.e. the dual space of A with the module multiplications

$$(af)(x) = f(xa), \qquad (fa)(x) = f(ax) \qquad (a, x \in A, \ f \in A') .$$

Given subsets E, F of A, A' respectively we denote by $E^{\perp}, F_{\perp}$ the sets given by

$$E^{\perp} = \{f \in A' : f(a) = 0 \; (a \in E)\}, \quad F_{\perp} = \{a \in A : f(a) = 0 \; (f \in F)\}.$$

Given $e \in A$, we shall as before write $A(1-e)$, $(1-e)A$ to denote the sets $\{a - ae : a \in A\}$, $\{a - ea : a \in A\}$ even when A does not have a unit element.

Definition 1. Given a subset F of A', an element $e \in A$ is a *left identity for F* if $ef = f \; (f \in F)$ and is a *right identity for F* if $fe = f \; (f \in F)$.

Lemma 2. (i) *Let u be a left identity for a non-zero right submodule F of A'. Then $F_{\perp}$ is a proper closed modular left ideal with right modular unit u.*

(ii) *Let u be a right modular unit for a proper modular left ideal E of A. Then $E^{\perp}$ is a non-zero weak* closed right submodule of A' with left identity u.*

Proof. (i) We have

$$f(a - au) = (f - uf)(a) = 0 \quad (a \in A, \; f \in F),$$
$$f(ax) = (fa)(x) = 0 \quad (a \in A, \; f \in F, \; x \in F_{\perp}).$$

Therefore $A(1-u) \subset F_{\perp}$, and $AF_{\perp} \subset F_{\perp}$. Also $F_{\perp} \neq A$ since $F \neq \{0\}$.

(ii) We have

$$(f - uf)(a) = f(a - au) = 0 \quad (a \in A, \; f \in E^{\perp}),$$
$$(fa)(x) = f(ax) = 0 \quad (a \in A, \; x \in E, \; f \in E^{\perp}).$$

Therefore $f - uf = 0 \; (f \in E^{\perp})$ and $E^{\perp}A \subset E^{\perp}$. ☐

Proposition 3. *The mapping $F \to F_{\perp}$ is a bijection of the set M of non-zero weak* closed right submodules of A' with left identity u onto the set of proper closed modular left ideals of A with right modular unit u. The mapping reverses inclusion, has inverse mapping $E \to E^{\perp}$, and maps the minimal elements of M onto the set of maximal modular left ideals with right modular unit u.*

Proof. Clear. ☐

It is clear from Proposition 3 and the fact that proper modular left ideals are contained in maximal ones that each element of M contains a minimal element of M. The next proposition and a compactness argument give an independent proof of this. Given $u \in A$ with $\|u\| \geq 1$, let

$$D(u) = \{f \in A' : \|f\| \leq 1 = f(u)\}.$$

Proposition 4. *Let* u *be a left identity for a non-zero weak* closed right submodule* F *of* A'. *Then* $F \cap D(u)$ *is not void.*

Proof. Let

$$\kappa = \sup \{|f(u)| : f \in F, \|f\| \leqslant 1\}.$$

We have $\kappa > 0$, for otherwise

$$f(a) = (u f)(a) = (f a)(u) = 0 \qquad (a \in A, f \in F),$$

giving $F = \{0\}$. Since the intersection of F with the unit ball of A' is weak* compact, the supremum κ is attained. Thus there exists $f_0 \in F$ with $\|f_0\| \leqslant 1$ and $f_0(u) = \kappa$. Given $a \in A$ with $\|a\| \leqslant 1$, we have $f_0 a \in F$ and $\|f_0 a\| \leqslant 1$. Therefore

$$|f_0(a)| = |(u f_0)(a)| = |(f_0 a)(u)| \leqslant \kappa,$$

and so $\|f_0\| \leqslant \kappa$, $\dfrac{1}{\kappa} f_0 \in D(u) \cap F$.   □

Proposition 5. *Let* F *be a minimal non-zero weak* closed right submodule of* A' *with a left identity* u. *Then there exists* $f_0 \in F$ *with* $f_0(u) \neq 0$; *and, for each such* f_0,

(i) $F_\perp = \{a \in A : a f_0 = 0\} = (f_0 A)_\perp$,

(ii) $A f_0$ *is an irreducible left* A-*module isomorphic to the regular left* A-*module* $A - F_\perp$.

Proof. By Proposition 4, there exists $f \in F \cap D(u)$, and so with $f(u) \neq 0$. Let $f_0 \in F$ with $f_0(u) \neq 0$. We have

$$u \notin \{a : a f_0 = 0\} = \{a : f_0(A a) = \{0\}\} = (f_0 A)_\perp \supset F_\perp.$$

The equality (i) therefore follows from the maximality of $F_\perp$.

Let $X = A - F_\perp$. Given $a_1, a_2 \in x \in X$, we have $a_1 - a_2 \in F_\perp$, and so $a_1 f_0 = a_2 f_0$, by (i). Thus we obtain a well-defined mapping T of X onto $A f_0$ given by

$$T x = a f_0 \qquad (a \in x \in X).$$

If $a \in x \in X$ and $T x = 0$, then $a f_0 = 0$, $a \in F_\perp$, $x = 0$. Thus T is a linear isomorphism of X onto $A f_0$. It is a module isomorphism; for if $b \in A$ and $a \in x \in X$, then $b a \in b x$, and so

$$T(b x) = (b a) f_0 = b(a f_0) = b T x.   □$$

Corollary 6. *Let* L *be a maximal modular left ideal of* A *with right modular unit* u. *Then there exists* $f_0 \in D(u) \cap L^\perp$ *such that* $f_0 u = f_0$, $L = (f_0 A)_\perp$, *and* $A f_0$ *is an irreducible left* A-*module isomorphic to the regular left* A-*module* $A - L$.

Proof. $L^{\perp}$ is a minimal non-zero weak* closed right submodule of A' with left identity u. By Proposition 4, there exists $f_0 \in D(u) \cap L^{\perp}$, and the rest follows from Proposition 5. □

It is now clear that every irreducible left A-module is isomorphic to a submodule $A f_0$ of A'.

We recall (from §11) that a bounded right approximate identity for A is a bounded net $\{e_\lambda\}_\Lambda$ in A such that

$$a e_\lambda \to a$$

for each $a \in A$.

Similar definitions apply to bounded left and two-sided approximate identities.

We recall also that the second dual A'' is a Banach algebra with respect to the Arens product (Example 9.13(v)), which is defined in terms of the A-bimodule A' by taking

$$(F f)(a) = F(f a) \qquad (a \in A, \ f \in A', \ F \in A''),$$
$$(F G)(f) = F(G f) \qquad (f \in A', \ F, G \in A'').$$

Proposition 7. *The Banach algebra A'' has a right identity if and only if A has a bounded right approximate identity.*

Proof. Let A have a bounded right approximate identity $\{e_\lambda\}_\Lambda$, and let $e_\lambda^\wedge$ denote the canonical image of e_λ in A''. Then $\{\|e_\lambda^\wedge\| : \lambda \in \Lambda\}$ is bounded and so has a weak* cluster point E. We have, for all $a \in A$, $f \in A'$,

$$e_\lambda^\wedge(f a) = (f a)(e_\lambda) = f(a e_\lambda) \to f(a),$$

and so $E(f a) = f(a)$. We have proved that $E f = f$ $(f \in A')$, and so for all $f \in A'$, $F \in A''$ we have

$$(F E)(f) = F(E f) = F(f).$$

Thus $F E = F$ $(F \in A'')$.

Suppose on the other hand that the Banach algebra A'' has a right identity E. Then E belongs to the weak* closure of some bounded subset of the canonical image $A^\wedge$ of A (see Dunford and Schwartz [130] Theorem V 4.5). Thus there exists a bounded net $\{e_\lambda\}_\Lambda$ in A such that $\{e_\lambda^\wedge\}_\Lambda$ converges to E in the weak* topology, i.e.

$$e_\lambda^\wedge(f) \to E(f) \qquad (f \in A').$$

Therefore

$$e_\lambda^\wedge(f a) \to E(f a) \qquad (a \in A, \ f \in A').$$

But $e_\lambda^\wedge(f a)=(f a)(e_\lambda)=f(a e_\lambda)$ and

$$E(f a)=(E f)(a)=a^\wedge(E f)=(a^\wedge E)(f)=a^\wedge(f)=f(a).$$

Therefore $f(a e_\lambda)\to f(a)$ $(a\in A,\ f\in A')$. Apply Proposition 11.4. $\square$

Corollary 8. *Let the Arens product in A'' be regular* (Example 9.13(v)). *Then A'' has a unit element if and only if A has a bounded two-sided approximate identity.*

Proof. Clear by Proposition 7 and Proposition 11.6. $\square$

§ 29. The Representation of Linear Functionals

Let A denote a Banach algebra over $\mathbb{F}$, and A' the dual space of A, which we shall also regard as the dual module of A. We consider representations of functionals $f\in A'$ in the form

$$f(a)=\langle a x_0,y_0\rangle \quad (a\in A),$$

where $(X,Y,\langle,\rangle)$ is a Banach A-module pairing and $x_0\in X,\ y_0\in Y$. We shall be concerned with representations of this kind which satisfy additional conditions, for example we may ask that the vectors x_0,y_0 be cyclic or that the modules be irreducible.

Given $f\in A'$, we shall use the following notation:

$$L_f=(f A)_\perp, \quad K_f=(A f)_\perp, \quad X_f=A-L_f, \quad Y_f=A-K_f.$$

Since $L_f=\{a\in A:Aa\subset\ker(f)\}$, L_f is a closed left ideal, and so X_f is a Banach left A-module. Similarly K_f is a closed right ideal and Y_f is a Banach right A-module. We define a bilinear form $\langle,\rangle_f$ on $X_f\times Y_f$ by

$$\langle x,y\rangle_f=f(b a) \quad (a\in x\in X_f, b\in y\in Y_f).$$

This is well-defined; for if $a_1,a_2\in x$ and $b_1,b_2\in y$, then $a_1-a_2\in L_f$, $b_1-b_2\in K_f$, and so

$$b_1 a_1-b_2 a_2=(b_1-b_2)a_1+b_2(a_1-a_2)\in\ker(f).$$

Also, for all $a\in x\in X_f$ and $b\in y\in Y_f$, we have

$$|\langle x,y\rangle_f|=|f(b a)|\leqslant\|f\|\,\|a\|\,\|b\|;$$

and so, by definition of the canonical norms on X_f,Y_f,

$$|\langle x,y\rangle_f|\leqslant\|f\|\,\|x\|\,\|y\|.$$

Thus $(X_f,Y_f,\langle,\rangle_f)$ is a Banach A-module pairing.

Definition 1. Let $f \in A'$, let $(X, Y, \langle, \rangle)$ be a Banach A-module pairing, and let $x_0 \in X$, $y_0 \in Y$. We say that f is *strictly represented* by $(X, Y, \langle, \rangle, x_0, y_0)$ if x_0 and y_0 are cyclic vectors, and

$$f(a) = \langle a x_0, y_0 \rangle \qquad (a \in A). \tag{1}$$

A vector $x_0 \in X$ is *weakly cyclic* if $A x_0$ is dense in X with respect to the weak topology $w(X, Y)$, and similarly for $y_0 \in Y$. We say that f is *weakly represented* by $(X, Y, \langle, \rangle, x_0, y_0)$ if x_0 and y_0 are weakly cyclic vectors and (1) holds. We say that f is *strictly (weakly) representable* if f is strictly (weakly) represented by some $(X, Y, \langle, \rangle, x_0, y_0)$.

Remark. In the theory of Banach star algebras the 'representable functionals' are defined in terms of the topologically cyclic vectors, i.e. those x_0 for which $A x_0$ is norm dense. This would also seem to be a natural concept in the present context, but we know of no positive results.

Proposition 2. *Let $f \in A' \setminus \{0\}$, let $u, v \in A$, and let*

$$u f = f v = f.$$

Then L_f is a proper closed modular left ideal with right modular unit u, K_f is a proper closed modular right ideal with left modular unit v, and f is strictly represented by $(X_f, Y_f, \langle, \rangle_f, x_0, y_0)$, where x_0 is the coset $u + L_f$ of u and y_0 is the coset $v + K_f$ of v.

Proof. Since u is a left identity for the right submodule $f A$, which is non-zero since $f = f v \in f A$, Lemma 28.2(i) shows that $L_f = (f A)_{\perp}$ is a proper closed modular left ideal with right modular unit u. A similar argument applies to K_f. Given $a \in x \in X_f$, we have $a u + L_f = a + L_f$, and so $x = a x_0$. Thus x_0 is a cyclic vector in X_f, and similarly y_0 is cyclic in Y_f.

We note that $L_f \subset \ker(f)$, since

$$L_f = (f A)_{\perp} \subset \{f v\}_{\perp} = \{f\}_{\perp} = \ker(f);$$

and similarly $K_f \subset \ker(f)$. We now have

$$a - v a u = (a - a u) + (a u - v a u) \in \ker(f)$$

and so

$$f(a) = f(v a u) = \langle a x_0, y_0 \rangle_f \qquad (a \in A). \quad \square$$

Corollary 3. *If A has a unit element then every non-zero continuous linear functional on A is strictly representable.*

Proposition 4. *Let A have a bounded two-sided approximate identity. Then every non-zero $f \in A'$ is weakly representable.*

Proof. Let $\{e_\lambda\}_A$ be the given bounded two-sided approximate identity, and let $f \in A'$. Let B denote the unitization $A + \mathbb{F}$ of A, and regard A as embedded in B. Let ξ be a cluster point of the bounded net $\{f(e_\lambda)\}_A$, and let g be defined on B by

$$g(\alpha + a) = \xi \alpha + f(a) \qquad (\alpha \in \mathbb{F}, a \in A).$$

Then $g \in B'$, and so g is strictly representable by $(X_g, Y_g, \langle , \rangle_g, x_0, y_0)$, with x_0, y_0 the cosets of 1 in X_g, Y_g. Since A is a subalgebra of $B, (X_g, Y_g, \langle , \rangle_g)$ is a Banach A-module pairing, and

$$f(a) = \langle a x_0, y_0 \rangle_g \qquad (a \in A).$$

It remains to prove that the vectors x_0, y_0, which are cyclic with respect to the algebra B, are also weakly cyclic with respect to the algebra A.

Given $x \in X$, there exist $\alpha \in \mathbb{F}$, $a \in A$ with $(\alpha + a)x_0 = x$. To prove that x_0 is weakly cyclic with respect to A, it is therefore enough to prove that x_0 belongs to the $w(X, Y)$ closure of $A x_0$. Let $\beta + b \in y \in Y_g$. Then

$$\begin{aligned}
\langle e_\lambda x_0 - x_0, y \rangle_g &= g((\beta + b)(e_\lambda 1 - 1)) \\
&= g(\beta e_\lambda + b e_\lambda - b - \beta) \\
&= f(\beta e_\lambda + b e_\lambda - b) - \beta \xi \\
&= f(b e_\lambda) - f(b) + \beta (f(e_\lambda) - \xi).
\end{aligned}$$

Since $b e_\lambda \to b$ weakly and ξ is a cluster point of the net $\{f(e_\lambda)\}_A$, it follows that $|\langle e_\lambda x_0 - x_0, y \rangle_g|$ is arbitrarily small for some arbitrarily large λ. Thus x_0 belongs to the $w(X, Y)$ closure of $A x_0$, and x_0 is weakly cyclic with respect to A. Similarly y_0 is weakly cyclic. $\quad\square$

Is the converse of Proposition 4 true? The next proposition takes us part of the way.

Proposition 5. *Let $f \in A'$ be weakly representable. Then*

$$f \in \overline{A f}^{w^*} \cap \overline{f A}^{w^*},$$

where $\overline{A f}^{w^*}, \overline{f A}^{w^*}$ denote the closure of $A f, f A$ in A' with respect to the weak topology.*

Proof. Suppose that $f \notin \overline{A f}^{w^*}$. Then there exists $a_0 \in A$ such that $f(a_0) = 1$, but $(A f)(a_0) = \{0\}$. Let f be weakly represented by $(X, Y, \langle , \rangle, x_0, y_0)$. Then $f(a) = \langle a x_0, y_0 \rangle = \langle x_0, y_0 a \rangle$ $(a \in A)$. Therefore

$$\langle A x_0, y_0 a_0 \rangle = \langle a_0 A x_0, y_0 \rangle = f(a_0 A) = (A f)(a_0) = \{0\}.$$

Since x_0 is weakly cyclic, this gives $y_0 a_0 = 0$. But this is absurd, since

$$\langle x_0, y_0 a_0 \rangle = f(a_0) = 1. \quad\square$$

We recall that, given an element x_0 of a left A-module A,

$$\ker(x_0) = \{a \in A : a x_0 = 0\},$$

and similarly for a right A-module.

Proposition 6. *Let $f \in A'$ be weakly representable by* $(X, Y, \langle\,,\,\rangle, x_0, y_0)$.
Then
(i) $L_f = \ker(x_0)$, $K_f = \ker(y_0)$;
(ii) *the mappings* U, V *defined by*

$$U x = a x_0 \quad (a \in x \in X_f), \qquad V y = y_0 b \quad (b \in y \in Y_f)$$

are continuous A-module monomorphisms of X_f into X, Y_f into Y respectively;
(iii) $\langle x, y \rangle_f = \langle U x, V y \rangle$ $(x \in X_f, y \in Y_f)$.
If x_0 and y_0 are cyclic vectors, then U, V are bicontinuous A-module isomorphisms of X_f onto X, Y_f onto Y respectively, and L_f, K_f are modular ideals.

Proof. (i) Since $y_0 A$ is $w(Y, X)$ dense in Y, we have

$$\ker(x_0) = \{a \in A : a x_0 = 0\} = \{a \in A : \langle a x_0, Y \rangle = \{0\}\}$$
$$= \{a \in A : \langle a x_0, y_0 A \rangle = \{0\}\}$$
$$= \{a \in A : \langle A a x_0, y_0 \rangle = \{0\}\}$$
$$= \{a \in A : f(A a) = \{0\}\} = L_f.$$

Similarly $\ker(y_0) = K_f$.

(ii) By (i), U is a well-defined mapping of X_f into X. Since X is a normed left A-module, there exists a positive constant κ such that $\|a x\| \leqslant \kappa \|a\| \|x\|$ $(a \in A, x \in X)$, and so $\|U x\| \leqslant \kappa \|a\| \|x_0\|$ $(a \in x \in X_f)$. Therefore, by definition of the canonical norm on X_f,

$$\|U x\| \leqslant \kappa \|x\| \|x_0\| \quad (x \in X_f).$$

Thus U is a continuous linear mapping of X_f into X. If $U x = 0$ and $a \in x$, we have $a x_0 = 0$, $a \in \ker(x_0) = L_f$, $x = 0$. Thus U has zero kernel. Given $b \in A$ and $a \in x \in X_f$, we have $b a \in b x$, and so

$$U b x = (b a) x_0 = b U x.$$

Thus U is a module monomorphism of X_f into X. Similarly for V.
(iii) Given $a \in x \in X_f$, $b \in y \in Y_f$, we have

$$\langle x, y \rangle_f = f(b a) = \langle b a x_0, y_0 \rangle = \langle a x_0, y_0 b \rangle = \langle U x, V y \rangle.$$

The final sentence in the proposition is an immediate consequence of Banach's isomorphism theorem and Proposition 24.4(ii). □

Corollary 7. *Let* $f \in A' \setminus \{0\}$. *The following statements are equivalent.*

(i) *There exist* $u, v \in A$ *with* $uf = fv = f$ *and* L_f, K_f *maximal modular left and right ideals respectively.*

(ii) f *is strictly representable by a dually irreducible Banach A-module pairing.*

Proof. Propositions 2 and 6. $\quad\square$

In the rest of this section we are concerned with geometric constructions for strictly representable functionals with one or both of X_f, Y_f irreducible.

Notation. We recall that $S(X)$ denotes the unit sphere of a normed linear space X, and that, given $e \in S(A)$, $D(e) = \{f \in S(A') : f(e) = 1\}$. We denote by $\Delta(e)$ the set

$$\Delta(e) = \{f \in D(e) : e f = f e = f\}.$$

The following proposition is stated to avoid repetition in later results.

Proposition 8. *Let* $e \in S(A)$ *and* $f \in \Delta(e)$. *Then* L_f *is a proper closed modular left ideal with right modular unit* e, K_f *is a proper modular right ideal with left modular unit* e, *and* f *is strictly represented by* $(X_f, Y_f, \langle, \rangle_f, x_0, y_0)$, *where* x_0, y_0 *are the cosets of* e *in* X_f, Y_f.

Proof. Proposition 2. $\quad\square$

Theorem 9. *Let* $e \in S(A) \cap \mathrm{q\text{-}Sing}(A)$. *Then*

(i) *there exists a maximal modular left ideal* L *with right modular unit* e, *and, for each such* L, *there exists* $f \in \Delta(e)$ *with* $L_f = L$;

(ii) *there exists a maximal modular right ideal* K *with left modular unit* e, *and, for each such* K, *there exists* $f \in \Delta(e)$ *with* $K_f = K$.

Proof. Suppose first that e does not have a left quasi-inverse, i.e. that $A(1-e) \neq A$. Then there exists a maximal modular left ideal L with right modular unit e. Let

$$E = D(e) \cap L^{\perp}.$$

Since $\|e\| = 1$ and e is a right modular unit for L, we have $d(e, L) = 1$, and so the Hahn-Banach theorem gives the existence of $f \in A'$ with $\|f\| = 1 = f(e)$ and $f(L) = \{0\}$; i.e. E is non-void. It is clear that E is a convex and weak* compact subset of A'. We prove next that $fe \in E$ whenever $f \in E$.

Given $f \in E$, we have

$$(f e)(L) = f(e L) \subset f(L) = \{0\},$$

since L is a left ideal. Also, since $e-e^2\in L$, we have

$$(fe)(e)=f(e^2)=f(e)=1\,.$$

Since $\|e\|=1$ we have $\|fe\|\leqslant\|f\|=1$. Therefore $\|fe\|=1$, and $fe\in E$.

We have now proved that the mapping $f\to fe$ sends E into E, and it is clearly weak* continuous. Therefore, by the Schauder-Tychonoff fixed point theorem (see Dunford and Schwartz [130, Theorem V. 10.5]), there exists $f\in E$ with $fe=f$. It follows that

$$f(a-ea)=0\qquad(a\in A)\,,$$

and so e has no right quasi-inverse.

Since $f(AL)\subset f(L)=\{0\}$, we have $L\subset L_f$; and, since $L_f\neq A$, this gives $L_f=L$. Since

$$A(1-e)\subset L\subset\ker(f)\,,$$

we have $ef=f$ and (i) is proved. Since e has no right quasi-inverse, a similar argument gives (ii).

We started with the assumption that e has no left quasi-inverse, but could equally well have started with the assumption that e has no right quasi-inverse. ☐

Remarks. (1) Given $u\in A$ with $r(u)=\|u\|\neq0$, and given that the scalar field is $\mathbb{C}$, a complex number λ can be chosen so that

$$\lambda u\in S(A)\cap\text{q-Sing}(A)\,.$$

(2) A. M. Sinclair has pointed out that the use of the Schauder-Tychonoff theorem in the proof of Theorem 9 can be avoided by using numerical range theory.

Corollary 10. *Let $e\in S(A)\cap\text{q-Sing}(A)$, and suppose that $\Delta(e)$ contains at most one element. Then*

(i) *$\Delta(e)$ contains exactly one element f;*

(ii) *there exists exactly one maximal modular left ideal with right modular unit e, namely L_f;*

(iii) *there exists exactly one maximal modular right ideal with left modular unit e, namely K_f;*

(iv) *f is strictly represented by the dually irreducible Banach A-module pairing $(X_f,Y_f,\langle,\rangle_f)$.*

Proof. Immediate. ☐

Remark. If $e\in S(A)\cap\text{q-Sing}(A)$ and the unit ball of A is smooth at e, i.e. $D(e)$ is a singleton $\{f\}$, then the conditions of Corollary 10 are obviously satisfied. In this case the proof would not need the appeal to the Schauder-Tychonoff fixed point theorem.

Corollary 11. *Let A be a Banach algebra over $\mathbb{C}$, let $e \in S(A) \cap q\text{-Sing}(A) \cap A^c$, and suppose that $\Delta(e)$ contains at most one element. Then $\Delta(e)$ contains exactly one element f, f is multiplicative, and $\ker(f) = (A(1-e))^-$.*

Proof. Since $a - ae = a - ea$, $A(1-e)$ is a bi-ideal contained in $\ker(f)$. Let J denote the closure of this ideal, let $B = A/J$, and let a' denote the canonical image of a in B. Since $f(e) = 1$, $e \notin J$, and so $B \neq \{0\}$, B has a unit element $1 = e'$.

Given $\phi \in D(1)$, let $g(a) = \phi(a')$ $(a \in A)$. Then $g \in D(e)$ and $eg = ge = g$. Thus $g \in \Delta(e)$, $g = f$. This proves that

$$\phi(b) = f(a) \quad (a \in b \in B),$$

and so $D(1)$ is a singleton, $\{\phi\}$ say. But, by the Bohnenblust-Karlin theorem [Corollary 10.15], $D(1)$ separates the points of B, and so $\ker(\phi) = \{0\}$. Therefore $\ker(f) = J = (A(1-e))^-$.

Finally, given $x, y \in A$, we have $y - f(y)e \in \ker(f)$, and, since J is a left ideal,

$$xy - f(y)xe \in \ker(f),$$

$$f(xy) = f(y)f(xe) = f(y)f(x). \quad \square$$

Proposition 12. *Let A be a complex $B^{\#}$-algebra (see Definition 26.12) in which the unit ball is smooth. Then A is dually semi-simple, i.e. the intersection of the kernels of the dually irreducible representations of A is the zero ideal.*

Proof. Let $a \in A$ with $\|a\| = 1$. We prove that there exists a dually irreducible representation π of A such that $a \notin \ker(\pi)$. Since $r(a^* a) = 1$, there exists $\lambda \in \mathbb{C}$ with $|\lambda| = 1$ such that $u = \lambda a^* a$ is quasi-singular. By Corollary 10, $\Delta(u) = \{f\}$, a singleton, and $(X_f, Y_f, \langle , \rangle_f)$ is a dually irreducible Banach A-module pairing. The corresponding representation π of A on X_f has kernel $L_f : A = P$ say. Then $a \notin P$; in fact, as in Proposition 26.13, we have $\|a^{\wedge}(P)\| = 1$, where $a^{\wedge}(P)$ is the coset of a in A/P. Thus A is dually semi-simple. $\quad \square$

In fact we have proved more. With Π_A^d denoting the set of kernels of dually irreducible representations of A, we have proved that A is isometrically isomorphic to a normed subdirect sum of $\{A/P : P \in \Pi_A^d\}$.

The concepts and results of §§ 27—29 are due to the authors [75, 76]. See also Fell [136].

Chapter IV. Minimal Ideals

§ 30. Algebraic Preliminaries

In this section A will denote an algebra. Definitions and results will usually be stated only for left ideals; with the obvious changes, they apply to right ideals.

Definition 1. *A minimal left ideal* of A is a left ideal $J \neq \{0\}$ such that $\{0\}$ and J are the only left ideals contained in J.

A *minimal idempotent* is a non-zero idempotent $e \in A$ such that eAe is a division algebra.

Obviously the minimal left ideals of A are the irreducible submodules of the regular left A-module A.

Lemma 2. *Let X be a minimal left ideal of A such that $X^2 \neq \{0\}$. Then there exists an idempotent $e \in X$ such that $Ae = X$; and each idempotent e with $Ae = X$ is minimal.*

Proof. There exists $x_0 \in X$ such that $X x_0 \neq \{0\}$. Then $X x_0$ is a non-zero submodule of X, and so $X x_0 = X$ and there exists $e \in X$ with $e x_0 = x_0$. By Proposition 24.4(i),

$$A(1-e) \subset \ker(x_0) = \{a \in A : a x_0 = 0\} .$$

Since $X \cap \ker(x_0)$ is a left ideal contained in X and X is not contained in $\ker(x_0)$, the minimality of X gives $X \cap \ker(x_0) = \{0\}$. Since $e \in X$, we have $e - e^2 \in X$. Also $e - e^2 \in A(1-e) \subset \ker(x_0)$, and so $e - e^2 = 0$. Ae is a left ideal contained in X and $Ae \neq \{0\}$, since $e \in Ae$. Therefore $Ae = X$.

Suppose now that e is any idempotent in A with $Ae = X$. Then $e = eee \in X$. Evidently eAe is an algebra with unit element e. Let $eae \neq 0$. Since $eae = e(eae) \in Aeae$, we have $\{0\} \neq Aeae \subset X$. Therefore $Aeae = X$ and there exists $b \in A$ with $beae = e$, and so with

$$(ebe)(eae) = e^2 = e .$$

We have proved that each non-zero element of eAe has a left inverse, and therefore eAe is a division algebra. $\square$

It is easy to see that $eAe=\mathfrak{D}e$, where $\mathfrak{D}$ is the division algebra
of Proposition 24.6. In fact, given $u\in eAe$, let T_u be defined on $X\ (=Ae)$
by $T_u x=xu$. Since $Au\subset Ae$, $T_u\in L(X)$. Also $aT_u x=axu=T_u(ax)$,
so that $T_u\in\mathfrak{D}$. $T_u e=eu=u$. Conversely, given $T\in\mathfrak{D}$, let $u=Te$.
Then $eu=T_,e^2=Te=u$, $u=eue\in eAe$.

Definition 3. An algebra A is *semi-prime* if $\{0\}$ is the only bi-ideal J
of A with $J^2=\{0\}$. Given a subset E of A, the *left annihilator* of E is
the set lan(E) and the *right annihilator* of E is the set ran(E) defined,
respectively, by

$$\text{lan}(E)=\{a\in A:aE=\{0\}\},\quad \text{ran}(E)=\{a\in A:Ea=\{0\}\}.$$

Lemma 4. *Let A be semi-prime, and let L be a left ideal of A with*
$L^2=\{0\}$. *Then* $L=\{0\}$.

Proof. lan(A) is a bi-ideal with $(\text{lan}(A))^2\subset(\text{lan}(A))A=\{0\}$, and so
lan$(A)=\{0\}$. [1]Since L is a left ideal, LA is a bi-ideal, and $(LA)(LA)\subset L^2 A$
$=\{0\}$. Therefore $LA=\{0\}$, $L\subset\text{lan}(A)=\{0\}$. ☐

Proposition 5. *Semi-simple algebras are semi-prime.*

Proof. Let J be a non-zero bi-ideal of A with $J^2=\{0\}$. For all $j\in J$,
$a\in A$, we have
$$(aj)\circ(-aj)=aj-aj-(aj)^2=0,$$
and so $aj\in\text{q-Inv}(A)$. Therefore, by Proposition 24.16, $J\subset\text{rad}(A)$. ☐

Proposition 6. *Let A be semi-prime. Then*
  (i) *L is a minimal left ideal of A if and only if $L=Ae$ where e is a minimal
idempotent in A;*
  (ii) *R is a minimal right ideal of A if and only if $R=eA$ where e is a
minimal idempotent in A.*

Proof. (i) If L is a minimal left ideal, then Lemma 2 shows that $L=Ae$
with e a minimal idempotent. Suppose then that e is a minimal idempotent
and let J be a left ideal with $\{0\}\neq J\subset Ae$. By Lemma 4, $J^2\neq\{0\}$ and
so there exist elements ae,be in J such that $aebe\neq0$. It follows that
$ebe\neq0$, and, since eAe is a division algebra, there exists $c\in eAe$ such
that $cebe=e$. Then $Ae=Acebe\subset Abe\subset J$, and so Ae is a minimal
left ideal.
  (ii) Similar. ☐

Lemma 7. *Let X be a minimal left ideal, and let $u\in A$. Then either*
$Xu=\{0\}$ *or Xu is a minimal left ideal.*

Proof. Suppose that $Xu\neq\{0\}$ and that J is a left ideal with
$$\{0\}\neq J\subset Xu.$$

---
[1] See footnote on page 124.

Let $K = \{x \in X : xu \in J\}$. Then K is a left ideal and $\{0\} \neq K \subset X$. Therefore $K = X$, $Xu \subset J$. □

Definition 8. If A has minimal left ideals, the smallest left ideal containing all of them is called the *left socle* of A. The *right socle* is similarly defined in terms of right ideals. If A has both minimal left and minimal right ideals, and if the left socle coincides with the right socle, it is called the *socle* of A and is denoted by $\operatorname{soc}(A)$. In these circumstances we say for brevity that $\operatorname{soc}(A)$ *exists*.

Lemma 9. *If A has minimal left ideals, then the left socle is a bi-ideal.*

Proof. Let A have minimal left ideals, and let S denote the left socle of A. Given a minimal left ideal X and $u \in A$, Lemma 7 shows that $Xu = \{0\}$ or Xu is a minimal left ideal. In both cases $Xu \subset S$. Therefore $Su \subset S$, and so S is a right ideal. By definition S is a left ideal. □

Proposition 10. *Let A be semi-prime and have minimal left ideals. Then $\operatorname{soc}(A)$ exists.*

Proof. By Proposition 6, A has minimal right ideals. By Lemma 9 the left socle of A is a bi-ideal which, by Proposition 6, contains all minimal idempotents. By Proposition 6 again, the left socle therefore contains the right socle. □

Proposition 11. *Let A be semi-prime and let e be a minimal idempotent in A. Then $A(1 - e)$ is a maximal modular left ideal and*

$$A = Ae \oplus A(1 - e).$$

Proof. By Proposition 6, Ae is an irreducible left A-module; so $\ker(e) = \{a : ae = 0\}$ is a maximal modular left ideal. Since e is an idempotent,

$$\{a : ae = 0\} = A(1 - e).$$

Since $Ae \not\subset \ker(e)$, we have $Ae \cap A(1 - e) = \{0\}$, and obviously $A = Ae + A(1 - e)$. □

§ 31. Minimal Ideals in Complex Banach Algebras

Throughout this section A will denote a complex Banach algebra. We develop some elementary properties of minimal ideals and minimal idempotents in A.

Lemma 1. *Let e be an idempotent in a normed algebra B. Then*
(i) $Be = \{b \in B : be = b\}$,
(ii) Be *is a closed left ideal.*

Proof. Clear. ☐

Proposition 2. *Let X be a minimal ideal in A with $X^2 \neq \{0\}$. Then X is closed.*

Proof. Lemma 30.2 and Lemma 1. ☐

Proposition 3. *Let e be a minimal idempotent in A. Then there exists $f \in A'$ such that $eae = f(a)e$ $(a \in A)$.*

Proof. By definition, eAe is a division algebra. It is also a normed algebra over $\mathbb{C}$, and so $eAe = \mathbb{C}e$. Thus there exists a linear functional f on A with $eae = f(a)e$ $(a \in A)$. Since $\|eae\| \leqslant \|e\|^2 \|a\|$, $f \in A'$. ☐

For the proof of Theorem 6, which is the main result in this section, we need some elementary properties of pairings. We shall use the bi-ideal $FBL(X, Y, \langle, \rangle)$ which we defined in Definition 27.10. We denote by δ_{ij} the Kronecker delta.

Lemma 4. *Let $(X, Y, \langle, \rangle)$ be a pairing and let $x_1, \ldots, x_n$ be linearly independent vectors in X. Then there exist $y_1, \ldots, y_n \in Y$ such that*

$$\langle x_i, y_j \rangle = \delta_{ij} \quad (1 \leqslant i, j \leqslant n).$$

Proof. Let X_0 be the subspace of X spanned by $\{x_1, \ldots, x_n\}$. Let T be the linear mapping of Y into X_0' defined by

$$(Ty)(x) = \langle x, y \rangle \quad (x \in X_0, y \in Y).$$

Then T is onto; for otherwise by the reflexivity of X_0 there exists non-zero x_0 with $\langle x_0, y \rangle = 0$ $(y \in Y)$. In particular, for each $i = 1, \ldots, n$, there exists $y_i \in Y$ such that

$$\langle x_i, y_j \rangle = \delta_{ij} \quad (i, j = 1, \ldots, n). \quad \text{☐}$$

Lemma 5. *Let $(X, Y, \langle, \rangle)$ be a normed pairing. Then $FBL(X, Y, \langle, \rangle)$ is the set of all operators of the form $T = \sum_{i=1}^{n} u_i \otimes v_i$ with $u_i \in X$, $v_i \in Y$.*

Proof. That all operators of the stated form belong to $FBL(X, Y, \langle, \rangle)$ has been noted already. Let $T \in FBL(X, Y, \langle, \rangle)$, and let $u_1, \ldots, u_n$ be a basis for TX. Then, for each $x \in X$, there exist $f_1(x), \ldots, f_n(x) \in \mathbb{F}$ with

$$Tx = f_1(x)u_1 + \cdots + f_n(x)u_n.$$

By Lemma 4, there exist $w_1, \ldots, w_n \in Y$ such that

$$\langle u_i, w_j \rangle = \delta_{ij} \quad (1 \leqslant i, j \leqslant n).$$

Take $v_j = T^* w_j$. Then, for all $x \in X$,

$$\langle x, v_j \rangle = \langle x, T^* w_j \rangle = \langle Tx, w_j \rangle = \sum_{i=1}^{n} \langle f_i(x) u_i, w_j \rangle = f_j(x),$$

$$Tx = \sum_{i=1}^{n} \langle x, v_i \rangle u_i.$$

Thus $T = \sum_{i=1}^{n} u_i \otimes v_i$. □

The next theorem is a collection of properties of semi-prime Banach algebras with minimal ideals.

Theorem 6. *Let A be semi-prime, let e be a minimal idempotent in A, and let f, $\langle, \rangle$ be defined by*

$$eae = f(a)e \quad (a \in A), \qquad \langle x, y \rangle e = yx \quad (x \in Ae, y \in eA).$$

Then:

(i) *$(Ae, eA, \langle, \rangle)$ is a dually irreducible Banach A-module pairing;*

(ii) *$f \in A'$ and is strictly represented by $(Ae, eA, \langle, \rangle, e, e)$;*

(iii) *if u is a minimal idempotent in A, then $\dim(uAe) \leq 1$;*

(iv) *$\mathrm{soc}(A)$ exists and*

$$\pi(\mathrm{soc}(A)) = FBL(Ae, eA, \langle, \rangle),$$

where π is the regular representation of A on Ae;

(v) *L_f is a maximal modular left ideal, and $L_f = A(1 - e)$;*

(vi) *K_f is a maximal modular right ideal, and $K_f = (1 - e)A$;*

(vii) *f is the only element g of A' with $eg = ge = g$ and $g(e) = 1$;*

(viii) *if also A is a primitive algebra, then π is a monomorphism.*

Proof. (i) By Proposition 30.6, Ae is a minimal left ideal, and, by Lemma 1 it is closed. Thus Ae is an irreducible Banach left A-module. Similarly eA is an irreducible Banach right A-module. If $x \in Ae$ and $\langle x, eA \rangle = \{0\}$, we have $eAx = \{0\}$. If $Ax \neq \{0\}$, the minimality of Ae gives $Ax = Ae$, and so $eAe = \{0\}$, $e = 0$, which is contradictory. Therefore $Ax = \{0\}$, and so $x = 0$, since A is semi-prime. Similarly $y = 0$ if $\langle Ae, y \rangle = \{0\}$, and so $(Ae, eA, \langle, \rangle)$ is a Banach pairing. Since $\langle ax, y \rangle e = yax = \langle x, ya \rangle e$, we have $\langle ax, y \rangle = \langle x, ya \rangle$ ($a \in A$, $x \in Ae$, $y \in eA$).

(ii) By Proposition 3, $f \in A'$. Also

$$f(a)e = eae = \langle ae, e \rangle e.$$

Thus $f(a) = \langle ae, e \rangle$ ($a \in A$), and also e is evidently a cyclic vector in Ae and in eA.

(iii) Let u be a minimal idempotent in A. Suppose that there exist $x_1, x_2 \in Ae$ with ux_1, ux_2 linearly independent. Then, by (i), there exists $a \in A$ with $aux_1 = ux_1$, $aux_2 = 0$. Therefore $uaux_1 = ux_1 \neq 0$, $uau \neq 0$, $Auau = Au$. But then

$$u x_2 \in A u x_2 = A u a u x_2 = \{0\},$$

which contradicts the linear independence of ux_1, ux_2.

(iv) The existence of soc(A) is given by Proposition 30.10. Given a minimal idempotent u, it follows from (iii) that $\pi(u)$ has rank at most one. By (i), we now have $\pi(u) \in FBL(Ae, eA, \langle, \rangle)$. For arbitrary $a \in A$, we have

$$\pi(au) = \pi(a)\pi(u) \in FBL(Ae, eA, \langle, \rangle),$$

and so π maps all minimal left ideals of A into $FBL(Ae, eA, \langle, \rangle)$. Therefore $\pi(\mathrm{soc}(A)) \subset FBL(Ae, eA, \langle, \rangle)$.

To prove the opposite inclusion, it is enough, by Lemma 5, to prove that $u \otimes v \in \pi(\mathrm{soc}(A))$ for all $u \in Ae$, $v \in eA$. For such u, v, we have $u = ue$, and so, for all $x \in Ae$,

$$(u \otimes v)(x) = \langle x, v \rangle u = \langle x, v \rangle ue = u \langle x, v \rangle e = uvx.$$

Thus $\pi(uv) = u \otimes v$.

(v) By Proposition 30.11, $A(1-e)$ is a maximal modular left ideal. But also $A(1-e) \subset L_f \neq A$.

(vi) is similar.

(vii) Let $g \in A'$ with $eg = ge = g$. If $a \in \ker(f)$, we have $eae = 0$, and so

$$a = a - ae + ae - e(ae) \in \ker(g).$$

Since $f(e) = g(e) = 1$, it follows that $g = f$.

(viii) If A is primitive, there exists an irreducible left A-module X with $\{a \in A : aX = \{0\}\} = \{0\}$. Then $eX \neq \{0\}$, $AeX = X$. If, for some $b \in A$, we have $bAe = \{0\}$, then $bX = bAeX = \{0\}$. $\quad\square$

In general, Banach algebras lack minimal ideals and minimal idempotents. In later sections of this chapter we shall be concerned with various algebraic and topological conditions for their existence. We end the present section by giving some sufficient geometric conditions.

Notation. We denote by A_1 the closed unit ball in A. We recall that A_1 is *smooth* at $e \in S(A)$ if $D(e)$ is a singleton, and that e is an *exposed point* of A_1 if there exists $g \in D(e)$ such that

$$\mathrm{Re}\, g(a) < 1 \qquad (a \in A_1 \setminus \{e\}).$$

We recall also that $\Delta(e) = \{f \in D(e) : ef = fe = f\}$.

Proposition 7. *Let $e \in S(A)$, let e be an idempotent, and let $\Delta(e)$ have at most one element. Then e is a minimal idempotent.*

Proof. Let $B = eAe$. Then B is a Banach algebra with unit element e. Let $\phi \in D(B, e) = \{\phi \in B' : \phi(e) = 1 = \|\phi\|\}$, and let f be defined on A by

$$f(a) = \phi(eae) \quad (a \in A).$$

Then $f \in D(A, e)$. Also

$$(ef)(a) = f(ae) = \phi(eae^2) = \phi(eae) = f(a),$$

so that $ef = f$. Similarly $fe = f$, and so $f \in \Delta(e)$. Therefore $D(B, e)$ is a singleton $\{\phi\}$ say. Since B is a complex unital Banach algebra, the Bohnenblust-Karlin theorem (Corollary 10.15) shows that $\ker(\phi) = \{0\}$, and so $B = \mathbb{C} e$. □

Corollary 8. *Let $e \in S(A)$, let e be an idempotent and let A_1 be smooth at e. Then e is a minimal idempotent.*

Proof. Clear. □

Corollary 9. *Let $e \in S(A) \cap \text{q-Sing}(A)$. Let A_1 be smooth at e, and let e be an exposed point of A_1. Then e is a minimal idempotent.*

Proof. Since $D(e) = \{f\}$, a singleton, Corollary 29.10 shows that $\Delta(e) = \{f\}$. Since e is exposed and f is the only element of $D(e)$, we have

$$\text{Re } f(a) < 1 \quad (a \in A_1 \setminus \{e\}).$$

But $e^2 \in A_1$ and $\text{Re } f(e^2) = \text{Re}(ef)(e) = \text{Re } f(e) = 1$. Therefore $e^2 = e$ and Corollary 8 applies. □

§ 32. Annihilator Algebras

We recall that, given a subset E of a Banach algebra A, the left and right annihilators of E are the sets $\text{lan}(E)$, $\text{ran}(E)$ given by

$$\text{lan}(E) = \{x \in A : xE = \{0\}\}, \quad \text{ran}(E) = \{x \in A : Ex = \{0\}\}.$$

Definition 1. A Banach algebra A is said to be an *annihilator algebra* if it satisfies the following axioms: for all closed left ideals L and all closed right ideals R,
  (i) $\text{ran}(L) = \{0\}$ if and only if $L = A$;
  (ii) $\text{lan}(R) = \{0\}$ if and only if $R = A$.

Notation. Throughout this section A will denote a complex Banach algebra that is also an annihilator algebra. $\mathscr{I}$ will denote the set of minimal idempotents in A.

We shall see eventually (Theorem 22) that the closure of the algebra $FBL(X)$, with X a reflexive Banach space, is an annihilator algebra and is the structure in terms of which semi-simple annihilator algebras are decomposed. We first develop in some detail the structure of semi-prime annihilator algebras.

Lemma 2. *Let* $u \in A$. *Then*

(i) $\operatorname{ran}(A(1-u)) = \{x \in A : ux = x\}$;

(ii) $\operatorname{lan}((1-u)A) = \{x \in A : xu = x\}$;

(iii) $\{x \in A : ux = x\} = \{0\}$ *if and only if* $A(1-u) = A$.

Proof. (i) $x \in \operatorname{ran}(A(1-u))$ if and only if $x - ux \in \operatorname{ran}(A) = \{0\}$.

(ii) Similar.

(iii) If $A(1-u) \neq A$, then $A(1-u)$ is contained in a proper closed left ideal, and so $\operatorname{ran}(A(1-u)) \neq \{0\}$. □

Theorem 3. *Let* M *be a maximal closed right ideal of* A *with* $\operatorname{lan}(M)$ *not contained in* $\operatorname{rad}(A)$. *Then* $\operatorname{lan}(M)$ *contains a minimal idempotent* e. *Also*

(i) $\operatorname{lan}(M) = Ae$ *and is a minimal left ideal;*

(ii) $M = (1-e)A$ *and is a maximal modular right ideal;*

(iii) eA *is a minimal right ideal and* $A(1-e)$ *is a maximal modular left ideal.*

Proof. By Proposition 24.16 (iii), $\operatorname{lan}(M)$ has an element e that has no left quasi-inverse. Thus $A(1-e) \neq A$ and, by Lemma 2 (iii), there exists $b \in A \setminus \{0\}$ with $eb = b$. We have $(e - e^2)M = \{0\}$, and so $M \subset \operatorname{ran}(\{e - e^2\})$. If $e - e^2 \neq 0$, we have $\operatorname{ran}(\{e - e^2\}) \neq A$, and so $M = \operatorname{ran}(\{e - e^2\})$. But, since $(e - e^2)b = 0$, this gives $b \in M$, and so $b = eb = 0$. Therefore $e - e^2 = 0$.

Since e is an idempotent, $ex = 0$ if and only if $x \in (1-e)A$. Therefore $A \neq (1-e)A = \operatorname{ran}(\{e\}) \supset M$, and so $M = (1-e)A$. Since M is modular and maximal closed, it is a maximal right ideal. Moreover, by Lemma 2,

$$\operatorname{lan}(M) = \operatorname{lan}((1-e)A) = \{x : xe = x\} = Ae.$$

Since $A = (1-e)A \oplus eA$ and $(1-e)A$ is a maximal right ideal, eA is a minimal right ideal. Let L be a maximal modular left ideal containing $A(1-e)$. Then

$$\{0\} \neq \operatorname{ran}(L) \subset \operatorname{ran}(A(1-e)) = eA,$$

and so $\operatorname{ran}(L) = eA$. Thus $L \subset \operatorname{lan}(eA) = A(1-e)$, and so $A(1-e)$ is a maximal modular left ideal. Since $A = A(1-e) \oplus Ae$, it follows that Ae is a minimal left ideal.

Finally, since Ae is a minimal left ideal and $(Ae)^2 \neq \{0\}$, Lemma 30.2 shows that e is a minimal idempotent. □

Lemma 4. *Let* A *be semi-prime and let* J *be a bi-ideal in* A. *Then* $\mathrm{lan}(J) = \mathrm{ran}(J)$, $J \cap \mathrm{lan}(J) = \{0\}$, *and* $(J \oplus \mathrm{lan}(J))^{-} = A$.

Proof. $\mathrm{lan}(J)$ is a bi-ideal, and $(J \cap \mathrm{lan}(J))^2 \subset \mathrm{lan}(J) J = \{0\}$. Since A is semi-prime, this gives $J \cap \mathrm{lan}(J) = \{0\}$. But then $J \mathrm{lan}(J) \subset J \cap \mathrm{lan}(J)$ $= \{0\}$, and so $\mathrm{lan}(J) \subset \mathrm{ran}(J)$. Similarly $\mathrm{ran}(J) \subset \mathrm{lan}(J)$. If $J \oplus \mathrm{lan}(J)$ is not dense in A, we have $\mathrm{lan}(J) \cap \mathrm{lan}(\mathrm{lan}(J)) \neq \{0\}$; but this is impossible since $\mathrm{lan}(J)$ is a bi-ideal. $\quad\square$

Proposition 5. *Let* A *be semi-prime and not a radical algebra. Then* $\mathrm{soc}(A)$ *exists, and*
  (i) $\mathrm{ran}(\mathrm{soc}(A)) = \mathrm{lan}(\mathrm{soc}(A)) = \mathrm{rad}(A)$,
  (ii) $(\mathrm{soc}(A))^{-} \cap \mathrm{rad}(A) = \{0\}$,
  (iii) $(\mathrm{soc}(A) \oplus \mathrm{rad}(A))^{-} = A$.

Proof. Since $\mathrm{rad}(A) \neq A$, there exist maximal modular right ideals. Let M be such an ideal. Then $\mathrm{lan}(M) \not\subset \mathrm{rad}(A)$, for we have $\mathrm{rad}(A) \subset M$, and so $\mathrm{lan}(M) \subset \mathrm{rad}(A)$ would give $(\mathrm{lan}(M))^2 = \{0\}$. Therefore, by Theorem 3, M is of the form $(1-e)A$ with $e \in \mathscr{I}$. By Proposition 30.10, $\mathrm{soc}(A)$ exists. Since the maximal modular right ideals coincide with the ideals of the form $(1-e)A$ with $e \in \mathscr{I}$, we have

$$\begin{aligned}
\mathrm{rad}(A) &= \bigcap \{(1-e)A : e \in \mathscr{I}\} \\
&= \bigcap \{\mathrm{ran}(Ae) : e \in \mathscr{I}\} \\
&= \mathrm{ran}(\mathrm{soc}(A)).
\end{aligned}$$

This proves (i), and it follows that $(\mathrm{soc}(A))^{-} \subset \mathrm{ran}(\mathrm{rad}(A))$. Therefore $(\mathrm{soc}(A))^{-} \cap \mathrm{rad}(A) \subset \mathrm{ran}(\mathrm{rad}(A)) \cap \mathrm{rad}(A) = \{0\}$.
  By (i) and Lemma 4, we have

$$(\mathrm{soc}(A) \oplus \mathrm{rad}(A))^{-} = (\mathrm{soc}(A)) \oplus \mathrm{ran}(\mathrm{soc}(A))^{-} = A. \quad\square$$

Corollary 6. A *is semi-simple if and only if* $A = (\mathrm{soc}(A))^{-}$.

Proof. If A is semi-simple, then A is semi-prime and $\mathrm{rad}(A) = \{0\}$. Therefore, by Proposition 5(iii), $A = (\mathrm{soc}(A))^{-}$.
  Suppose that $A = (\mathrm{soc}(A))^{-}$ and that J is a closed bi-ideal with $J^2 = \{0\}$. It is enough to prove that $J = \{0\}$, for then A is semi-prime and Proposition 5(ii) gives $\mathrm{rad}(A) = \{0\}$.
  Given a minimal left ideal L we have either $L \cap J = L$ or $L \cap J = \{0\}$. In both cases we have $JL = \{0\}$. Therefore $\mathrm{soc}(A) \subset \mathrm{ran}(J)$, $A \subset \mathrm{ran}(J)$, $J = \{0\}$. $\quad\square$

Corollary 7. *Let* A *be semi-prime, and let* L *be a left ideal not contained in* $\mathrm{rad}(A)$. *Then* L *contains a minimal idempotent.*

Proof. Suppose that $L \cap \mathscr{I} = \emptyset$. Then $A e \cap L = \{0\}$ $(e \in \mathscr{I})$. Given $a \in A$ and $e \in \mathscr{I}$, either $A e a$ is a minimal left ideal for $A e a = \{0\}$ (Lemma 30.7). In both cases $A\,e\,a\cap L=\{0\}$. Thus $e\,A\cap L=\{0\}$, and so

$$e\,A\,L\subset e\,A\cap L=\{0\}.$$

We have now proved that $\operatorname{soc}(A)L=\{0\}$, $L\subset\operatorname{ran}(\operatorname{soc}(A))=\operatorname{rad}(A)$.   □

Lemma 8. *Let A be semi-prime, let B be a closed bi-ideal of A, and let L be a closed left ideal of B. Then:*
   (i) *B is a semi-prime Banach algebra;*
   (ii) *L is a closed left ideal of A;*
   (iii) *if $\operatorname{ran}(L)\cap B=\{0\}$, then $BA\subset L$;*
   (iv) *$\operatorname{rad}(B)=B\cap\operatorname{rad}(A)$.*

Proof. By Lemma 4, $(B\oplus\operatorname{lan}(B))^{-}=A$, and so, since $(B+\operatorname{lan}(B))L=BL\subset L$, it follows that $AL\subset L$. Thus L is a left ideal of A which is obviously closed in A. Since A is semi-prime, we now have $L=\{0\}$ if $L^{2}=\{0\}$, and so B is semi-prime.

Let $I=L\oplus\operatorname{lan}(B)$. Then I is a left ideal of A and

$$BI\subset L. \tag{1}$$

If $\operatorname{ran}(L)\cap B=\{0\}$, then $\operatorname{ran}(I)\cap B=\{0\}$; and so we have in turn $\operatorname{ran}(I)B=\{0\}$, $\operatorname{ran}(I)\subset\operatorname{lan}(B)\subset I$, $(\operatorname{ran}(I))^{2}=\{0\}$, $\operatorname{ran}(I)=\{0\}$, $I^{-}=A$. Therefore, by (1), $BA\subset L$.

For (iv) see Corollary 24.20.   □

Theorem 9. *Let A be semi-prime and let B be a closed bi-ideal with $(A\,B)^{-}=(B\,A)^{-}=B$. Then B is a semi-prime annihilator Banach algebra. If also $B\cap\operatorname{rad}(A)=\{0\}$, then B is semi-simple.*

Proof. Let L be a closed left ideal of B with $L\neq B$. By Lemma 8, we have $\operatorname{ran}(L)\cap B\neq\{0\}$, for otherwise $BA\subset L$, and so $B=(BA)^{-}\subset L$. Thus L has a non-zero right annihilator relative to B. Likewise, proper closed right ideals of B have non-zero left annihilators in B. Since A is semi-prime, B has zero annihilator relative to B. Finally if $B\cap\operatorname{rad}(A)=\{0\}$, then, by Lemma 8(iv), $\operatorname{rad}(B)=\{0\}$.   □

Corollary 10. *Let A be semi-prime. Then $(\operatorname{soc}(A))^{-}$ is a semi-simple annihilator Banach algebra.*

Proof. Let $B=(\operatorname{soc}(A))^{-}$. For all $e\in\mathscr{I}$, we have $e=e^{2}\in A\operatorname{soc}(A)$. Thus $\operatorname{soc}(A)\subset A\operatorname{soc}(A)$, $B\subset(AB)^{-}$. Similarly $B=(BA)^{-}$, and so Theorem 9 applies.   □

Definition 11. A *minimal closed bi-ideal* is a closed bi-ideal $J\neq\{0\}$ that contains no closed bi-ideals other than $\{0\}$ and J.

Corollary 12. *Let A be semi-prime, and let B be a minimal closed bi-ideal of A not contained in $\operatorname{rad}(A)$. Then B is a semi-simple annihilator Banach algebra.*

Proof. $(AB)^-$ is a closed bi-ideal of A contained in B and not zero. Therefore $(AB)^- = B$, and similarly $(BA)^- = B$. Also $B \cap \mathrm{rad}(A) = \{0\}$, by minimality of B. Thus Theorem 9 applies. □

Even when A is semi-prime, we do not know that $\mathrm{rad}(A)$ is an annihilator algebra. The following proposition takes us part of the way in this direction.

Proposition 13. *Let A be semi-prime, let $R = \mathrm{rad}(A)$, and [1]let $B = (R^2)^-$. Then*
   (i) *B is a semi-prime radical annihilator Banach algebra;*
   (ii) *$A = (\mathrm{soc}(A) \oplus B)^-$ if $R \neq A$;*
   (iii) *R is an annihilator algebra if and only if $R = B$.*

Proof. (i) We note first that

$$\mathrm{ran}(R^n) \cap R = \{0\} \qquad (n \in \mathbb{N}). \tag{2}$$

For if $x \in \mathrm{ran}(R^n) \cap R$, then $R^{n-1} x \subset \mathrm{ran}(R) \cap R = \{0\}$, so that $x \in \mathrm{ran}(R^{n-1}) \cap R$. Finally $x \in \mathrm{ran}(R) \cap R = \{0\}$.
   Since $\mathrm{ran}(R^4) \cap R = \{0\}$, Lemma 8 (iii) with $L = (R^4)^-$ gives $RA \subset (R^4)^-$ and so $B \subset (B^2)^-$. Therefore $B = (AB)^- = (BA)^-$, and Theorem 9 shows that B is a semi-prime annihilator Banach algebra. To see that B is a radical algebra, note that any closed subalgebra of a radical Banach algebra is a radical algebra.
   (ii) By (2), we have $\mathrm{ran}(R^2) \cap R = \{0\}$, and so $\mathrm{ran}(B) \cap R = \{0\}$. By Proposition 5 (i), this gives $\mathrm{ran}(B) \cap \mathrm{ran}(\mathrm{soc}(A)) = \{0\}$. So $\mathrm{ran}(\mathrm{soc}(A) \oplus B) = \{0\}$, $(\mathrm{soc}(A) \oplus B)^- = A$.
   (iii) We have observed in (ii) that $\mathrm{ran}(B) \cap R = \{0\}$. If R is an annihilator algebra, this implies that $B = R$. The converse is immediate from (i). □

It is natural to ask when we have $A = (\mathrm{soc}(A))^- \oplus \mathrm{rad}(A)$. Sufficient conditions are obtained by taking $B = (\mathrm{soc}(A))^-$ in the following proposition.

Proposition 14. *Let A be semi-prime, let B be a closed bi-ideal of A, and suppose that the left regular representation of B on B is a homeomorphism. Then*

$$A = B \oplus \mathrm{lan}(B) = B \oplus \mathrm{ran}(B).$$

Proof. Let B_1 denote the closed unit ball of B. Our hypothesis gives the existence of a positive constant κ such that

$$\sup \{ \|bx\| : x \in B_1 \} \geq \kappa \|b\| \qquad (b \in B).$$

---

[1] See footnote on page 124.

Given $b \in B$, $x \in B_1$, $j \in \text{lan}(B)$, we have $jx = 0$, and so

$$\|bx\| = \|(b-j)x\| \leqslant \|b-j\|.$$

Therefore

$$\kappa \|b\| \leqslant d(b, \text{lan}(B)) \quad (b \in B). \tag{3}$$

Given $a \in A$, Lemma 4 gives the existence of $b_n \in B$, $j_n \in \text{lan}(B)$ such that $\lim_{n \to \infty} (b_n + j_n) = a$. By (3) it follows that $\{b_n\}$ is a Cauchy sequence in B, and so $a \in B + \text{lan}(B)$. Finally, by Lemma 4, $\text{ran}(B) = \text{lan}(B)$.   □

An alternative reduction of a semi-prime annihilator algebra to a semi-simple annihilator algebra is obtained by dividing out the radical, as follows.

Proposition 15. *Let A be semi-prime, and let J be a closed bi-ideal such that*

$$J = \{a \in A : aA \subset J\} = \{a \in A : Aa \subset J\}.$$

Then A/J is an annihilator Banach algebra. In particular $A/\text{rad}(A)$ is a semi-simple annihilator Banach algebra.

Proof. Let $B = A/J$, let π be the canonical mapping of A onto B, and let L be a proper closed left ideal of B. Then $\pi^{-1}(L)$ is a closed left ideal of A with $J \subset \pi^{-1}(L) \neq A$. Then $\text{ran}(\pi^{-1}(L))$ is not contained in J, since otherwise $\text{ran}(\pi^{-1}(L)) \subset \pi^{-1}(L)$, which is impossible with A semi-prime. Therefore $\pi(\text{ran}(\pi^{-1}(L))) \neq \{0\}$, $\text{ran}(L) \neq \{0\}$.

Suppose that $Bb = \{0\}$. Then $b = \pi(a)$ for some $a \in A$, and $Aa \subset J$. Therefore $a \in J$, $b = 0$. It is now clear that B is an annihilator algebra. If $Aa \subset \text{rad}(A)$, then, by Corollary 24.19, $a \in \text{rad}(A)$.   □

The reduction of a semi-simple annihilator algebra is carried out in terms of its minimal closed bi-ideals, the existence of which is established by the following proposition.

Proposition 16. *Let A be semi-prime, and let J be a closed bi-ideal not contained in $\text{rad}(A)$. Then J contains a minimal closed bi-ideal K with $K \cap \text{rad}(A) = \{0\}$.*

Proof. By Corollary 7, J contains a minimal idempotent e. Let K be the intersection of the closed bi-ideals containing e, and let I be a closed bi-ideal with $I \subset K$. If $e \in I$, then $I = K$. Suppose that $e \notin I$. Then

$$IAe \subset Ae \cap I = \{0\},$$

and so e belongs to the closed bi-ideal $\text{ran}(I)$. Therefore $I \subset K \subset \text{ran}(I)$, $I^2 = \{0\}$, $I = \{0\}$; K is a minimal closed bi-ideal. Since $e \in K$, we have $K \not\subset \text{rad}(A)$, and so $K \cap \text{rad}(A) = \{0\}$ by the minimality of K.   □

In the next proposition we collect together the preceding results with the simplifications that occur when A is semi-simple.

Proposition 17. *Let A be semi-simple. Then each non-zero left ideal contains a minimal idempotent, each non-zero closed bi-ideal contains a minimal closed bi-ideal, each minimal closed bi-ideal B is a semi-simple annihilator Banach algebra with $\{0\}$ and B as its only closed bi-ideals, $A = (\mathrm{soc}(A))^-$, and A is the closure of the direct sum of the minimal closed bi-ideals of A.*

Proof. Everything has been proved except the last statement. Plainly minimal closed bi-ideals have zero intersection with each other. If their direct sum is not dense in A, its annihilator contains a minimal closed bi-ideal, which then annihilates itself. □

Except for questions concerning the topology and the norm the study of semi-simple annihilator algebras has been reduced to the study of annihilator algebras with no proper minimal closed bi-ideals, to which we now turn.

Definition 18. A Banach algebra B is *topologically simple* if its only closed bi-ideals are $\{0\}$ and B.

Lemma 19. *Let A be semi-simple and topologically simple, let e be a minimal idempotent, and, given $y \in eA$, let ϕ_y be the linear functional defined on Ae by*
$$\phi_y(x)e = yx \quad (x \in Ae).$$
Then the mapping $y \to \phi_y$ is a bicontinuous linear isomorphism of eA onto $(Ae)'$.

Proof. Since $\mathrm{lan}(Ae)$ is a proper closed bi-ideal and A is topologically simple, we have
$$\mathrm{lan}(Ae) = \{0\}. \tag{4}$$
It is now clear that the mapping $y \to \phi_y$ is a continuous linear injection of eA into $(Ae)'$.

Let $\phi \in (Ae)' \setminus \{0\}$, let $N = \ker(\phi)$, choose $u \in Ae$ with $\phi(u)=1$, and let g denote the linear functional on A defined by
$$g(a) = \phi(ae) \quad (a \in A).$$
Then $g \in A'$, and $K_g = \{a \in A : g(aA) = \{0\}\} \neq A$. For, if $K_g = A$, we have
$$\phi(Ae) = \phi(Aee) \subset \phi(AAe) = g(AA) = \{0\},$$
$\phi = 0$. Therefore K_g is a proper closed right ideal of A, and we may choose $b \in \mathrm{lan}(K_g) \setminus \{0\}$. We have $N = Ne = NeAe = NAe$, and so
$$g(NA) = \phi(NAe) = \phi(N) = \{0\}.$$

Therefore $N \subset K_g$, and so $bN = \{0\}$. Since $x - \phi(\hat{x})u \in N$, this gives

$$bx = \phi(x)bu \qquad (x \in Ae).$$

By (4), we now have $bu \neq \{0\}$. Therefore $Abu \neq \{0\}$, $Abu = Ae$. Choose $c \in A$ with $cbu = e$. Then

$$ecbx = \phi(x)ecbu = \phi(x)e \qquad (x \in Ae),$$

$\phi = \phi_{ecb}$. We have proved that the mapping $y \to \phi_y$ is a continuous linear bijection of eA onto $(Ae)'$, and Banach's isomorphism theorem completes the proof. □

Theorem 20. *Let A be semi-simple and topologically simple, let e be a minimal idempotent, and let π denote the left regular representation of A on Ae. Then Ae is a reflexive Banach space X, $\pi(\mathrm{soc}(A)) = FBL(X)$, and $\pi(A)$ is contained in $(FBL(X))^-$, the closure of $FBL(X)$ in $BL(X)$ with respect to the operator norm.*

Proof. Let ϕ_y be defined as in Lemma 19, and, given $x \in Ae$, let ψ_x be defined on eA by

$$\psi_x(y)e = yx \qquad (y \in eA).$$

Then $x \to \psi_x$ is a bicontinuous linear bijection of Ae onto $(eA)'$, and $\psi_x(y) = \phi_y(x) = \hat{x}(\phi_y)$, where $\hat{x}$ is the canonical image of x in $(Ae)''$.

Given $\Phi \in (Ae)''$, let ψ be defined on eA by

$$\psi(y) = \Phi(\phi_y) \qquad (y \in eA).$$

Then $\psi \in (eA)'$, and so there exists $x \in Ae$ with $\psi = \psi_x$. Thus

$$\Phi(\phi_y) = \psi_x(y) = \hat{x}(\phi_y) \qquad (y \in eA),$$

and, since $\{\phi_y : y \in eA\} = (Ae)'$, this proves that $\Phi = \hat{x}$. Thus Ae is a reflexive Banach space.

Each bounded linear operator on Ae of rank 1 is of the form $u \otimes \phi$ with $u \in Ae$ and $\phi \in (Ae)'$, and is therefore of the form $u \otimes \phi_y$ for some $y \in eA$. We have, for $x \in Ae$,

$$(u \otimes \phi_y)(x) = \phi_y(x)u = u\phi_y(x)e = uyx = \pi(uy)x.$$

Thus $u \otimes \phi_y = \pi(uy) \in \pi(\mathrm{soc}(A))$. The rest is straightforward. □

In Theorem 22 we give a converse of Theorem 20. As a preliminary we develop some elementary properties of certain algebras of operators on Banach spaces.

Lemma 21. *Let X be a Banach space, let $|\cdot|$ be the operator norm on $BL(X)$, and let B be a subalgebra of $BL(X)$ that contains $FBL(X)$ and is a Banach algebra with respect to an algebra-norm $\|\cdot\|$. Suppose* *further that* $\|b\| \geqslant |b|$ ($b \in B$) *and that* $F BL(X)$ *is dense in* B *with respect to* $\|\cdot\|$. *Let* K *be a closed right ideal of* B, *and let* $[K X]$ *denote the closed linear span of* $\{k x: k \in K, x \in X\}$. *Then*

(i) *for each* $\phi \in X'$, *the mapping* $x \to x \otimes \phi$ *is a continuous linear mapping of* X *into* B (*with the norm* $\|\cdot\|$),

(ii) $\operatorname{ran}(\operatorname{lan}(K)) = \{b \in B: b X \subset [K X]\}$;

(iii) $K \cap F BL(X) = \{b \in F BL(X): b X \subset [K X]\}$;

(iv) $(\operatorname{ran}(\operatorname{lan}(K))) B \subset K$.

Proof. (i) Since $F BL(X) \subset B$, X is an irreducible left B-module, and so B is semi-simple. Let $\phi \in X' \backslash \{0\}$. Then $X \otimes \phi$ is a minimal left ideal of B, for, if $x_0 \in X \backslash \{0\}$, we have

$$B(x_0 \otimes \phi) = (B x_0) \otimes \phi = X \otimes \phi.$$

Since B is semi-simple, it follows that $X \otimes \phi$ is a closed left ideal of B (Proposition 31.2). The mapping $x \to x \otimes \phi$ is a linear bijection of X onto $X \otimes \phi$, and

$$\|x \otimes \phi\| \geqslant |x \otimes \phi| = \|x\| \cdot \|\phi\|.$$

Therefore, by Banach's isomorphism theorem, this mapping is continuous.

(ii) Clearly, $\operatorname{lan}(K) = \{a \in B: a[K X] = \{0\}\}$. If $b \in B$ and $b X \subset [K X]$, we have $\operatorname{lan}(K) b X \subset \operatorname{lan}(K)[K X] = \{0\}$, and so $b \in \operatorname{ran}(\operatorname{lan}(K))$. Suppose on the other hand that $b X \not\subset [K X]$, and choose $x_0 \in X$ with $b x_0 \notin [K X]$. By the Hahn-Banach theorem, we may choose $\phi \in X'$ with $\phi(b x_0) = 1$ and $\phi([K X]) = \{0\}$. Then $x_0 \otimes \phi \in \operatorname{lan}(K)$, but

$$((x_0 \otimes \phi) b)(x_0) = \phi(b x_0) x_0 = x_0 \neq 0,$$

and so $b \notin \operatorname{ran}(\operatorname{lan}(K))$.

(iii) Given $k \in K$, $u \in X \backslash \{0\}$, $\phi \in X'$, we have

$$(k u) \otimes \phi = k(u \otimes \phi) \in K.$$

Since X is irreducible and K is a right ideal, we have $K X = K u$, $[K X] = (K u)^-$. Thus, given $y \in [K X]$, there exist $k_n \in K$ with $y = \lim_{n \to \infty} k_n u$. By (i), $k_n u \otimes \phi \to y \otimes \phi$ as $n \to \infty$, with respect to the norm $\|\cdot\|$. Therefore $y \otimes \phi \in K$. We have proved that K contains all operators of the form $y \otimes \phi$ with $y \in [K X]$ and $\phi \in X'$.

Let $b \in F BL(X)$ with $b X \subset [K X]$. Then $b = \sum_{i=1}^{n} u_i \otimes \phi_i$ with $u_i \in X$, $\phi_i \in X'$. With the natural bilinear form X, X' is a pairing. We may suppose that $\phi_1, \ldots, \phi_n$ are linearly independent, and then, by Lemma 31.4, there exist $x_1, \ldots, x_n \in X$ with $\phi_j(x_i) = \delta_{ij}$ ($1 \leqslant i, j \leqslant n$). Then $u_i = b x_i \in [K X]$, and so $u_i \otimes \phi_i \in K$, $b \in K$. We have now proved that $\{b \in F BL(X): b X \subset [K X]\} \subset K \cap F BL(X)$, and the opposite inclusion is obvious from (ii).

(iv) Let $a \in \operatorname{ran}(\operatorname{lan}(K))$, $b_0 \in FBL(X)$. Then $ab_0 \in FBL(X)$, and $ab_0 X \subset aX \subset [KX]$, by (ii). So $ab_0 \in K$. Since $FBL(X)$ is dense in B (with respect to $\|\cdot\|$) and K is closed, we have $aB \subset K$. □

Theorem 22. *Let X be a reflexive Banach space, let $|\cdot|$ be the operator norm on $BL(X)$, and let B be a subalgebra of $BL(X)$ that contains $FBL(X)$ and is a Banach algebra with respect to an algebra-norm $\|\cdot\|$. Suppose further that $\|b\| \geqslant |b|$ ($b \in B$), and that $FBL(X)$ is dense in B with respect to $\|\cdot\|$. Then $(B, \|\cdot\|)$ is a semi-simple and topologically simple annihilator Banach algebra. If $\|\cdot\| = |\cdot|$ on B, then B has the $B^{\#}$-property.*

Proof. It is obvious that $\operatorname{lan}(B) = \{0\}$. Let K be a closed right ideal with $\operatorname{lan}(K) = \{0\}$. Then $\operatorname{ran}(\operatorname{lan}(K)) = B$, and, by Lemma 21 (ii), $X = BX \subset [KX]$. Therefore, by Lemma 21 (iii), $K \cap FBL(X) = FBL(X)$; and, since $FBL(X)$ is dense in B, $B \subset K$.

Since X is a reflexive Banach space, the mapping $T \to T^*$ maps $FBL(X)$ onto $FBL(X')$, and therefore maps B onto a subalgebra B^* of $BL(X')$ which, with the norm transferred from B, satisfies the conditions of the theorem. Since this mapping reverses products, it is now clear that B is an annihilator algebra. It is also clear that B is a semi-simple Banach algebra.

Let J be a non-zero closed bi-ideal of B. Then $JX \neq \{0\}$, and so $JX \supset BJX = X$. Therefore, by Lemma 21 (iii), $J \cap FBL(X) = FBL(X)$. Using again the density of $FBL(X)$ in B, we have $B \subset J$, B is topologically simple.

Suppose now that $\|\cdot\| = |\cdot|$, and let $b \in S(B)$. As an operator norm limit of finite rank operators, b is a compact linear operator. Therefore, since the unit ball of X is weakly compact, b attains its operator norm, i.e. there exists $u \in S(X)$ with $\|bu\| = 1$. Choose $\phi \in S(X')$ with $\phi(bu) = 1$, and let $c = u \otimes \phi$. Then $c \in S(B)$ and $(cb)(u) = \phi(bu)u = u$. This shows that 1 is an eigenvalue of the operator cb, and so $r(cb) \geqslant 1 = \|c\| \|b\|$. Take $b^{\#} = c$. We have proved that B has the $B^{\#}$-property. □

We return to the study of the abstract annihilator algebra A, and show that the results can be substantially improved if the norm of A has the $B^{\#}$ property.

Lemma 23. *Let p be an algebra-norm on A. Then*

$$\lim_{n \to \infty} p(a^n)^{\frac{1}{n}} = r(a) \qquad (a \in A).$$

Proof. Let B denote the completion of A with respect to the norm p, let $a \in A$, and let $\lambda \in \operatorname{Sp}(A, a) \backslash \{0\}$. Then $\lambda^{-1} a$ is quasi-singular. Suppose that it has no left quasi-inverse. Then $A(1 - \lambda^{-1} a)$ is contained in a proper closed left ideal of A. Choose $u \in \operatorname{ran}(A(1 - \lambda^{-1} a)) \backslash \{0\}$.

Then $au = \lambda u$, and so $\lambda \in \mathrm{Sp}(B, a)$. It is now clear that $\mathrm{Sp}(A, a) \setminus \{0\}$ $= \mathrm{Sp}(B, a) \setminus \{0\}$. □

Proposition 24. *Let A have the $B^\#$-property. Then the norm of A is minimal in the sense that if p is an algebra-norm on A with $p(a) \leqslant \|a\|$ $(a \in A)$, then $p(a) = \|a\|$ $(a \in A)$.*

Proof. Let p be an algebra-norm on A with $p(a) \leqslant \|a\|$ $(a \in A)$. Given $a \in A$, there exists $a^\# \in A$ with $\|a^\#\| = \|a\|$ and $\|a^\#\| \, \|a\| = r(a^\# a)$. Then, by Lemma 23,

$$\|a^\#\| \, \|a\| = r(a^\# a) = \lim_{n \to \infty} p((a^\# a)^n)^{\frac{1}{n}} \leqslant p(a^\# a) \leqslant p(a^\#) p(a).$$

Since $p(a^\#) \leqslant \|a^\#\|$ and $p(a) \leqslant \|a\|$, it now follows that $p(a) = \|a\|$. □

Definition 25. Given a family $\{A_\lambda : \lambda \in \Lambda\}$ of Banach algebras, the $B(\infty)$-*sum* of the family is the subalgebra $\sum_0 \{A_\lambda : \lambda \in \Lambda\}$ of the normed full direct sum $\sum \{A_\lambda : \lambda \in \Lambda\}$ (see Definition 26.10) consisting of those function $f \in \sum \{A_\lambda : \lambda \in \Lambda\}$ such that, for every $\varepsilon > 0$, the set $\{\lambda \in \Lambda : \|f(\lambda)\| \geqslant \varepsilon\}$ is finite.

It is obvious that the $B(\infty)$-sum is a normed subdirect sum in the sense of Definition 26.10, and it is easy to prove that it is a Banach algebra.

Theorem 26. *Let A have the $B^\#$-property, and let $\{M_\lambda : \lambda \in \Lambda\}$ denote its set of minimal closed bi-ideals. Then A is isometrically isomorphic to the $B(\infty)$-sum $\sum_0 \{M_\lambda : \lambda \in \Lambda\}$, and each M_λ is isometrically isomorphic to $(F\,BL(X_\lambda))^-$ for a reflexive Banach space X_λ.*

Proof. Given $a \in A$ and $\lambda \in \Lambda$, let $|a|_\lambda = \sup \{\|ax\| : x \in M_\lambda, \|x\| \leqslant 1\}$, and let $|a| = \sup \{|a|_\lambda : \lambda \in \Lambda\}$. We know that A is semi-simple (Proposition 26.13), and therefore, by Proposition 17, $A = B^-$, where B denotes the direct sum of the ideals M_λ. Therefore $|\cdot|$ is an algebra-norm on A. Since $|a| \leqslant \|a\|$ $(a \in A)$, Proposition 24 gives

$$|a| = \|a\| \qquad (a \in A). \tag{5}$$

Let $a \in B$. Then $a = m_1 + \cdots + m_n$ with $m_k \in M_{\lambda_k}$ $(k = 1, \ldots, n)$. Since the M_λ annihilate each other, we have

$$|a|_\lambda = |m_k| \ (\lambda = \lambda_k), \qquad |a|_\lambda = 0 \ (\lambda \notin \{\lambda_1, \ldots, \lambda_n\}).$$

Therefore, by (5),

$$\|a\| = \max(\|m_k\| : 1 \leqslant k \leqslant n).$$

Let $a(\cdot)$ denote the function defined on Λ by taking $a(\lambda_k) = m_k$ $(k = 1, \ldots, n)$, $a(\lambda) = 0$ $(\lambda \notin \{\lambda_1, \ldots, \lambda_n\})$. Then $a(\cdot) \in \sum_0 \{M_\lambda : \lambda \in \Lambda\}$, and $\|a(\cdot)\|_\infty = \|a\|$.

It is now clear that the mapping $a \to a(\cdot)$ is an isometric isomorphism of the algebra B onto a dense subalgebra of $\sum_0 \{M_\lambda : \lambda \in \Lambda\}$. Therefore it extends by continuity to an isometric isomorphism of A onto $\sum_0 \{M_\lambda : \lambda \in \Lambda\}$. We denote this extended isomorphism also by $a \to a(\cdot)$.

We know already that each M_λ is a semi-simple topologically simple annihilator Banach algebra. We prove that it satisfies the $B^\#$-condition. Let $\mu \in \Lambda$ and let $m \in M_\mu$ with $\|m\| = 1$. Then there exists $m^\# \in A$ with $\|m^\#\| = 1$ and $\|(m^\# m)^n\| = 1$ $(n \in \mathbb{N})$. Since $m^\# m \in M_\mu$, we have $(m^\# m)(\lambda) = 0$ $(\lambda \neq \mu)$, and $(m^\# m)(\mu) = m^\#(\mu)m$. Therefore

$$(m^\# m)(\lambda) = (m^\#(\mu)m)(\lambda) \qquad (\lambda \in \Lambda),$$

and, since $a \to a(\cdot)$ is an isomorphism, this gives $m^\# m = m^\#(\mu)m$. Also $\|m^\#(\mu)\| \leqslant \|m^\#\| = 1$, but $\|m^\#(\mu)\| \|m\| \geqslant \|m^\#(\mu)m\| = \|m^\# m\| = 1$. Thus $\|m^\#(\mu)\| = 1$, and $\|(m^\#(\mu)m)^n\| = \|(m^\# m)^n\| = 1$ $(n \in \mathbb{N})$. This proves that M_μ has the $B^\#$ property.

Let $M = M_\lambda$ for some λ, let e be a minimal idempotent in M, and let π denote the left regular representation of M on Me. We have $\|\pi(a)\| = \sup\{\|ax\| : x \in Me, \|x\| \leqslant 1\} \leqslant \|a\|$, and π is injective. Let $p(a) = \|\pi(a)\|$ $(a \in M)$. By Proposition 24, $p(a) = \|a\|$ $(a \in A)$, and therefore π is an isometry. Theorem 20 completes the proof.   □

Definition 27. A Banach algebra is a *dual algebra* if for each closed left ideal L and each closed right ideal K,

$$\operatorname{lan}(\operatorname{ran}(L)) = L, \qquad \operatorname{ran}(\operatorname{lan}(K)) = K.$$

It is easy to see that a dual algebra is an annihilator algebra. The converse is not in general true even for semi-simple commutative algebras (see Johnson [210]).

Proposition 28. *Let A be semi-simple and topologically simple. Then A is a dual algebra if and only if*

$$a \in (aA)^- \cap (Aa)^- \qquad (a \in A). \tag{6}$$

Proof. Suppose that A is a dual algebra, let $a \in A$ and let $J = (aA)^-$. Then $(\operatorname{lan}(J))aA = \{0\}$, $(\operatorname{lan}(J))a = \{0\}$, $a \in \operatorname{ran}(\operatorname{lan}(J)) = J$. Similarly $a \in (Aa)^-$.

Suppose on the other hand that A satisfies condition (6). Let e be a minimal idempotent, let $X = Ae$, let π be the left regular representation of A on X, and let $B = \pi(A)$. We define a norm $\|\cdot\|$ on B by taking $\|\pi(b)\| = \|b\|$. Then B is a subalgebra of $BL(X)$ satisfying the conditions of Lemma 21, and π is an isometric isomorphism of A onto B.

Let K be a closed right ideal of B, and let $b \in \operatorname{ran}(\operatorname{lan}(K))$. By Lemma 21 (iv), $bB \subset K$, and so $b \in (bB)^- \subset K$. Thus $\operatorname{ran}(\operatorname{lan}(K)) = K$; and the same holds for closed right ideals of A. Since the conditions on A are symmetrical, it follows that also $\operatorname{lan}(\operatorname{ran}(L)) = L$ for closed left ideals L of A.   ☐

Definition 29. A Banach space X has the *approximation property* if, for every compact subset E of X and every $\varepsilon > 0$, there exists $T \in F BL(X)$ such that $\|Tx - x\| < \varepsilon$ $(x \in E)$.

Corollary 30. *Let X be a reflexive Banach space with the approximation property, and let B denote the closure of $F BL(X)$ in $BL(X)$ with respect to the operator norm. Then B is a dual Banach algebra.*

Proof. By Theorem 22, B is a semi-simple topologically simple annihilator Banach algebra. Let $b \in B$, and let U denote the closed unit ball of X. Then b is a compact linear operator, and so bU is contained in a compact subset of X. Given $\varepsilon > 0$, there exists $t \in F BL(X)$ with $\|tx - x\| < \varepsilon$ $(x \in bU)$, and so $\|tb - b\| \leqslant \varepsilon$. Thus $b \in (Bb)^-$.

Since X is reflexive, its dual space X' has the approximation property (see Grothendieck [178] Proposition 36, p. 167) and so the above argument applies to the image B^* of B in $BL(X')$. Therefore $b \in (bB)^-$, and Proposition 28 applies.   ☐

The theory in the present section had its origins in the theory of dual Banach algebras developed by Kaplansky in [232, 233] with the main emphasis on dual B^*-algebras. Semi-simple topologically simple dual Banach algebras can be expressed in terms of $(F BL(X))^-$ for some reflexive Banach space X, but it was not clear that $(F BL(X))^-$ is always a dual Banach algebra. The weaker axioms of an annihilator algebra were introduced by Bonsall and Goldie [80, 63] to fill this gap. The question whether $(F BL(X))^-$ is dual for every reflexive Banach space remained open until very recently when Davie [110], gave a counter-example, using Enflo's example of a reflexive Banach space without the approximation property [133].

The theory of annihilator algebras has been effectively generalized by several authors by weakening the annihilator axioms to apply only to a specified class of the closed ideals, for example to modular ideals or to closed right ideals. A full bibliography of these developments is given in Barnes [43].

§ 33. Compact Action on Banach Algebras

We have seen in Theorem 31.6 that, if e is a minimal idempotent in a semi-prime Banach algebra A, then the elements of $\operatorname{soc}(A)$ act by left multiplication on Ae as bounded linear operators of finite rank, and therefore as compact linear operators. Conversely, the existence of non-zero elements of a Banach algebra A that act as compact linear operators on suitable subalgebras of A provides a source of minimal idempotent. In this section we study the association between Banach algebras and compact linear operators from a somewhat wider standpoint.

Notation. Given a Banach space X, the closed unit ball of X is denoted by X_1, and the set of all compact linear operators on X by $KL(X)$. Thus $KL(X)$ is the set of all $t \in BL(X)$ such that $t X_1$ is contained in a compact subset of X.

We take as known certain elementary properties of compact linear operators. $KL(X)$ is a closed bi-ideal of $BL(X)$ containing $FBL(X)$. If $\{\lambda_n\}$ is a sequence of distinct eigenvalues of $t \in KL(X)$, then $\lim_{n \to \infty} \lambda_n = 0$. Let $t \in KL(X)$, let λ be a non-zero eigenvalue of t, and let $N_k = \ker((\lambda - t)^k)$. Then for each $k \in \mathbb{N}$, $\dim(N_k) < \infty$, and there exist positive integers k with $N_{k+1} = N_k$. The least such k is called the *ascent* of λ. All this involves only the most elementary normed linear space theory. The remaining, less trivial parts of the Riesz-Schauder theory of compact linear operators will be developed in the present section.

Compactness of a given linear operator is nearly always a consequence of the Ascoli-Arzelà theorem in one of its forms. For our purposes Theorem 2 below will suffice.

Notation. Let E be compact topological space, F a metric space with distance function d, $C(E,F)$ the set of continuous mappings of E into F, τ_p the topology of pointwise convergences, and τ_u the topology of uniform convergence on $C(E,F)$. We recall that a subset $\mathscr{E}$ of $C(E,F)$ is equicontinuous if for every $x_0 \in E$ and $\varepsilon > 0$, there exists a neighbourhood U of x_0 such that

$$d(f(x), f(x_0)) < \varepsilon \quad (x \in U, f \in \mathscr{E}).$$

Lemma 1. *Let $\mathscr{E}$ be an equicontinuous subset of $C(E,F)$. Then τ_u coincides with τ_u on $\mathscr{E}$.*

Proof. Let $\varepsilon > 0$. Each point $a \in E$ has an open neighbourhood $U(a)$ such that

$$d(f(x), f(a)) < \tfrac{1}{3}\varepsilon \quad (x \in U(a), f \in \mathscr{E}).$$

We choose $a_1, \ldots, a_n \in E$ such that $E \subset \bigcup_{k=1}^{n} U(a_k)$. Given $f_0 \in \mathscr{E}$, let V be the τ_p neighbourhood of f_0 defined by

$$V = \{ f \in \mathscr{E} : d(f(a_k), f_0(a_k)) < \tfrac{1}{3}\varepsilon \ (k = 1, \ldots, n) \}.$$

If $f \in V$ and $x \in E$, then $x \in U(a_k)$ for some $k \in \{1,\dots,n\}$, and so $d(f(x), f_0(x)) < \varepsilon$. Thus $\sup \{d(f(x), f_0(x)): x \in E\} \leqslant \varepsilon$. The rest is clear. □

Theorem 2 (Ascoli-Arzelà). *Let $\mathscr{E}$ be an equicontinuous subset of $C(E,F)$ such that, for each $x \in E$, the set $\{f(x): f \in \mathscr{E}\}$ is contained in a compact subset of F. Then the τ_u closure of $\mathscr{E}$ in $C(E,F)$ is τ_u compact.*

Proof. For each $x \in E$, the set $\{f(x): f \in \mathscr{E}\}$ is contained in a compact subset Y_x of F. Let $\mathscr{F}$ denote the closure of $\mathscr{E}$ in the product space $\Pi\{Y_x: x \in E\}$. We prove that $\mathscr{F}$ is an equicontinuous subset of $C(E,F)$. Let $x_0 \in E$, $\varepsilon > 0$. Then there exists a neighbourhood U of x_0 such that

$$d(f(x), f(x_0)) < \tfrac{1}{3}\varepsilon \qquad (x \in U, f \in \mathscr{E}).$$

Let $x_1 \in U$ and $g \in \mathscr{F}$. By definition of the product topology, there exists $f \in \mathscr{E}$ such that $d(f(x_i), g(x_i)) < \tfrac{1}{3}\varepsilon$ $(i = 0,1)$. Then $d(g(x_1), g(x_0)) < \varepsilon$. We have proved that

$$d(g(x), g(x_0)) < \varepsilon \qquad (x \in U, g \in \mathscr{F}),$$

and so $\mathscr{F}$ is equicontinuous. By Tychonoff's theorem and Lemma 1, it follows that $\mathscr{F}$ is compact in the topology τ_u. Lemma 1 also shows that $\mathscr{E}$ is τ_u dense in $\mathscr{F}$, and so $\mathscr{F}$ is the τ_u closure of $\mathscr{E}$ in $C(E,F)$. □

Notation. Let X be a Banach space over $\mathbb{F}$, let $B = BL(X)$, and given $s, t \in B$, let $T_{s,t}$ be the mapping defined on B by

$$T_{s,t}a = sat \qquad (a \in B).$$

Evidently $T_{s,t} \in BL(B)$. We recall that for $t \in B$, $\{t\}^c = \{a \in B: at = ta\}$.

Theorem 3. *Let $s, t \in K L(X)$. Then*
(i) $T_{s,t} \in K L(B)$,
(ii) *the mapping $a \to at$ is a compact linear operator on $\{t\}^c$.*

Proof. (i) $E = (t X_1)^-$ is compact, $\mathscr{E} = s B_1$ is an equicontinuous subset of $C(E,X)$, and, for each $x \in E$, $\{s B_1 x\}^-$ is compact. Therefore, by Theorem 2, $\mathscr{E}$ is contained in a uniformly compact subset of $C(E,X)$. It follows that, given $b_n \in B_1$ $(n \in \mathbb{N})$, there exists a subsequence $\{b_{n_k}\}$ such that $\{s b_{n_k}\}$ converges uniformly on E. Then $\{s b_{n_k} t\}$ converges uniformly on X_1, i.e. converges in operator norm. Therefore $T_{s,t} \in K L(B)$.

(ii) Let $A = \{t\}^c$, and let E be as in (i). Then $A_1|_E$ is an equicontinuous subset of $C(E,X)$. Given $a \in A_1$, we have $at X_1 = ta X_1 \subset t X_1$, and so $aE \subset E$. Therefore, for each $x \in E$, $A_1 x$ is contained in the compact set E. It now follows by Theorem 2 that $A_1|_E$ is contained in a uniformly compact subset of $C(E,X)$. As in the proof of (i) we now see that, given $a_n \in A_1$ $(n \in \mathbb{N})$, there exists a subsequence $\{a_{n_k}\}$ such that $\{a_{n_k} t\}$ converges in operator norm. □

Corollary 4 (Schauder [336]). *Let* $t \in BL(X)$. *Then* $t \in KL(X)$ *if and only if* $t^* \in KL(X')$.

Proof. Given $u \in X$, $\phi \in X'$, $s, t \in BL(X)$, we have

$$T_{s,t}(u \otimes \phi) = s u \otimes t^* \phi. \tag{1}$$

Fix $u \in X$, $s \in KL(X)$ with $\|u\| = \|su\| = 1$. Then $\|u \otimes \phi\| = \|\phi\|$ and $\|su \otimes t^* \phi\| = \|t^* \phi\|$. If $t \in KL(X)$, we have $T_{s,t} \in KL(B)$, and so $t^* \in KL(X')$. If $t^* \in KL(X')$, we now have $t^{**} \in KL(X'')$, and so $t \in KL(X)$.   ☐

Corollary 5. *Let* $s, t \in BL(X) \backslash \{0\}$, *and let* $T_{s,t} \in KL(B)$. *Then* $s, t \in KL(X)$.

Proof. First fix ϕ with $t^* \phi \neq 0$. Then, by (1) and the compactness of $T_{s,t}$, s is compact. Next fix u with $su \neq 0$. Then, by (1) and the compactness of $T_{s,t}$, t^* is compact. Hence, by Corollary 4, t is compact.   ☐

Notation. For the rest of this section, A will denote a complex Banach algebra, and X will denote a complex Banach space.

Lemma 6. *Let* $t \in A$, *and let the mapping* $a \to ta$ $(a \in A)$ *be a compact linear operator on* A. *Then*

(i) 0 *is the only possible point of accumulation of* $\mathrm{Sp}(t)$;

(ii) *each* $\lambda \in \mathrm{Sp}(t) \backslash \{0\}$ *is an eigenvalue, i. e. there exists* $u \in A \backslash \{0\}$ *with* $tu = \lambda u$.

Proof. Suppose first that $\lambda \in \partial \mathrm{Sp}(t) \backslash \{0\}$. By Proposition 3.9, there exist $x_n \in S(A)$ such that $\lim_{n \to \infty} t x_n - \lambda x_n = 0$. By the compactness of the mapping $a \to ta$, there exists a convergent subsequence $\{t x_{n_k}\}$. Let $u = \lim_{k \to \infty} t x_{n_k}$. Then $u = \lim_{k \to \infty} \lambda x_{n_k}$, $\|u\| = |\lambda| \neq 0$, $tu = \lim_{k \to \infty} \lambda t x_{n_k} = \lambda u$.

Since 0 is the only possible point of accumulation of eigenvalues of a compact operator, we have now proved that 0 is the only possible point of accumulation of $\partial \mathrm{Sp}(t)$. Let $\varepsilon > 0$ and $E = \{\lambda \in \mathrm{Sp}(t) : |\lambda| = \varepsilon\}$. If $\lambda \in E$, then $\alpha \lambda \in \partial \mathrm{Sp}(t)$ for some real $\alpha \geqslant 1$. Therefore if E were infinite, so would be the set $\{\lambda \in \partial \mathrm{Sp}(t) : |\lambda| \geqslant \varepsilon\}$, and $\partial \mathrm{Sp}(t)$ would have a non-zero point of accumulation. Therefore E is finite, and therefore $\mathrm{Sp}(t) = \partial \mathrm{Sp}(t)$.   ☐

Proposition 7. *Let* A *be semi-prime, let* $t \in A$, *let the mapping* $a \to ta$ $(a \in A)$ *be a compact linear operator on* A, *and let* $\lambda \in \mathrm{Sp}(t) \backslash \{0\}$. *Then there exists a minimal idempotent* e *of* A *with* $te = \lambda e$.

Proof. Let $J = \{a \in A : ta = \lambda a\}$. Then J is a non-zero closed right ideal with finite dimension. Therefore J contains a minimal right ideal M, and, since A is semi-prime, $M = Ae$ for some minimal idempotent e.   ☐

Theorem 8. *Let t be a central element of A such that the mapping $a \to ta$ $(a \in A)$ is a compact linear operator on A, let $\lambda \in \mathrm{Sp}(t) \setminus \{0\}$, let v be the ascent of the eigenvalue λ, and let*

$$N = \{a \in A : (t - \lambda)^v a = 0\}, \quad R = (t - \lambda)^v A.$$

Then

(i) *N and R are closed bi-ideals with $NR = RN = N \cap R = \{0\}$;*

(ii) *$A = N \oplus R$;*

(iii) *there exists a non-zero central idempotent p such that*

$$N = pA, \quad R = (1 - p)A;$$

(iv) *$\lambda^{-1} t(1 - p) \in \mathrm{q\text{-}Inv}(R)$.*

Proof. Since t belongs to the centre of A, N and R are bi-ideals. If $a \in N \cap R$, then $a = (t - \lambda)^v b$ for some $b \in A$, and $(t - \lambda)^{2v} b = (t - \lambda)^v a = 0$. Therefore, v being the ascent of λ, we have $(t - \lambda)^v b = 0$, $a = 0$. We have proved that $N \cap R = \{0\}$, and it follows that $NR = RN = \{0\}$.

By Lemma 6, N is a non-zero closed bi-ideal. Let $B = A/N$, and let $a \to a'$ denote the canonical mapping of A onto B. It is easy to check that the mapping $x \to t'x$ $(x \in B)$ is compact, and so Lemma 6 applies to B, t' in place of A, t. It follows that $\lambda \notin \mathrm{Sp}(t')$. For otherwise there exists $x \in B \setminus \{0\}$ with $t'x = \lambda x$. We have $x = a'$ for some $a \in A$, and $(t - \lambda)a \in N$. But then we have in turn $(t - \lambda)^{v+1} a = 0$, $(t - \lambda)^v a = 0$, $a \in N$, $x = 0$. Therefore

$$\lambda^{-1} t' \in \mathrm{q\text{-}Inv}(B), \tag{2}$$

and so $(\lambda - t')B = B$. It follows that $(\lambda - t')^v B = B$. Therefore, given $a \in A$, there exists $x \in A$ with $a' = (\lambda - t')^v x'$, i.e. $a - (\lambda - t)^v x \in N$. This proves that $A = N \oplus R$.

Since $A = N \oplus R$, there exist $u \in N$, $v \in R$ with

$$\lambda^v - (\lambda - t)^v = u + v.$$

Given $x \in N$, we have $(\lambda - t)^v x = 0$ and $vx = 0$, and so $\lambda^v x = ux$. Let $p = \lambda^{-v} u$. Then $p \in N$ and $px = x$ $(x \in N)$. Thus $p^2 = p$, and $p \neq 0$ since $N \neq \{0\}$. The element $\lambda^v - (\lambda - t)^v$ is central, and so, for all $a \in A$,

$$au - ua + av - va = 0.$$

Since $au - ua \in N$ and $av - va \in R$, this gives $au - ua = 0$ $(a \in A)$, and so p is central. That $N = pA$ is now clear, and it is easy to check that $R = \{x \in A : px = 0\} = (1 - p)A$. Therefore R is closed.

By (2), there exists $a \in A$ with $\lambda^{-1} t' + a' - \lambda^{-1} t'a' = 0$. Therefore $\lambda^{-1} t + a - \lambda^{-1} ta \in N$, $(\lambda^{-1} t + a - \lambda^{-1} ta)(1 - p) = 0$, $\lambda^{-1} t(1 - p) \in \mathrm{q\text{-}Inv}(R)$. □

Remark. The principal results of the Riesz-Schauder theory for compact linear operators are easily deduced from Theorem 8. Let

$t \in KL(X)$, and let A be the smallest closed subalgebra of $BL(X)$ containing $(1, t)$. By Theorem 3 (ii), the mapping $a \to ta$ is a compact linear operator on A, and so Lemma 6 applies. Therefore $\mathrm{Sp}(A, t) = \partial \, \mathrm{Sp}(A, t)$, and so $\mathrm{Sp}(A, t) = \mathrm{Sp}(BL(X), t) = \mathrm{Sp}(t)$. Thus 0 is the only possible point of accumulation of $\mathrm{Sp}(t)$, and each $\lambda \in \mathrm{Sp}(t) \backslash \{0\}$ is an eigenvalue of t, as an operator on X. For there exists $a \in A \backslash \{0\}$ with $ta = \lambda a$, and any $x \in a X \backslash \{0\}$ will serve as an eigenvector in X.

Suppose now that $\lambda \in \mathrm{Sp}(t) \backslash \{0\}$, let v, p be as in Theorem 8, and let $q = 1 - p$. We have $(t - \lambda)^v A = A q$, and so there exist $a, b \in A$ with $q = (t - \lambda)^v a$, $(t - \lambda)^v = b q$. Therefore $\ker(t - \lambda)^v = \ker(q) = p X$, and $(t - \lambda)^v X = q X$. Thus

$$X = \ker((t - \lambda)^v) \oplus (t - \lambda)^v X,$$

and it is now easy to check that v is the ascent of λ with t regarded as an operator on X.

It is also of interest to note that these results also hold for any $t \in BL(X)$ such that the mapping $t \to ta$ is a compact linear operator on A, since the hypothesis $t \in KL(X)$ was needed only to establish this property of t.

Definition 9. A *compact Banach algebra* is a complex Banach algebra A such that for each $t \in A$ the mapping $a \to tat$ is a compact linear operator on A.

Example 10. $KL(X)$ is a compact Banach algebra. This is immediate from Theorem 3 (i).

Proposition 11. *Let A be a compact Banach algebra, B a closed subalgebra of A, and I a closed bi-ideal of A. Then B and A/I are compact Banach algebras.*

Proof. Straightforward. □

Lemma 12. *Suppose that there exists a dense subset E of A such that the mapping $a \to tat$ is a compact linear operator on A for each $t \in E$. Then A is a compact Banach algebra.*

Proof. Let $t \in A$, $t_n \in E$, $\lim_{n \to \infty} t_n = t$, and let T, T_n be the operators on A defined by $Ta = tat$, $T_n a = t_n a t_n$ $(a \in A)$. Then

$$\|Ta - T_n a\| = \|(t - t_n) at + t_n a(t - t_n)\| \leqslant \|a\| \, (\|t\| + \|t_n\|) \, \|t - t_n\|,$$

and so $T \in KL(A)$. □

Remark. It follows from Lemma 12 that any semi-prime Banach algebra with dense socle is a compact Banach algebra. In particular, any semi-simple annihilator Banach algebra is a compact Banach algebra.

Example 10 provides an indication of the distinction between annihilator Banach algebras and compact Banach algebras since, by Enflo [133], there exist reflexive Banach spaces X with $KL(X) \neq (FBL(X))^-$.

Proposition 13. *Let* $\{A_\lambda : \lambda \in \Lambda\}$ *be a set of compact Banach algebras. Then their* $B(\infty)$-*sum* $\sum_0 \{A_\lambda : \lambda \in \Lambda\}$ *is a compact Banach algebra.*

Proof. Let $A = \sum_0 \{A_\lambda : \lambda \in \Lambda\}$, and let E be the set of all $f \in A$ such that $\{\lambda \in \Lambda : f(\lambda) \neq 0\}$ is finite. Lemma 12 applies. $\square$

Theorem 14. *Let A be a compact Banach algebra. Then*
(i) *each left ideal J of A that is not contained in* $\mathrm{rad}(A)$ *contains a non-zero idempotent,*
(ii) *for each idempotent $e \in A$ the subalgebra eAe has finite dimension.*

Proof. (i) Let J be a left ideal of A that is not contained in $\mathrm{rad}(A)$. Then there exists $t \in J$ with $1 \in \mathrm{Sp}(t)$, and we also have $1 \in \mathrm{Sp}(t^2)$. Let $A(t)$ denote the least closed subalgebra of A containing t. We have

$$t^2 a = tat \quad (a \in A(t)),$$

and so the mapping $a \to t^2 a$ is a compact linear operator on $A(t)$. We have $1 \in \mathrm{Sp}(A(t), t^2)$, and so, by Theorem 8, $A(t)$ contains a non-zero idempotent p with $p(1-t^2)^\nu = 0$. Then $p = p(1-(1-t^2)^\nu) \in J$, since J is a left ideal and $t \in J$.

(ii) Let e be a non-zero idempotent in A. Then eAe is a closed subalgebra with e as its unit element. By Proposition 11, eAe is a compact Banach algebra; and the mapping $x \to exe$ $(x \in eAe)$ is the identity operator. Since this mapping is compact, the unit ball of eAe is compact, eAe has finite dimension. $\square$

Theorem 15. *Let A be a semi-prime compact Banach algebra. Then each left ideal of A that is not contained in* $\mathrm{rad}(A)$ *contains a minimal idempotent. If* $A \neq \mathrm{rad}(A)$, $\mathrm{soc}(A)$ *exists.*

Proof. Let J be a left ideal of A with $J \not\subset \mathrm{rad}(A)$. By Theorem 14, J contains a non-zero idempotent u and the subalgebra uAu has finite dimension. Let $B = uAu$. Then B is semi-prime; for if L is a left ideal of B with $L^2 = \{0\}$, we have $Lu = uL = L$, and so

$$(AL)^2 = (ALu)(AuL) = ALBL \subset AL^2 = \{0\},$$

$AL = \{0\}$, $L = \{0\}$.

Since B has finite dimension it has minimal left ideals, and therefore, since it is also semi-prime, there exists an idempotent $e \in B$ with $eBe = \mathbb{C}e$. We have $eu = ue = e$, and so $e \in J$ and

$$eAe = euAue = eBe = \mathbb{C}e.$$

Thus e is a minimal idempotent with respect to the algebra A.

Finally, if $A \neq \mathrm{rad}(A)$, there exists a left ideal J with $J \not\subset \mathrm{rad}(A)$. □

The next theorem shows that a primitive compact Banach algebra is faithfully and irreducibly represented by an algebra of compact operators. Such an algebra is semi-simple and therefore has minimal idempotents, by Theorem 15.

Theorem 16. *Let A be a primitive compact Banach algebra, let e be a minimal idempotent in A, let π be the left regular representation of A on Ae, and let $\langle,\rangle$ be defined by $\langle x,y\rangle e = yx$ $(x \in Ae, y \in eA)$. Then π is a norm reducing monomorphism and*

$$F\,BL(Ae,eA,\langle,\rangle) \subset \pi(A) \subset KL(Ae) \cap BL(Ae,eA,\langle,\rangle).$$

Proof. By Theorem 31.6, π is a norm reducing monomorphism, and

$$F\,BL(Ae,eA,\langle,\rangle) = \pi(\mathrm{soc}(A)) \subset \pi(A) \subset BL(Ae,eA,\langle,\rangle).$$

Given $u \in Ae$, $v \in eA$, we have $\pi(uv) = u \otimes v$, and therefore

$$\|uv\| \geqslant \|u \otimes v\| = \|u\| \sup\{|\langle x,v\rangle|: x \in Ae, \|x\| \leqslant 1\}.$$

Let $t \in A\backslash\{0\}$ and choose $v \in eA$ with $vt \neq 0$. Then

$$\sup\{|\langle x,vt\rangle|: x \in Ae, \|x\| \leqslant 1\} = \kappa > 0,$$

and

$$\|uvt\| \geqslant \kappa\|u\| \qquad (u \in Ae). \tag{3}$$

Let $x_n \in Ae$ with $\|x_n\| \leqslant 1$ $(n \in \mathbb{N})$. Then there exists a subsequence $\{x_{n_k}\}$ such that $\{tx_{n_k}vt\}$ converges. Therefore, by (3), $\{tx_{n_k}\}$ converges, and so $\pi(t) \in KL(Ae)$. □

Corollary 17. *$KL(X)$ is the greatest subalgebra A of $BL(X)$ such that (i) X is irreducible with respect to A, (ii) A is a compact Banach algebra with respect to some algebra-norm majorizing the operator norm on A.*

Proof. $KL(X)$ has the properties (i) and (ii). Suppose on the other hand that A is a non-zero subalgebra of $BL(X)$ satisfying (i) and (ii). Then A is a primitive Banach algebra and has a minimal idempotent e. Choose $x_0 \in eX\backslash\{0\}$. Then $aex_0 = ax_0$, and so $Aex_0 = Ax_0 = X$. Also, if $ae \neq 0$, then $Aae = Ae$, and so $aex_0 \neq 0$. Thus the mapping $u \rightarrow ux_0$ is a continuous linear isomorphism of Ae onto X. By Banach's isomorphism theorem, the inverse mapping is also continuous, and so $A \subset KL(X)$, by Theorem 16. □

Theorem 18. *Let A be a compact Banach algebra. Then the structure space Π_A is discrete.*

Proof. The canonical mapping of A onto $A/\mathrm{rad}\,A$ induces a homeomorphism between the structure spaces of these algebras, and $A/\mathrm{rad}\,A$ is a compact Banach algebra. We therefore assume, without losing generality, that A is semi-simple.

Let $P_0 \in \Pi_A$. Our aim is to prove that the singleton $\{P_0\}$ is an open set, i.e. that $\Pi_A \backslash \{P_0\}$ is closed. Let $J = \cap\{P \in \Pi_A : P \neq P_0\}$. By definition of the hull-kernel topology it suffices to prove that

$$P_0 \not\supseteq J. \tag{4}$$

We first prove the existence of an idempotent $p \in A \backslash P_0$. Let $a \to a'$ denote the canonical mapping of A onto A/P_0. Since A/P_0 is a primitive compact Banach algebra, it contains a minimal idempotent q. Choose $t \in A$ with $t' = q$, and let $A(t)$ denote the least closed subalgebra of A containing t. Since the canonical mapping is a homomorphism and $(t^2)' = q^2 = q$, we have $1 \in \mathrm{Sp}(A, t^2)$, and therefore $1 \in \mathrm{Sp}(A(t), t^2)$. Also the mapping $a \to t^2 a$ $(a \in A(t))$ is compact, and so, by Theorem 8, $A(t)$ contains a non-zero idempotent p such that

$$(1-p)A(t) = R = (t^2 - 1)^v A(t).$$

Since $(t')^n = q$ $(n \in \mathbb{N})$, we have $(A(t))' = \mathbb{C}q$. Thus $R' = (q-1)^v \mathbb{C}q = \{0\}$. Therefore $(1-p')\mathbb{C}q = R' = \{0\}$, $p'q = q \neq 0$, $p' \neq 0$, $p \notin P_0$.

Let $B = pAp$. By Theorem 14(ii), B has finite dimension, and by Theorem 26.14, the mapping $P \to P \cap B$ is a homeomorphism of $\Gamma = \Pi_A \backslash \mathrm{hul}(B)$ onto Π_B. It follows that B is a semi-simple finite dimensional algebra, and its structure space is therefore discrete (by Proposition 26.7). Since $p \notin P_0$, $P_0 \in \Gamma$ and $Q_0 = P_0 \cap B \in \Pi_B$. Therefore

$$J \cap B = \cap\{P \cap B : P \in \Pi_A, P \neq P_0\}$$

$$= \cap\{Q : Q \in \Pi_B, Q \neq Q_0\} \not\subseteq Q_0 = P_0 \cap B.$$

Thus (4) is proved. □

The crucial Theorem 3(i) and Corollary 5 are due to Vala [391]. Compact Banach algebras were introduced and developed by Alexander in [8, 9], where further results will be found. Banach algebras A with the stronger assumption that the mapping $a \to ta$ $(a \in A)$ is compact were studied under the name completely continuous algebras by Kaplansky [233] and Freundlich [138]. For the other results in this section and some related ideas see Bonsall [69, 70].

§ 34. *H*-Algebras*

In this section A will denote a complex Banach algebra. Our main interest will be the theory of H^*-algebras, but we start with a wider class of algebras. We consider Banach algebras that are equipped with a bilinear form through which all the continuous linear functionals on the algebra are represented by elements of the algebra.

Definition 1. We say that A has an *inner dual module* if there exist a bi-ideal B of A and a non-degenerate bilinear form $\langle,\rangle$ on $A \times B$ such that the following axioms hold:

(i) $\langle a_1 a_2, b \rangle = \langle a_1, a_2 b \rangle = \langle a_2, b a_1 \rangle$ $(a_1, a_2 \in A, b \in B)$;

(ii) for each $b \in B$ the functional $b^\wedge$ defined by

$$b^\wedge(a) = \langle a, b \rangle \quad (a \in A)$$

belongs to A', and $A' = \{b^\wedge : b \in B\}$.

Since $\langle,\rangle$ is non-degenerate, the mapping $b \to b^\wedge$ is injective, and so the axioms (i) and (ii) are equivalent to the statement that the mapping $b \to b^\wedge$ is an A-module isomorphism of the bi-ideal B onto the dual module A'.

Example 2. Let A be the group algebra $L^1(G)$ of a compact group G, and take $B = L^\infty(G)$, $\langle a, b \rangle = \int_G a(t) b(t^{-1}) d\mu(t)$ $(a \in A, b \in B)$. Since $\langle a, b \rangle = (a * b)(e)$, where e is the identity of the group, the first equality in (i) is clear. The second equality depends on the inverse invariance of Haar measure on a compact group. The well known identification of the dual space of $L^1(G)$ with $L^\infty(G)$ gives (ii), and so A has an inner dual module.

Notation. Given that A has an inner dual module, we denote by B and $\langle,\rangle$ a bi-ideal and form as in Definition 1. Then, given a subset E of A, we denote by $E_\perp$ the set $\{b \in B : \langle E, b \rangle = \{0\}\}$.

Lemma 3. *Let A have an inner dual module, and let L be a left ideal of A. Then*

(i) $L_\perp$ *is a right ideal of A;*

(ii) $\mathrm{ran}(L) \cap B = (AL)_\perp$;

(iii) $\mathrm{ran}(L) \cap B = \{0\}$ *if and only* [1] *if* $(AL)^- = A$.

The corresponding statements hold for right ideals.

Proof. Let $x \in L_\perp$ and $a \in A$. Then $xa \in B$ and

$$\langle L, xa \rangle = \langle aL, x \rangle \subset \langle L, x \rangle = \{0\}.$$

Thus $xa \in L_\perp$, and $L_\perp$ is a right ideal of A.

---

[1] See footnote on page 124.

Let $b \in (AL)_\perp$, then $b \in B$ and

$$\langle A, Lb \rangle = \langle AL, b \rangle = \{0\},$$

and so $Lb = \{0\}$ by the non-degeneracy of $\langle , \rangle$. Thus $b \in \mathrm{ran}(L) \cap B$. Plainly this argument reverses.

By the Hahn-Banach theorem, $(AL)^- \neq A$ if and only if there exists $f \in A' \backslash \{0\}$ with $f(AL) = \{0\}$. Since $A' = \{b^\wedge : b \in B\}$, it follows that $(AL)^- \neq A$ if and only if $(AL)_\perp \neq \{0\}$. Thus (iii) follows from (ii).

Given a right ideal R, we prove similarly that $R_\perp$ is a left ideal, $\mathrm{lan}(R) \cap B = (RA)_\perp$, and $\mathrm{lan}(R) \cap B = \{0\}$ if and only if $(RA)^- = A$. □

Proposition 4. *Let A have an inner dual module and let A be semi-prime. Then A is an annihilator algebra.*

Proof. Since A is semi-prime, $\mathrm{ran}(A) = \mathrm{lan}(A) = \{0\}$, and so, by Lemma 3, A is an annihilator algebra. □

Remarks. (1) We do not know an example in which A has an inner dual module and is semi-prime but is not semi-simple.

(2) Which annihilator algebras have an inner dual module?

Proposition 5. *Let A have an inner dual module, and suppose that*

$$a \in (aA)^- \cap (Aa)^- \qquad (a \in A).$$

Then A is a dual algebra.

Proof. Let L be a closed left ideal of A and let $a_0 \in A \backslash L$. By the Hahn-Banach theorem, there exists $f \in A'$ with $f(L) = \{0\}$ and $f(a_0) = 1$. Then $f = b^\wedge$ for some $b \in B$, and we have $\langle L, b \rangle = \{0\}$, $\langle a_0, b \rangle = 1$. Thus $b \in L_\perp \subset (AL)_\perp = \mathrm{ran}(L) \cap B$. Also $\langle Aa_0, b \rangle \neq \{0\}$, for otherwise, by continuity of $b^\wedge$, $\langle a_0, b \rangle \in \langle (Aa_0)^-, b \rangle = \{0\}$. Therefore $\langle A, a_0 b \rangle \neq \{0\}$, $a_0 b \neq 0$. We have proved that $b \in \mathrm{ran}(L)$, and so $a_0 \notin \mathrm{lan}(\mathrm{ran}(L))$. Thus $\mathrm{lan}(\mathrm{ran}(L)) = L$, and a similar result holds for closed right ideals. □

Remark. Let G be a compact group. Then the group algebra $L^1(G)$ has a two-sided approximate identity and so, by Proposition 5, it is a dual algebra.

Definition 6. An *H*-algebra* is a Banach star algebra $(A, \|\cdot\|, *)$ on which is defined an inner product $(,)$ such that the following axioms hold, for all $a, x, y \in A$,

(i) $\|x\|^2 = (x, x)$;
(ii) $(ax, y) = (x, a^* y)$, $(xa, y) = (x, ya^*)$.

Thus an H^*-algebra is a complex Hilbert space that is also a Banach algebra with respect to the Hilbert space norm and has an involution which corresponds to taking adjoints of the left and right multiplication operators.

Remark. The original definition of H^*-algebras, due to Ambrose [18], is more general than Definition 6. Instead of assuming the existence of an involution on A satisfying (ii), it is assumed that for each $a \in A$ there is a non-void subset $\mathfrak{A}(a)$ of A such that, for all $x, y \in A$,

$$(ax, y) = (x, by), \qquad (xa, y) = (x, yb) \qquad (b \in \mathfrak{A}(a)).$$

For details see Ambrose [18].

Example 7. Let G be a compact group. Then the Hilbert space $L^2(G)$, with products defined by convolution and with the involution $f^*(s) = (f(s^{-1}))^*$, is an H^*-algebra.

Notation. For the rest of this section, H will denote an H^*-algebra, and given a subset E of H, $E^\perp$ will denote its Hilbert space orthogonal complement in H.

Proposition 8. H *has an inner dual module, and, if* L *is a left ideal,* $L^\perp$ *is a closed left ideal, and*

$$\operatorname{ran}(L) = ((HL)^\perp)^*.$$

Proof. Take $B = H$ and $\langle a, b \rangle = (a, b^*)$ $(a, b \in H)$. Since $(,)$ is sesquilinear, $\langle, \rangle$ is bilinear. The Riesz representation theorem for Hilbert spaces together with the identities

$$\langle a_1 a_2, b \rangle = (a_1 a_2, b^*) = (a_2, a_1^* b^*) = (a_2, (b a_1)^*) = \langle a_2, b a_1 \rangle,$$

$$\langle a_1 a_2, b \rangle = (a_1 a_2, b^*) = (a_1, b^* a_2^*) = (a_1, (a_2 b)^*) = \langle a_1, a_2 b \rangle,$$

prove that H has an inner dual module. For any subset E of H, we have $E^\perp = (E_\perp)^*$, and so the rest follows from Lemma 3. $\quad\Box$

Lemma 9. $\operatorname{lan}(H) = \operatorname{ran}(H) = (H^2)^\perp = \{x \in H : x^* x = 0\} = \operatorname{rad}(H).$

Proof. By Proposition 8 with $L = H$, and the corresponding statement for right ideals, we have

$$\operatorname{ran}(H) = ((H^2)^\perp)^* = \operatorname{lan}(H). \tag{1}$$

Since $\operatorname{lan}(H)H = \{0\}$, we have $H(\operatorname{lan}(H))^* = H^*(\operatorname{lan}(H))^* = \{0\}$, and so $(\operatorname{lan}(H))^* \subset \operatorname{ran}(H) = \operatorname{lan}(H)$. Thus $\operatorname{lan}(H) = (\operatorname{lan}(H))^{**} \subset (\operatorname{lan}(H))^*$, and (1) now gives $(H^2)^\perp = (\operatorname{lan}(H))^* = \operatorname{lan}(H)$.

If $x \in \operatorname{ran}(H)$, we obviously have $x^* x = 0$. On the other hand, if $x^* x = 0$, we have $(xH, xH) = (H, x^* xH) = \{0\}$, and so $xH = \{0\}$, $x \in \operatorname{lan}(H)$.

It is easy to see that $\mathrm{lan}(H)\subset\mathrm{rad}(H)$. We prove that $\mathrm{rad}(H)\subset\mathrm{lan}(H)$. Given $a\in H$, let T_a denote the operator on H defined by $T_a x=ax\,(x\in H)$, and let

$$|a| = \|T_a\| = \sup\{\|ax\|: x\in H, \|x\|\leqslant 1\}.$$

The mapping $a\to T_a$ is a star homomorphism of H into $BL(H)$, and so $T_{a^*a}=(T_a)^* T_a$, a positive self-adjoint operator on H. For such operators the norm is equal to the spectral radius, and therefore

$$|a^* a| = \|T_{a^*a}\| = \lim_{n\to\infty} \|(T_{a^*a})^n\|^{\frac{1}{n}} = \lim_{n\to\infty} |(a^* a)^n|^{\frac{1}{n}} \leqslant \lim_{n\to\infty} \|(a^* a)^n\|^{\frac{1}{n}} = r(a^* a).$$

If $a\in\mathrm{rad}(H)$, we have $r(a^* a)=0$, $|a^* a|=0$, $a^* a H=\{0\}$, $(aH,aH) = (H,a^* a H)=\{0\}$, $aH=\{0\}$, $a\in\mathrm{lan}(H)$. □

The following theorem is the 'first Wedderburn structure theorem' for H^*-algebras.

Theorem 10. *Let $A=(H^2)^-$. Then $H=A\oplus A^\perp$, A is an H^*-algebra with zero annihilator in itself, and $A^\perp=\mathrm{rad}(H)$.*

Proof. That $H=A\oplus A^\perp$ is clear, since A is a closed linear subspace of the Hilbert space H. That $A^\perp=\mathrm{rad}(H)$ is given by Lemma 9. It is not evident that A is an H^*-algebra, since we do not know that A is a star ideal of H, though this would obviously be the case if the involution were continuous. However we show that we can construct an involution on A with respect to which it is an H^*-algebra.

Given $b\in A$, let $b^\dagger$ denote the projection of b^* on A, i.e.

$$b^* = b^\dagger + c$$

with $b^\dagger\in A$ and $c\in A^\perp$. Since $(A^\perp)^*=A^\perp$, we have

$$b-b^{\dagger *} = (b^*-b^\dagger)^* = c^*\in A^\perp,$$

and so $b-b^{\dagger\dagger}=(b-b^{\dagger *})+(b^{\dagger *}-b^{\dagger\dagger})\in A\cap A^\perp=\{0\}$. It is now clear that † is a linear involution on A. By Lemma 9, $A^\perp=\mathrm{lan}(H)=\mathrm{ran}(H)$, and so $ab^*=ab^\dagger$, $b^* a=b^\dagger a$ $(a\in A)$. Therefore

$$(ab)^\dagger - b^\dagger a^\dagger = (ab)^\dagger - b^* a^* = (ab)^\dagger -(ab)^*\in A\cap A^\perp = \{0\},$$

and † is an involution on A. Also

$$(bx,y) = (x,b^* y) = (x,b^\dagger y), \qquad (xb,y) = (x,yb^*) = (x,yb^\dagger),$$

and so A with the involution † is an H^*-algebra.

Suppose finally that $b\in A\cap\mathrm{lan}(A)$. Then $bH^2=\{0\}$, $(bH,bH) = (bH^2,b)=\{0\}$, $bH=\{0\}$, $b\in\mathrm{lan}(H)=A^\perp$, $b=0$. Therefore the left annihilator of A in A is zero. □

Notation. We assume for the rest of this section that $\text{lan}(H)=\{0\}$.

Theorem 11. *H is a semi-simple dual algebra, and*

$$a\in(aH)^-\cap(Ha)^- \qquad (a\in H).$$

Proof. Let $a\in H$, and let $L=(Ha)^-$. Then $a=b+c$ with $b\in L$ and $c\in L^\perp$. Let $x\in A$. Then $xc=xa-xb\in L$, and also $xc\in L^\perp$, since $L^\perp$ is a left ideal. Therefore $xc=0$, $c\in\text{ran}(H)=\text{lan}(H)=\{0\}$, $c=0$, $a\in L$. Similarly $a\in(aH)^-$. By Propositions 8 and 5, it now follows that H is a dual algebra, and it is semi-simple by Lemma 9. ☐

Lemma 12. *Let L be a closed left ideal and J a closed bi-ideal of H. Then* $L^\perp=(\text{ran}(L))^*$, $J^\perp=\text{lan}(J)=\text{ran}(J)$, $J^*=J$.

Proof. Let $a\in(HL)^\perp$. Then $(Ha,L)=(a,HL)=\{0\}$. Therefore $Ha\subset L^\perp$, and, by Theorem 11, $a\in(Ha)^-\subset L^\perp$. Since we obviously have $L^\perp\subset(HL)^\perp$, this proves that $L^\perp=(HL)^\perp$. By Proposition 8, we now have $L^\perp=(\text{ran}(L))^*$.

$J^\perp$ is a closed bi-ideal, and so $J^\perp J\subset J^\perp\cap J=\{0\}$. Therefore $J\subset\text{ran}(J^\perp)=(J^{\perp\perp})^*=J^*$. Therefore also $J^*\subset J^{**}=J$, $J^*=J$. Since $\text{ran}(J)$ is a closed bi-ideal it follows that

$$\text{ran}(J)=(\text{ran}(J))^*=J^\perp.$$

The proof is easily completed. ☐

Theorem 13. *H is the orthogonal sum of the minimal closed bi-ideals of H, each of which is a topologically simple H*-algebra with zero annihilator in itself.*

Proof. Let $\{M_\alpha:\alpha\in\Lambda\}$ denote the set of minimal closed bi-ideals of H. Since H is a semi-simple annihilator algebra, we know, by Proposition 32.17, that H is the closure of the (algebraic) direct sum of $\{M_\alpha:\alpha\in\Lambda\}$, and each M_α is topologically simple. By Lemma 12, $M_\alpha^*=M_\alpha$, and so that each M_α is an H^*-algebra. Also, if $\alpha\neq\beta$, $M_\alpha M_\beta\subset M_\alpha\cap M_\beta=\{0\}$ and so $M_\alpha\subset\text{lan}(M_\beta)=M_\beta^\perp$. Thus the subspaces M_α are mutually orthogonal and so H is their orthogonal sum. Finally $M_\alpha\cap\text{lan}(M_\alpha)=M_\alpha\cap M_\alpha^\perp=\{0\}$. ☐

Lemma 14. *Each minimal left ideal of H is of the form He with e a self-adjoint idempotent.*

Proof. Each minimal left ideal of H is of the form Hu with u a minimal idempotent. Then $uu^*u=\lambda u$ for some $\lambda\in\mathbb{C}$, and so $(u^*u)^2=\lambda u^*u$. We have $(u^*u)^*(u^*u)\neq0$, for otherwise, by Lemma 9 we have in turn $u^*u=0$, $u=0$. Therefore λ is real and non-zero. Let $e=\lambda^{-1}u^*u$. Then $e^2=e=e^*$ and $He=Hu$. ☐

Lemma 15. *Let* e, f *be self-adjoint minimal idempotents in* H. *Then the following statements are equivalent:*

(i) e, f *are orthogonal;* (ii) $ef = 0$, (iii) He, Hf *are orthogonal subspaces.*

Proof. $(e, f) = (e^2, f^2) = (ef^*, e^* f) = (ef, ef) = \|ef\|^2$,

$$(He, Hf) = (Hef^*, H) = (Hef, H). \quad \square$$

Theorem 16. *Let* H *be topologically simple, and let* $\{e_\alpha : \alpha \in \Lambda\}$ *be a maximal set of mutually orthogonal self-adjoint minimal idempotents in* H. *Then:*

(i) $e_\alpha e_\beta = 0 \ (\alpha \neq \beta)$;

(ii) *the minimal left ideals* He_α *are mutually orthogonal and* H *is their orthogonal sum;*

(iii) $\dim(e_\alpha H e_\beta) = 1 \ (\alpha, \beta \in \Lambda)$;

(iv) *the one-dimensional subspaces* $e_\alpha H e_\beta$ *are mutually orthogonal and* H *is their orthogonal sum.*

Proof. (i) and (ii) are clear from Lemmas 15 and 14, for if the orthogonal sum L of $\{He_\alpha : \alpha \in \Lambda\}$ is not H, then $L^\perp$ is a non-zero left ideal and therefore contains a minimal left ideal (Corollary 32.7) and hence a self-adjoint minimal idempotent orthogonal to all e_α.

(iii) By Theorem 31.6 (iii), $\dim(e_\alpha H e_\beta) \leq 1$. But also, since H is topologically simple, $\operatorname{lan}(He_\beta) = \{0\}$, and so $e_\alpha H e_\beta \neq \{0\}$.

(iv) $$(e_\alpha H e_\beta, e_\gamma H e_\delta) = (He_\beta, e_\alpha e_\gamma H e_\delta) = \{0\} \text{ if } \alpha \neq \gamma,$$
$$(e_\alpha H e_\beta, e_\gamma H e_\delta) = (e_\alpha H, e_\gamma H e_\delta e_\beta) = \{0\} \text{ if } \delta \neq \beta.$$

Thus $e_\alpha H e_\beta$ and $e_\gamma H e_\delta$ are orthogonal unless $\alpha = \gamma$ and $\beta = \delta$. With the sums representing orthogonal sums we have in turn $\sum_{\beta \in \Lambda}^\oplus He_\beta = H$, $\sum_{\beta \in \Lambda}^\oplus e_\alpha H e_\beta = e_\alpha H$, $\sum_{\alpha, \beta \in \Lambda}^\oplus e_\alpha H e_\beta = \sum_{\alpha \in \Lambda}^\oplus e_\alpha H = H$. $\quad \square$

Corollary 17. *With* H, $\{e_\alpha : \alpha \in \Lambda\}$ *as in Theorem 16, there exist non-zero elements* $e_{\alpha\beta}$ *of the one-dimensional subspace* $e_\alpha H e_\beta$ *such that, for all* $\alpha, \beta, \gamma, \delta \in \Lambda$,

$$(e_{\alpha\beta})^* = e_{\beta\alpha}, \quad e_{\alpha\alpha} = e_\alpha, \quad e_{\alpha\beta} e_{\gamma\delta} = 0 \ (\beta \neq \gamma), \quad e_{\alpha\beta} e_{\beta\gamma} = e_{\alpha\gamma}.$$

Proof. We have $(e_\alpha A e_\beta)^* = e_\beta A e_\alpha$, $(e_\alpha A e_\beta)(e_\gamma A e_\delta) = \{0\}$ if $\beta = \gamma$, $(e_\alpha A e_\beta)(e_\beta A e_\gamma) = e_\alpha A e_\gamma$. Fix $\kappa \in \Lambda$ and choose $e_{\kappa\alpha} \in e_\kappa A e_\alpha$ with $e_{\kappa\alpha}(e_{\kappa\alpha})^* = e_\kappa = e_{\kappa\kappa}$. Then $(e_{\kappa\alpha})^* e_{\kappa\alpha}$ is a self-adjoint idempotent in $e_\alpha A e_\alpha$, and so $(e_{\kappa\alpha})^* e_{\kappa\alpha} = e_\alpha$. Now take $e_{\alpha\kappa} = (e_{\kappa\alpha})^*$ and $e_{\alpha\beta} = e_{\alpha\kappa} e_{\kappa\beta}$. $\quad \square$

The theory of H^*-algebras has been generalized to several classes of complemented algebras by Saworotnow [329], Tomiuk [386], Alexander [6], and others, see Barnes [43] for a full bibliography.

Chapter V. Star Algebras

§ 35. Commutative Banach Star Algebras

Recall from § 12 that a Banach star algebra is a complex Banach algebra A with an involution * satisfying, for all $x, y \in A$, $\alpha \in \mathbb{C}$,

(i) $(x+y)^* = x^* + y^*$,
(ii) $(\alpha x)^* = \alpha^* x^*$,
(iii) $x^{**} = x$,
(iv) $(xy)^* = y^* x^*$.

Sym(A) denotes the set of self-adjoint elements of A.

Definition 1. Given a Banach star algebra A and a linear functional f on A, the linear functional f^* is defined on A by

$$f^*(a) = (f(a^*))^* \qquad (a \in A).$$

Then $f \to f^*$ is a linear involution on the algebraic dual of A, and A' is an invariant subspace if the involution on A is continuous (and conversely, by a closed graph argument). A linear functional f on A is *self-adjoint* if $f^* = f$. Self-adjoint functionals are real valued on Sym(A). Conversely, given a real linear functional f_0 on Sym(A) we may define a self-adjoint linear functional f on A by

$$f(a) = f_0\left(\frac{1}{2}(a+a^*)\right) + i f_0\left(\frac{1}{2i}(a-a^*)\right) \qquad (a \in A).$$

The set of all continuous self-adjoint linear functionals on A is denoted by Sym(A').

Notation. For the remainder of this section A denotes a commutative Banach star algebra, with carrier space Φ_A, Shilov boundary $\check{S}(A)$, and Gelfand representation $a \to a^\wedge$.

Proposition 2. *The mapping* $\phi \to \phi^*$ *($\phi \in \Phi_A$) is a homeomorphism of* Φ_A *onto itself and maps* $\check{S}(A)$ *onto* $\check{S}(A)$.

Proof. The linear involution $f \to f^*$ maps Φ_A into Φ_A, and hence the mapping $\phi \to \phi^*$ ($\phi \in \Phi_A$) is a bijection. The continuity is clear.

For $a \in A$,

$$\max\{|a^\wedge(\phi^*)|: \phi \in \check{S}(A)\} = \max\{|a^{*\wedge}(\phi)|: \phi \in \check{S}(A)\}$$
$$= r(a^*) = r(a),$$

and hence $\check{S}(A) \subset \check{S}(A)^*$. Then $\check{S}(A)^* \subset \check{S}(A)^{**} = \check{S}(A)$. □

Proposition 2 shows that a necessary condition for the existence of an involution on a complex commutative Banach algebra is the existence of a homeomorphism of period 2 on its carrier space. Civin and Yood [99] give an example of a semi-simple complex commutative Banach algebra which does not admit an involution.

Theorem 3. *The following conditions on A are equivalent:*
(i) $\mathrm{Sp}(h) \subset \mathbb{R}$ $(h \in \mathrm{Sym}(A))$;
(ii) $a \to a^\wedge$ *is a star homomorphism;*
(iii) $r(a^* a) = r(a)^2$ $(a \in A)$.

Proof. (i) $\Rightarrow$ (ii). Let condition (i) hold, and let $a = h + ik$, where $h, k \in \mathrm{Sym}(A)$. Then $h^\wedge$, $k^\wedge$ are real valued on Φ_A, and thus, for $a \in A$, $\phi \in \Phi_A$,

$$a^{*\wedge}(\phi) = (h - ik)^\wedge(\phi)$$
$$= h^\wedge(\phi) - i k^\wedge(\phi)$$
$$= (h^\wedge(\phi) + i k^\wedge(\phi))^*$$
$$= (a^\wedge(\phi))^*.$$

(ii) $\Rightarrow$ (iii). Let $a \to a^\wedge$ be a star homomorphism. Then, for $a \in A$,

$$r(a^* a) = \max\{|(a^* a)^\wedge(\phi)|: \phi \in \Phi_A\}$$
$$= \max\{(a^\wedge(\phi))^* a^\wedge(\phi): \phi \in \Phi_A\}$$
$$= \max\{|a^\wedge(\phi)|^2: \phi \in \Phi_A\}$$
$$= r(a)^2.$$

(iii) $\Rightarrow$ (i). Let condition (iii) hold, and suppose that condition (i) does not hold. Then there exist $h \in \mathrm{Sym}(A)$, $\phi \in \Phi_A$ such that $\phi(h) = \alpha + i$, where $\alpha \in \mathbb{R}$. Choose $u \in A$ such that $\phi(u) = 1$ and let $b = (h - \alpha + ni)^m u$, where $m, n \in \mathbb{N}$. Since $\phi(b) = i^m(1 + n)^m$, we have

$$(1 + n)^{2m} \leqslant r(b)^2 = r(b^* b)$$
$$= r([(h - \alpha)^2 + n^2]^m u^* u)$$
$$\leqslant [(r(h) + |\alpha|)^2 + n^2]^m r(u^* u).$$

It follows in turn that, for $n \in \mathbb{N}$,

$$(1 + n)^2 \leqslant [(r(h) + |\alpha|)^2 + n^2](r(u^* u))^{\frac{1}{m}} \quad (m \in \mathbb{N}),$$
$$(1 + n)^2 \leqslant (r(h) + |\alpha|)^2 + n^2,$$
$$1 + 2n \leqslant (r(h) + |\alpha|)^2.$$

This contradiction completes the proof.   □

Note that Theorem 3 characterizes those commutative Banach star algebras for which the spectral semi-norm satisfies the B^* condition (Definition 12.16).

Theorem 4. *Let A be a commutative B^*-algebra with unit. Then the Gelfand representation is an isometric star isomorphism of A onto $C(\Phi_A)$.*

Proof. For $a \in A$ we have

$$\|a^{*n}\| \, \|a^n\| = \|a^{*n} \, a^n\| = \|(a^* a)^n\| \qquad (n \in \mathbb{N}),$$

and hence $r(a)^2 = r(a^*)r(a) = r(a^* a)$. By Theorem 3 the Gelfand representation is a star homomorphism of A into $C(\Phi_A)$. Given $h \in \mathrm{Sym}(A)$ we have

$$\|h^{2^n}\| = \|h\|^{2^n} \quad (n \in \mathbb{N}),$$

and hence $r(h) = \|h\|$. It follows that

$$r(a)^2 = r(a^* a) = \|a^* a\| = \|a\|^2 \qquad (a \in A).$$

Since $A^{\wedge}$ contains the constants, separates the points of Φ_A, and is closed in $C(\Phi_A)$, the Stone-Weierstrass theorem completes the proof.   □

Remarks. (1) An alternative proof that the Gelfand representation is a star homomorphism is provided by Propositions 12.20, 10.6 and the trivial part of Theorem 3.

(2) An elaboration of the above arguments shows that if A has a unit and satisfies the weaker condition

$$\|a^*\| \, \|a\| \leqslant \kappa \|a^* a\| \qquad (a \in A),$$

for some positive constant κ, then the Gelfand representation is a homeomorphic star isomorphism of A onto $C(\Phi_A)$. This last conclusion is also obtained under the conditions $1 \in A$, $\mathrm{Sp}(h) \subset \mathbb{R}$ $(h \in \mathrm{Sym}(A))$ and

$$\|h\| \leqslant \kappa \, r(h) \qquad (h \in \mathrm{Sym}(A))$$

for some positive constant κ.

Theorem 5. *Let A be a commutative B^*-algebra without unit. Then the Gelfand representation is an isometric star isomorphism of A onto $C_0(\Phi_A)$.*

Proof. Essentially as for Theorem 4.   □

The above theorems are due to Gelfand and Naimark [152].

For star algebras there is a natural modification of the concept of commutative subsets which is due to Civin and Yood [99].

Definition 6. Let B be a star algebra. A subset E of B is *normal* if $E \cup E^*$ is a commutative subset of B. An element x of B is *normal* if $x x^* = x^* x$.

Proposition 7. *Each normal subset of a star algebra B is contained in a maximal normal subset of B. Each maximal normal subset is a commutative star subalgebra of B, contains 1 if A has a unit element, and is closed if B is a normed algebra.*

Proof. Argue as in Proposition 15.3, noting that a subset E of B is normal if and only if $E \subset (E \cup E^*)^c$. ☐

Theorem 8. *Let B be a star algebra with unit, let a be a normal element of B and let M be a maximal normal subset of B containing $\{a\}$. Then*
  (i) $(z-a)^{-1} \in M$ $(z \in \mathbb{C} \backslash \mathrm{Sp}(B,a))$,
  (ii) $\mathrm{Sp}(M,a) = \mathrm{Sp}(B,a)$.

Proof. Let $z \in \mathbb{C} \backslash \mathrm{Sp}(B,a)$. As in Theorem 15.4, $M \cup \{(z-a)^{-1}\}$ is a commutative subset of B. Since a is normal, $(z-a)^{-1}$ is normal, and so $M \cup \{(z-a)^{-1}\}$ is a normal subset of B. By maximality of $M, (z-a)^{-1} \in M$, $z \in \mathbb{C} \backslash \mathrm{Sp}(M,a)$. ☐

§ 36. Continuity of the Involution

Throughout this section A denotes a Banach star algebra with involution $*$. We consider various conditions under which the involution $*$ is continuous; in the process we develop some elementary representation theory for Banach star algebras with continuous involution. There are very few known examples of discontinuous involutions; the examples at the end of the section give some indication of the pathology involved in discontinuous involutions.

We shall make frequent use of the linear involution $f \to f^*$ introduced in Definition 35.1. Recall that $\mathrm{Sym}(A')$ denotes the set of all continuous self-adjoint linear functionals on A.

Proposition 1. *The following statements are equivalent:*
  (i) $*$ *is continuous on A;*
  (ii) $\mathrm{Sym}(A')$ *is separating on A;*
  (iii) $\mathrm{Sym}(A)$ *is closed in A.*

Proof. (i) $\Rightarrow$ (ii). Let $*$ be continuous on A, and let $h \in \mathrm{Sym}(A) \backslash \{0\}$. By the Hahn-Banach theorem, there exists a continuous real linear functional f_0 on $\mathrm{Sym}(A)$ with $f_0(h)=1$. Let f be the associated self-adjoint complex linear functional on A defined by

$$f(x)=f_0\left(\frac{1}{2}(x+x^*)\right)+if_0\left(\frac{1}{2i}(x-x^*)\right) \qquad (x\in A).$$

Then f is continuous, since $*$ is continuous, $f\in\mathrm{Sym}(A')$, $f(h)=1$. Since $A=\mathrm{Sym}(A)+i\,\mathrm{Sym}(A)$, and elements of $\mathrm{Sym}(A')$ are real valued on $\mathrm{Sym}(A)$, it follows that $\mathrm{Sym}(A')$ is separating on A.

(ii) $\Rightarrow$ (iii). Let condition (ii) hold, and let $\{h_n\}\subset\mathrm{Sym}(A)$, $\lim_{n\to\infty} h_n=p+iq$, where $p,q\in\mathrm{Sym}(A)$. For $f\in\mathrm{Sym}(A')$ we have

$$if(q)=f(iq)=\lim_{n\to\infty} f(h_n-p).$$

Since f is real valued on $\mathrm{Sym}(A)$, it follows that $f(q)=0$ for each $f\in\mathrm{Sym}(A')$, $q=0$. Thus $\mathrm{Sym}(A)$ is closed.

(iii) $\Rightarrow$ (i). Let condition (iii) hold, and let $\lim_{n\to\infty} x_n=x$, $\lim_{n\to\infty} x_n^*=y$. Then

$$x+y=\lim_{n\to\infty}(x_n+x_n^*)\in\mathrm{Sym}(A),$$

and similarly $i(x-y)\in\mathrm{Sym}(A)$. Therefore

$$x+y=x^*+y^*, \qquad x-y=-x^*+y^*,$$

and so $y=x^*$. Apply the closed graph theorem.   □

Note that Proposition 1 remains true if A is only a Banach space with linear involution.

Theorem 2. *Let A be semi-simple. Then $*$ is continuous.*

Proof. A second Banach algebra norm $|\cdot|$ on A is obtained by taking

$$|a|=\|a^*\| \qquad (a\in A).$$

Apply Theorem 25.9.   □

Let X be a complex vector space. A *conjugate bilinear form* (or *sesquilinear form*) on X is a mapping $\langle,\rangle\colon X\times X\to\mathbb{C}$ such that
(i) $x\to\langle x,y\rangle$ $(x\in X)$ is linear for each $y\in X$;
(ii) $y\to\langle x,y\rangle$ $(y\in X)$ is conjugate linear for each $x\in X$, i.e.

$$\langle x,\alpha y+\beta z\rangle=\alpha^*\langle x,y\rangle+\beta^*\langle x,z\rangle.$$

Definition 3. A *self-dual space* is a pair $(X,\langle,\rangle)$ where X is a complex vector space and $\langle,\rangle$ is a non-degenerate conjugate bilinear form on X such that

$$\langle x,y\rangle=\langle y,x\rangle^* \qquad (x,y\in X).$$

A *normed self-dual space* is a self-dual space in which X is a normed space and $\langle , \rangle$ is continuous, i.e. there exists a positive constant κ such that

$$|\langle x, y \rangle| \leqslant \kappa \|x\| \|y\| \quad (x, y \in X).$$

A *Banach self-dual space* is a normed self-dual space in which X is a Banach space.

Note that if X is a normed self-dual space there exists a continuous conjugate linear monomorphism of X onto a total subspace of X'.

Example 4. Let Δ be a compact Hausdorff space and let $X = C(\Delta)$. Let μ be a positive finite regular Borel measure on Δ with support Δ and let

$$\langle x, y \rangle = \int_\Delta x y^* d\mu \quad (x, y \in X).$$

Then $(X, \langle , \rangle)$ is a Banach self-dual space.

Definition 5. Let $(X, \langle , \rangle)$ be a normed self-dual space. Operators $S, T \in BL(X)$ are said to be *adjoint* with respect to $\langle , \rangle$ if

$$\langle Sx, y \rangle = \langle x, Ty \rangle \quad (x, y \in X).$$

As in Definition 27.4, to each $S \in BL(X)$ there corresponds at most one $T \in BL(X)$ such that S, T are adjoint with respect to $\langle , \rangle$; the operator T, if it exists, is called the *adjoint of S with respect to $\langle , \rangle$*, and is denoted by S^*. The set of operators $S \in BL(X)$ that have adjoints with respect to $\langle , \rangle$ is denoted by $BL(X, \langle , \rangle)$, and the norm $\|\cdot\|_d$ is defined on $BL(X, \langle , \rangle)$ by

$$\|S\|_d = \max \{\|S\|, \|S^*\|\}.$$

It is clear that $BL(X, \langle , \rangle)$ is a subalgebra of $BL(X)$ and that $S \to S^*$ is an isometric involution on $BL(X, \langle , \rangle)$.

Proposition 6. *Let $(X, \langle , \rangle)$ be a Banach self-dual space.*
(i) *$BL(X, \langle , \rangle)$ with the norm $\|\cdot\|_d$ is a Banach star-normed algebra.*
(ii) *$BL(X, \langle , \rangle)$ is a closed subalgebra of $BL(X)$ if and only if the involution $S \to S^*$ on $BL(X, \langle , \rangle)$ is continuous with respect to the operator norm on X.*

Proof. (i) As in Lemma 27.5.
(ii) Let $S \to S^*$ be continuous with respect to $\|\cdot\|$, the operator norm on X. Then $\|\cdot\|$ and $\|\cdot\|_d$ are equivalent norms on $BL(X, \langle , \rangle)$ and (i) applies. Let $BL(X, \langle , \rangle)$ be closed in $BL(X)$. $BL(X, \langle , \rangle)$ is primitive, since it contains the finite rank operators given by $\langle , \rangle$, and hence $\|\cdot\|, \|\cdot\|_d$ are equivalent by Theorem 25.9. ☐

Theorem 9 below is a generalization of Proposition 6 (ii), and the proof of Theorem 9 is more elementary.

Definition 7. Let $(X, \langle , \rangle)$ be a normed self-dual space. A *star representation* of A on X is a star homomorphism of A into $BL(X, \langle , \rangle)$.

The construction used in Theorem 8 below (due to Schatz [335]) is a slight modification of the construction given in § 29. A further modification of the construction is given in the next section.

Theorem 8. *Let the involution on A be continuous. Then there exists a star representation $a \to \pi(a)$ of A on a Banach self-dual space X which is a homeomorphism of A into $BL(X, \langle , \rangle)$. If A is star-normed, then π is an isometry on* $\mathrm{Sym}(A)$.

Proof. We may suppose without loss that A has a unit; otherwise we embed A in its unitization. We consider the case in which A is star-normed; the general case follows by considering the equivalent norm $|\cdot|$ on A defined by

$$|a| = \max \{\|a\|, \|a^*\|\} \qquad (a \in A).$$

Let $\varLambda = \{f \in \mathrm{Sym}(A') : \|f\| = 1\}$, and given $f \in \varLambda$ let

$$L_f = \{a \in A : f(A a) = \{0\}\}, \qquad X_f = A - L_f,$$

$$\langle x, y \rangle_f = f(b^* a) \qquad (a \in x \in X_f, \ b \in y \in X_f).$$

Then X_f is a Banach self-dual space, and the left regular representation $a \to \pi_f(a)$ on X_f is a norm-decreasing star representation (cf. § 29).

Let $X = \sum_{f \in \varLambda}^{\oplus} X_f$, i.e. let X be the set of all functions ϕ belonging to the Cartesian product $\varPi_{f \in \varLambda} X_f$ such that

$$\|\phi\|^2 = \sum_{f \in \varLambda} \|\phi(f)\|^2 < \infty.$$

Then X is a Banach self-dual space with respect to the form $\langle , \rangle$ defined by

$$\langle \phi, \psi \rangle = \sum_{f \in \varLambda} \langle \phi(f), \psi(f) \rangle_f.$$

Let π be defined on A by

$$(\pi(a) \phi)(f) = \pi_f(a)(\phi(f)) \qquad (f \in \varLambda, \ \phi \in X, \ a \in A).$$

Then π is a norm-decreasing star representation of A on X.

Given $h \in \mathrm{Sym}(A)$, by the Hahn-Banach theorem, there exists $f_0 \in A'$ such that $f_0(h) = \|h\|$, $\|f_0\| = 1$. Let f be the self-adjoint functional on A defined for $x \in A$ by $f(x) = \frac{1}{2} f_0(x) + \frac{1}{2}(f(x^*))^*$. Since A is star-normed, it follows that $f \in \varLambda$. Also

$$\|h\| = f(h) = f(h + a) \leqslant \|h + a\| \qquad (a \in L_f),$$

and so $\|h\| = \|h'\|$, where h' denotes the coset $h + L_f$. Let ϕ be the element of X defined by

$$\phi(f) = 1', \qquad \phi(g) = 0 \qquad (g \in \Lambda \backslash \{f\}) \, .$$

Then $\|\phi\| = 1$, $\|\pi(h)\phi\| = \|h'\|$. This shows that π is an isometry on $\mathrm{Sym}(A)$. Finally, given $a = h + ik$ where $h, k \in \mathrm{Sym}(A)$, let f and ϕ be as above. Then

$$\|h\| = |\phi(f)(h)| \leqslant |\phi(f)(a)| \leqslant \|\pi(a)\| \, ,$$

and similarly $\|k\| \leqslant \|\pi(a)\|$. Therefore

$$\|a\| \leqslant \|h\| + \|k\| \leqslant 2 \|\pi(a)\| \, . \quad \square$$

Theorem 9. *If there exists a continuous star representation of A on a normed self-dual space, with zero kernel, then the involution on A is continuous.*

Proof. Let π be a continuous star representation of A on X, with zero kernel. Let $\{h_n\} \subset \mathrm{Sym}(A)$ and let $\lim_{n \to \infty} h_n = b$. Then for $x, y \in X$,

$$\langle \pi(b)x, y \rangle = \lim_{n \to \infty} \langle \pi(h_n)x, y \rangle = \lim_{n \to \infty} \langle x, \pi(h_n)y \rangle = \langle x, \pi(b)y \rangle \, .$$

Therefore $\pi(b) = (\pi(b))^* = \pi(b^*)$, $b = b^* \in \mathrm{Sym}(A)$. Apply Proposition 1. $\square$

Example 10. Let B be an infinite dimensional complex Banach algebra in which $B^2 = \{0\}$. Let $\{u_n\} \cup \{v_\lambda\}$ be a Hamel basis for B consisting of elements of norm 1. Let

$$v_\lambda^* = v_\lambda, \qquad u_{2n}^* = n u_{2n-1}, \qquad u_{2n-1}^* = \frac{1}{n} u_{2n} \, .$$

Then * extends by conjugate linearity to a discontinuous involution on B.

Example 11. Let B be a complex Banach algebra with respect to two non-equivalent norms $\|\cdot\|_1$, $\|\cdot\|_2$. Let * be an involution on B that is continuous with respect to $\|\cdot\|_1$. Let C denote the direct sum $B \oplus B$ with the norm

$$\|(a, b)\| = \max \{\|a\|_1, \|b\|_2\} \, ,$$

with the product

$$(a_1, b_1)(a_2, b_2) = (a_1 a_2, b_1 b_2) \, ,$$

and with the involution

$$(a, b)^* = (b^*, a^*) \, .$$

There exists a positive constant κ such that

$$\|a^*\|_1 \leqslant \kappa \|a\|_1 \qquad (a \in B).$$

Let $\{a_n\}$ be a sequence in B such that $\lim\limits_{n \to \infty} \|a_n\|_1 = 0$ but $\lim\limits_{n \to \infty} \|a_n\|_2 = 1$. Then

$$\|(0, a_n)\| = \|a_n\|_2, \qquad \|(0, a_n)^*\| = \|(a_n^*, 0)\| = \|a_n^*\|_1 \leqslant \kappa \|a_n\|_1.$$

Thus C is a Banach star algebra with discontinuous involution.

§ 37. Star Representations and Positive Functionals

Throughout this section A denotes a Banach star algebra. We shall relate star representations of A on Hilbert spaces to positive linear functionals on A. The main theorems of the section are concerned with the automatic continuity of star representations and positive linear functionals.

Definition 1. A *Hilbert A-module* is a Hilbert space $(H, \langle, \rangle)$ which is a Banach left A-module such that

$$\langle ax, y \rangle = \langle x, a^* y \rangle \qquad (x, y \in H, \ a \in A).$$

We recall from Definition 9.12 that there exists a positive constant κ for which

$$\|ax\| \leqslant \kappa \|a\| \|x\| \qquad (x \in H, \ a \in A). \tag{1}$$

A star representation of A on a Hilbert space H is a star homomorphism π of A into $BL(H)$. A Hilbert A-module H gives a corresponding (continuous) star representation π where

$$\pi(a)x = ax \qquad (x \in H, \ a \in A). \tag{2}$$

Conversely, a continuous star representation π of A on H gives a corresponding Hilbert A-module structure on H defined by (2).

Theorem 3 below shows that every star representation of A on a Hilbert space is automatically continuous. It follows that condition (1) in the definition of a Hilbert A-module is redundant.

Notation. Throughout this section H denotes a Hilbert space with inner product $\langle, \rangle$, and $|T|$ denotes the operator norm of $T \in BL(H)$.

Lemma 2. *Let C be a star subalgebra of $BL(H)$.*
 (i) *C is semi-simple.*
 (ii) *If C is a Banach algebra with respect to some norm, then*

$$|a|^2 \leqslant r(a^* a) \qquad (a \in C).$$

Proof. (i) The standard iteration (employed in the proof of Theoren 35.4) gives

$$|h| = \lim_{n \to \infty} |h^n|^{\frac{1}{n}} \quad (h \in \mathrm{Sym}(C)). \tag{3}$$

Given $q \in \mathrm{rad}(C)$, we have $q^* q \in \mathrm{rad}(C)$, and (3) then gives

$$|q|^2 = |q^* q| = 0 .$$

(ii) By (3), Theorem 5.7 and Theorem 5.8 we have, for $h \in \mathrm{Sym}(C)$,

$$|h| = \lim_{n \to \infty} |h^n|^{\frac{1}{n}} \leqslant \sup\{|z| : z \in \mathrm{Sp}(h)\} = r(h) .$$

For $a \in C$,

$$|a|^2 = |a^* a| \leqslant r(a^* a) . \quad \square$$

Theorem 3. *Every star representation of A on a Hilbert space is continuous.*

Proof. Let π be a star representation of A on H with kernel J and image C. C is semi-simple by Lemma 2, and so J is closed, by Proposition 25.10. The quotient norm on A/J induces a Banach algebra norm $\|\cdot\|$ on C under the natural algebra isomorphism of A/J onto C. It is now sufficient to show that

$$|a| \leqslant \kappa \|a\| \quad (a \in C)$$

for some positive constant κ.

Let $a_n, b, c \in C$ with

$$\lim_{n \to \infty} \|a_n - b\| = 0, \quad \lim_{n \to \infty} \|a_n^* - c\| = 0 .$$

By Lemma 2,

$$|a_n^* - c|^2 \leqslant r((a_n^* - c)^* (a_n^* - c))$$
$$\leqslant \|a_n - c^*\| \, \|a_n^* - c\| .$$

Since $\{\|a_n\|\}$ is bounded, it follows that $\lim\limits_{n \to \infty} |a_n^* - c| = 0$, and similarly $\lim\limits_{n \to \infty} |a_n^* - b^*| = 0$. Therefore $b^* = c$, and the closed graph theorem gives a positive constant γ such that

$$\|a^*\| \leqslant \gamma \|a\| \quad (a \in C).$$

Finally,

$$|a|^2 \leqslant r(a^* a) \leqslant \|a^*\| \, \|a\| \leqslant \gamma \|a\|^2 \quad (a \in C). \quad \square$$

Definition 4. A linear functional f on A is *positive* if

$$f(a^* a) \geqslant 0 \quad (a \in A).$$

Example 5. Let π be a star representation of A on H, let $x \in H$, and let f be the linear functional on A defined by

$$f(a) = \langle \pi(a)x, x \rangle \quad (a \in A).$$

Then, for $a \in A$,

$$f(a^* a) = \langle \pi(a^*)\pi(a)x, x \rangle = \langle \pi(a)x, \pi(a)x \rangle \geqslant 0,$$

and so f is positive.

Notation. Given a linear functional f on A, and given $b \in A$, we define the linear functional f_b on A by

$$f_b(a) = f(b^* a b) \quad (a \in A).$$

Clearly f_b is positive if f is positive.

Lemma 6. *Let f be a positive linear functional on A. Then, for $a, b \in A$, $h \in \mathrm{Sym}(A)$,*
  (i) $f(b^* a) = (f(a^* b))^*$;
  (ii) $|f(b^* a)|^2 \leqslant f(a^* a) f(b^* b)$;
  (iii) $|f_b(h)| \leqslant f(b^* b) r(h)$;
  (iv) $|f_b(a)| \leqslant f(b^* b)(r(a^* a))^{\frac{1}{2}}$.

Proof. Let $\lambda, \mu \in \mathbb{C}$ and let $w = \lambda a + \mu b$. Then $f(w^* w) \geqslant 0$, and so

$$|\lambda|^2 f(a^* a) + \lambda^* \mu f(a^* b) + \lambda \mu^* f(ab^*) + |\mu|^2 f(b^* b) \geqslant 0.$$

Parts (i) and (ii) follow by suitable choice of λ and μ.
  (iii) Given $h \in \mathrm{Sym}(A)$ with $r(h) < 1$, by Proposition 12.11, there exist $p, q \in \mathrm{Sym}(A)$ such that $p \circ p = h, q \circ q = -h$. Let $u = b - pb, v = b - qb$. Then

$$u^* u = b^*(1 - p)^2 b = b^*(1 - h)b,$$

$$v^* v = b^*(1 - q)^2 b = b^*(1 + h)b.$$

This gives

$$f(b^*(1 - h)b) \geqslant 0, \qquad f(b^*(1 + h)b) \geqslant 0,$$

and hence $|f_b(h)| \leqslant f(b^* b)$.
  (iv) By (ii) and (iii),

$$|f_b(a)|^2 = |f(b^*(ab))|^2 \leqslant f(b^* b) f(b^* a^* ab),$$

$$f_b(a^* a) \leqslant f(b^* b) r(a^* a). \quad \square$$

Given a star representation of A on H, Example 5 gives associated positive linear functionals on A. Conversely, let f be a positive linear functional on A and, as in Theorem 36.8, let

$$L_f = \{a \in A : f(Aa) = \{0\}\}, \quad X_f = A - L_f,$$

$$\langle x, y \rangle_f = f(b^* a) \quad (a \in x \in X_f, \; b \in y \in X_f).$$

It follows from Lemma 6 that $L_f = \{a : f(a^* a) = 0\}$ and that $\langle, \rangle_f$ is an inner product on X_f. Let H_f be the Hilbert space completion of X_f. Given $a \in A$, $b \in x \in X_f$, we have, by Lemma 6,

$$\|\pi_f(a)x\|^2 = \langle ax, ax \rangle_f = f(b^* a^* ab) \leqslant r(a^* a)\, f(b^* b),$$

and hence $\|\pi_f(a)x\| \leqslant (r(a^* a))^{\frac{1}{2}}\|x\|$. Therefore $\pi_f(a)$ extends to a bounded linear operator on H_f; and the extension, which we also denote by $\pi_f(a)$, satisfies

$$|\pi_f(a)| \leqslant (r(a^* a))^{\frac{1}{2}} \quad (a \in A).$$

Thus π_f is a star representation of A on H_f and is continuous by Theorem 3. The associated module multiplication is given by

$$ax = \pi_f(a)x \quad (a \in A, \; x \in H_f).$$

Proposition 7. *For each positive linear functional f on A, H_f is a Hilbert A-module.*

Proof. Above. ☐

Theorem 8. *Let f be a positive linear functional on A and let $b \in A$. Then f_b is continuous.*

Proof. Let $b \in x \in X_f$. Then

$$f_b(a) = \langle \pi_f(a)x, x \rangle_f \quad (a \in A).$$

Apply Proposition 7. ☐

Corollary 9. *Let A have a unit. Then every positive linear functional on A is continuous.*

Proof. Put $b = 1$ in Theorem 8. ☐

Remarks. Let A have a unit and let f be a positive linear functional on A. Lemma 6 shows that f is self-adjoint and also

$$|f(a)| \leqslant f(1)(r(a^* a))^{\frac{1}{2}} \quad (a \in A).$$

If A is commutative or star-normed, it follows that $\|f\| \leqslant f(1)$. If, further, A is unital, then $\|f\| = f(1)$. We shall see in the next section that for unital B^*-algebras every continuous linear functional f with $\|f\| = f(1)$ is positive.

Definition 10. A non-zero positive linear functional f on A is *representable* if there exists a star representation π of A on H and a topologically cyclic vector $x \in H$ such that

$$f(a) = \langle \pi(a)x, x \rangle \quad (a \in A).$$

A representable functional is automatically continuous, by Theorem 3. If A has a unit, it is clear from the proof of Theorem 8 that every positive linear functional on A is representable.

Theorem 11. *A non-zero positive linear functional f on A is representable if and only if there exists $\kappa > 0$ such that*

$$|f(a)|^2 \leqslant \kappa f(a^* a) \quad (a \in A).$$

Proof. Let f be representable, say

$$f(a) = \langle \pi(a)x, x \rangle \quad (a \in A).$$

Then

$$|f(a)|^2 \leqslant \|\pi(a)x\|^2 \|x\|^2 = \langle \pi(a^* a)x, x \rangle \|x\|^2 = \|x\|^2 f(a^* a).$$

Suppose now that $|f(a)|^2 \leqslant \kappa f(a^* a)$ $(a \in A)$. Let $B = A + \mathbb{C}$ be the unitization of A, and extend f to g on B by

$$g(a + \lambda) = f(a) + \kappa \lambda \quad (a + \lambda \in B).$$

Then, for $a + \lambda \in B$,

$$
\begin{aligned}
g((a + \lambda)^*(a + \lambda)) &= f(a^* a) + \lambda^* f(a) + \lambda f(a)^* + \kappa |\lambda|^2 \\
&\geqslant f(a^* a) - 2|\lambda| |f(a)| + \kappa |\lambda|^2 \\
&\geqslant f(a^* a) - 2|\lambda| \kappa^{\frac{1}{2}} (f(a^* a))^{\frac{1}{2}} + \kappa |\lambda|^2 \\
&= ((f(a^* a))^{\frac{1}{2}} - \kappa^{\frac{1}{2}} |\lambda|)^2 .
\end{aligned}
$$

Thus g is positive and so representable, say

$$g(b) = \langle \pi(b)x, x \rangle \quad (b \in B),$$

where x is a topologically cyclic vector of H with respect to B. Let

$$N = \{y \in H : \pi(a)y = 0 \ (a \in A)\}.$$

Since $f \neq 0$, $x \notin N$. Let $x = u + v$ where $u \in N^{\perp}$, $v \in N$; then $u \neq 0$. Since $\pi(A)N \subset N$ and $\pi(A)N^{\perp} \subset N^{\perp}$, we have

$$f(a) = \langle \pi(a)u, u \rangle \quad (a \in A).$$

Let $K=(\pi(A)u)^-$. It is now sufficient to show that $u\in K$. Let $u=p+q$ where $p\in K$, $q\in Y=K^\perp\cap N^\perp$. Since $\pi(A)K\subset K$ and $\pi(A)Y\subset Y$, it follows that

$$\pi(a)q\in K\cap K^\perp=\{0\}\qquad (a\in A).$$

Then $q\in N\cap N^\perp=\{0\}$, $u\in K$. □

Corollary 9 gives the automatic continuity of all positive linear functionals under the condition that A have a unit. It is clear that some algebraic condition on A is required to ensure the continuity of all positive linear functionals; indeed if A is an infinite dimensional Banach star algebra with $A^2=\{0\}$, then any linear functional on A is positive. The next three theorems are due to Murphy [283] and Varopoulos [394].

Definition 12. Given positive linear functionals f,g on A, we say that f *dominates* g if $f-g$ is positive, and we then write $f\geqslant g$.

Theorem 13. [1]*Let $A^2=A$, and let every non-zero positive linear functional on A dominate a non-zero continuous positive linear functional. Then every positive linear functional on A is continuous.*

Proof. Each $a\in A$ is of the form $a=\sum_{j=1}^n u_j v_j$, and, for $u,v\in A$, we have

$$4uv=(v+u^*)^*(v+u^*)-(v-u^*)^*(v-u^*)$$
$$+i(v+iu^*)^*(v+iu^*)-i(v-iu^*)^*(v-iu^*).$$

Therefore $\{a^* a: a\in A\}$ spans A, and so $\geqslant$ is a partial order on the set of positive linear functionals on A.

Let f be a non-zero positive linear functional on A, let

$$D=\{g\in A'\backslash\{0\}: g \text{ is positive, } f\geqslant g\},$$

and let C be a chain in the partially ordered set $(D,\geqslant)$. Each $a\in A$ is of the form $a=\sum_{j=1}^m \alpha_j b_j^* b_j$, and since

$$g(b^* b)\leqslant f(b^* b)\qquad (b\in A,\ g\in C),$$

we may define a linear functional ϕ on A by

$$\phi(a)=\lim_{g\in C} g(a)\qquad (a\in A).$$

Clearly ϕ is positive and $\phi\leqslant f$. Moreover for $g\in C$, $a\in A$,

$$|g(a)|\leqslant \sum_{j=1}^m |\alpha_j| g(b_j^* b_j)\leqslant \sum_{j=1}^m |\alpha_j| f(b_j^* b_j).$$

By the uniform boundedness theorem, there exists $\kappa>0$ such that

$$\|g\|\leqslant\kappa\qquad (g\in C),$$

-----
[1] See footnote on page 124.

and hence $|\phi(a)| \leqslant \kappa \|a\|$ $(a \in A)$, ϕ is continuous. By Zorn's Lemma, D has a maximal element, say g_0.

If $f - g_0 \neq 0$, there exists a non-zero continuous positive linear functional g_1 such that $f - g_0 - g_1$ is positive. Then $g_0 + g_1 \in D$, $g_0 + g_1 \geqslant g_0$, and so, by maximality, $g_1 = 0$. This contradiction establishes that $f = g_0$, f is continuous. $\square$

Theorem 14. *Let* $A^2 = A$, *and let* A *be commutative. Then every positive linear functional on* A *is continuous.*

Proof. Let f be a non-zero positive linear functional on A. By Theorem 8, f_b is a continuous positive linear functional for each $b \in A$. If $f_b = 0$ $(b \in A)$, then, by Lemma 6 (ii),

$$|f(b^* u^* v)|^2 \leqslant f_b(u^* u) f(v^* v) = 0 \qquad (b, u, v \in A),$$

and so $f(A^3) = \{0\}$, $f = 0$. Choose $b \in A$ such that $f_b \neq 0$ and $\|b^* b\| < 1$. By Proposition 12.11 there exists $p \in \mathrm{Sym}(A)$ such that $p \circ p = b^* b$, and then, for $a \in A$,

$$a^* (1 - b^* b) a = v^* v$$

where $v = a - p a$. Therefore, since A is commutative,

$$(f - f_b)(a^* a) = f(v^* v) \geqslant 0 \qquad (a \in A),$$

and so $f \geqslant f_b$. Apply Theorem 13. $\square$

Theorem 15. *Let* A *have a bounded two-sided approximate identity. Then every positive linear functional on* A *is continuous.*

Proof. Let f be a positive linear functional on A. Then, by Theorem 8, f_b is continuous for each $b \in A$. For $a, b, c \in A$,

$$4 f(b a c) = f_{c+b^*}(a) - f_{c-b^*}(a) + i f_{c+ib^*}(a) - i f_{c-ib^*}(a),$$

and so the linear functional $a \to f(b a c)$ is continuous.

Let $\{a_n\} \subset A$, $\lim_{n \to \infty} a_n = 0$. By Corollary 11.12 (and its analogue), there exist $b, c, w_n \in A$ such that

$$a_n = b w_n c, \qquad \lim_{n \to \infty} w_n = 0.$$

Then $\lim_{n \to \infty} f(a_n) = \lim_{n \to \infty} f(b w_n c) = 0$, f is continuous. $\square$

We shall exploit in the following sections the relation between positive linear functionals on A and Hilbert A-modules. The example below gives a large class of Banach star algebras for which the positive linear functionals are essentially trivial, i.e. $H_f = \{0\}$ for each positive linear functional.

Example 16. Let B be a Banach star algebra and let C denote the direct sum $B \oplus B$ with the norm

$$\|(a, b)\| = \max \{\|a\|, \|b\|\},$$

with the product

$$(a_1, b_1)(a_2, b_2) = (a_1 a_2, b_1 b_2),$$

and with the involution

$$(a, b)^* = (b^*, a^*).$$

Let f be a linear functional on C. Then there exist linear functionals ϕ, ψ on B such that

$$f(a, b) = \phi(a) + \psi(b) \qquad (a, b \in B).$$

Suppose that f is positive. Then

$$\phi(b^* a) + \psi(a^* b) \geqslant 0 \qquad (a, b \in B).$$

Multiplication by appropriate scalars gives

$$\phi(b^* a) = \psi(a b^*) = 0 \qquad (a, b \in B)$$

and so $f(u^* u) = 0$ $(u \in C)$.

Given $f \in \text{Sym}(A')$, the Schatz construction (see Theorem 36.8) gives a star representation of A on $X_f = A - L_f$, where X_f has the quotient norm. When f is positive, this quotient norm does not in general coincide with the inner product norm on X_f and hence the Schatz construction need not produce Hilbert A-modules. Very recently Lance [252] has repaired this deficiency by modifying the Schatz construction in terms of balanced self-dual Banach spaces as follows.

Definition 17. Let $(X, \langle, \rangle)$ be a self-dual space. A norm p on X is *admissible with respect to* $\langle, \rangle$ if

$$|\langle x, y \rangle| \leqslant p(x) p(y) \qquad (x, y \in X).$$

Given an admissible norm p on X the norm p^* is defined on X by

$$p^*(x) = \sup \{|\langle x, y \rangle| : p(y) \leqslant 1\} \qquad (x \in X).$$

We say that p is *balanced* if $p^* = p$. A Banach space $(X, \|\cdot\|)$ is *balanced* if there is a bilinear form $\langle, \rangle$ such that $(X, \langle, \rangle)$ is self-dual and $\|\cdot\|$ is balanced with respect to $\langle, \rangle$.

A Hilbert space is clearly balanced with respect to the given inner product. Lance [252] shows that l^1 is balanced, but l^p is not balanced for $1 < p < 2$, $2 < p < \infty$, and c_0 is not balanced.

Lemma 18. *Let $(X, \langle , \rangle)$ be a self-dual space and let p be an admissible norm on X. Then*

   (i) $|\langle x, y \rangle| \leqslant p^*(x) p(y)$ $(x, y \in X)$;

   (ii) *if q is a norm on X and $q \leqslant p$, then $p^* \leqslant q^*$;*

   (iii) $p^{**} \leqslant p$, *and p^{**} is admissible;*

   (iv) $(\frac{1}{2}(p + p^*))^* \leqslant \frac{1}{2}(p + p^*)$.

Proof. (i) and (ii) are clear.

(iii) For $x \in X$,

$$p^{**}(x) = \sup\{|\langle x, y \rangle| : p^*(y) \leqslant 1\}$$
$$= \sup\{|\langle y, x \rangle| : p^*(y) \leqslant 1\}$$
$$\leqslant p(x).$$

Since p is admissible, $p^* \leqslant p$, and (ii) gives $p^* \leqslant p^{**}$. Hence

$$|\langle x, y \rangle| \leqslant p^{**}(x) p^*(y) \leqslant p^{**}(x) p^{**}(y) \qquad (x, y \in X).$$

(iv) Let $x, y \in X$ with $y \neq 0$, $\frac{1}{2}p(y) + \frac{1}{2}p^*(y) \leqslant 1$. We have

$$|\langle x, y \rangle| \leqslant p^*(x) p(y), \qquad |\langle x, y \rangle| \leqslant p^{**}(x) p^*(y).$$

Let $t = \frac{1}{2}p(y)$, so that $0 < t < 1$, $\frac{1}{2}p^*(y) \leqslant 1 - t$. Then

$$|\langle x, y \rangle| \leqslant \frac{1}{4 t(1 - t)} |\langle x, y \rangle|$$
$$\leqslant \frac{1}{2}\left(\frac{1}{p(y)} + \frac{1}{p^*(y)}\right)|\langle x, y \rangle|$$
$$\leqslant \frac{1}{2}(p^*(x) + p^{**}(x))$$
$$\leqslant \frac{1}{2}(p(x) + p^*(x)). \quad \square$$

Theorem 19. *Let p be an admissible norm on a self-dual space $(X, \langle , \rangle)$. Then there exists a balanced norm $\tilde{p}$ on X with $\tilde{p} \leqslant p$.*

Proof. Let $p_1 = \frac{1}{2}(p + p^*)$. By Lemma 18 (iv), $p_1^* \leqslant p_1$ and p_1 is admissible. Also $p^* \leqslant p_1^* \leqslant p_1 \leqslant p$ and so

$$p_1(x) - p_1^*(x) \leqslant \frac{1}{2}(p(x) - p^*(x)) \qquad (x \in X).$$

We now define inductively a sequence of admissible norms $\{p_n\}$ by

$$p_{n+1} = \frac{1}{2}(p_n + p_n^*) \qquad (n \in \mathbb{N}).$$

We have $p_n^* \leqslant p_{n+1}^* \leqslant p_{n+1} \leqslant p_n$ and

$$p_n(x) - p_n^*(x) \leqslant 2^{-n}(p(x) - p^*(x)) \qquad (x \in X).$$

Let $\tilde{p}(x) = \inf_n p_n(x) = \sup_n p_n^*(x)$ $(x \in X)$. Then $\tilde{p}$ is a norm on X,

$$p_n^* \leqslant \tilde{p} \leqslant p_n, \qquad p_n^* \leqslant \tilde{p}^* \leqslant p_n^{**} \leqslant p_n \qquad (n \in \mathbb{N}).$$

Therefore $\tilde{p}^* = \tilde{p} \leqslant p$. $\quad \square$

Example 20. Let $(X, \langle,\rangle)$ be an inner product space and let p be an admissible norm on X. Then $\tilde{p}(x)=\langle x,x\rangle^{\frac{1}{2}}$ $(x\in X)$. To see this note that $\langle x,x\rangle\leqslant\tilde{p}(x)^2$, since $\tilde{p}$ is admissible. Apply Lemma 18 (ii).

Lemma 21. *Let p be an admissible norm on a self-dual space $(X,\langle,\rangle)$, and let $T\in BL(X,\langle,\rangle,p)$. [1] Then $T\in BL(X,\langle,\rangle,\tilde{p})$ and $\|T\|_{\tilde{p}}\leqslant\|T\|_p$.*

Proof. For $x\in X$,

$$\begin{aligned}
p^*(Tx) &= \sup\{|\langle Tx,y\rangle|: p(y)\leqslant 1\} \\
&= \sup\{|\langle x, T^*y\rangle|: p(y)\leqslant 1\} \\
&\leqslant p^*(x)\sup\{p(T^*y): p(y)\leqslant 1\} \\
&= |T^*|_p p^*(x) \\
&\leqslant \|T\|_p p^*(x),
\end{aligned}$$

and similarly $p^*(T^*x)\leqslant|T|_p p^*(x)\leqslant\|T\|_p p^*(x)$. Thus $T\in BL(X,\langle,\rangle,p^*)$, and $\|T\|_{p^*}\leqslant\|T\|_p$. In the notations of the proof of Theorem 19, we now obtain $\|T\|_{p_1}\leqslant\|T\|_p$, and, in turn, $\|T\|_{p_n}\leqslant\|T\|_p$, $\|T\|_{\tilde{p}}\leqslant\|T\|_p$. $\quad\square$

We are now ready to modify the Schatz construction of Theorem 36.8. Let A be star-normed and let

$$\Lambda=\{f\in\mathrm{Sym}(A'): \|f\|=1\}\,.$$

Given $f\in\Lambda$ let p denote the quotient norm on the self-dual space $(X_f,\langle,\rangle_f)$. Then p is admissible and by Theorem 19 there is a balanced norm $\tilde{p}$ on X_f with $\tilde{p}\leqslant p$. (Note that if f is positive then $\tilde{p}$ is the inner product norm on X_f, by Example 20.) Then $a\to\pi_f(a)$ is a star representation on X_f, and, by Lemma 21,

$$\|\pi_f(a)\|_{\tilde{p}}\leqslant\|a\| \qquad (a\in A)\,.$$

Let E_f be the completion of X_f with respect to $\tilde{p}$. Then $\langle,\rangle_f$ extends to a form on E_f, so that E_f is a balanced Banach self-dual space. Moreover $a\to\pi_f(a)$ has an extension by continuity to a star representation on E_f. Let $E=\sum_{f\in\Lambda}^{\oplus}E_f$, so that, in the obvious manner, E becomes a balanced Banach self-dual space. Finally let

$$\Psi(a)(x)(f)=\pi_f(a)x(f) \qquad (a\in A,\ x\in E,\ f\in\Lambda)\,.$$

Then Ψ is a norm-decreasing star representation of A on E which is an isometry on $\mathrm{Sym}(A)$. Finally we give a condition which makes Ψ an isometry.

Definition 22. A *D^*-algebra* is a Banach star-normed algebra such that

$$2\|a\|=\sup\{\|x^*a+a^*x\|: x\in S(A)\} \qquad (a\in A)\,.$$

---

[1] Given a linear-norm q on X, $|T|_q$ denotes the corresponding operator norm and $\|T\|_q=\max\{|T|_q,|T^*|_q\}$.

Theorem 23. *A D*-algebra A is isometrically star isomorphic to a closed self-adjoint subalgebra of* $BL(E, \langle , \rangle)$ *for some balanced Banach space E.*

Proof. It is sufficient to prove that the representation Ψ above is norm increasing. Since Ψ is isometric on $\mathrm{Sym}(A)$ we have, for $a \in A$,

$$\|a\| = \tfrac{1}{2} \sup \{ \|x^* a + a^* x\| : \|x\| \leqslant 1 \}$$
$$= \tfrac{1}{2} \sup \{ \|\Psi(x^* a + a^* x)\| : \|x\| \leqslant 1 \}$$
$$\leqslant \tfrac{1}{2} (\|\Psi(a)\| + \|\Psi(a^*)\|)$$
$$= \|\Psi(a)\| . \quad \square$$

Lance [252] shows that a closed self-adjoint unital subalgebra A of $BL(E, \langle \, \rangle)$ is D^* if $A \supset F\,BL(E, \langle , \rangle)$.

§ 38. Characterizations of C*-Algebras

A C^*-algebra is a closed self-adjoint subalgebra of $BL(H)$ for some Hilbert space H. A B^*-algebra is a Banach star algebra A such that

$$\|a^* a\| = \|a\|^2 \quad (a \in A).$$

We observed in § 12 that a C^*-algebra is a B^*-algebra. We prove here the Gelfand-Naimark theorem, which states that a B^*-algebra is isometrically star isomorphic to some C^*-algebra.

Let A be a complex unital Banach algebra. An element $h \in A$ is Hermitian (Definition 10.12) if it has real numerical range, i.e. if $f(h) \in \mathbb{R}$ whenever $f \in A'$, $f(1) = 1 = \|f\|$. We shall make significant use of the fact (Corollary 10.13) that h is Hermitian if and only if

$$\|\exp(ith)\| = 1 \quad (t \in \mathbb{R}).$$

We denote the set of all Hermitian elements of A by $\mathrm{Her}(A)$.

Definition 1. A *V-algebra* is a complex unital Banach algebra A such that $A = \mathrm{Her}(A) + i\,\mathrm{Her}(A)$, i.e. each $a \in A$ is of the form $a = h + ik$ with $h, k \in \mathrm{Her}(A)$.

If A is a unital B^*-algebra, then $\mathrm{Her}(A) = \mathrm{Sym}(A)$, by Proposition 12.20, and so A is a V-algebra. We also prove here the Vidav-Palmer theorem which states that a V-algebra, with involution defined by $(h + ik)^* = h - ik$, is a B^*-algebra.

Notation. In Lemmas 2, 3, 4 below, A denotes a complex unital Banach algebra. Given $E \subset \mathbb{C}$, $\mathrm{co}\,E$ denotes the convex hull of E, i.e. the intersection of all convex subsets of $\mathbb{C}$ that contain E.

Lemma 2. Her(A) *is a real Banach space and* $i(hk-kh) \in$ Her(A) *whenever* $h, k \in$ Her(A).

Proof. It is clear from the definition that Her(A) is a closed real linear subspace of A. Let $h, k \in$ Her(A), let $t \in \mathbb{R}$, and let

$$w = \exp(ith)\exp(itk)\exp(-ith)\exp(-itk).$$

Then $\|w\| \leqslant 1$ and $\|w^{-1}\| \leqslant 1$. Therefore $\|w\| = 1$, and so

$$\|1 - t^2(hk-kh)\| = 1 + 0(t^3) \quad (t \to 0).$$

Interchanging h and k gives

$$\|1 + t^2(hk-kh)\| = 1 + 0(t^3) \quad (t \to 0),$$

and so

$$\|1 + t(hk-kh)\| = 1 + o(t) \quad (t \to 0).$$

Therefore

$$\lim_{t \to 0} \frac{1}{t} \{\|1 + itx\| - 1\} = 0,$$

where $x = i(hk-kh)$. Apply Theorem 10.10. □

Lemma 3. $V(h) = \mathrm{co}\, \mathrm{Sp}(h)$ $(h \in \mathrm{Her}(A))$.

Proof. By Propositions 10.4 and 10.6, $\mathrm{co}\, \mathrm{Sp}(h) \subset V(h) \subset \mathbb{R}$ and, by Theorem 10.17, $r(h - t) = v(h - t)$ $(t \in \mathbb{R})$, where

$$v(x) = \max\{|\lambda| : \lambda \in V(x)\}. \quad \square$$

Lemma 4. *Let* $a = h + ik$; *where* $h, k \in \mathrm{Her}(A)$, $hk = kh$. *Then* $V(a) = \mathrm{co}\, \mathrm{Sp}(a)$.

Proof. Since the conditions of the lemma hold for a replaced by $\lambda + \mu a$ $(\lambda, \mu \in \mathbb{C})$, since $\mathrm{co}\, \mathrm{Sp}(a)$ is the intersection of the closed half-planes that contain $\mathrm{Sp}(a)$, and since $V(a)$ is compact convex, it is enough to prove that $\max \mathrm{Re}\, V(a) = \max \mathrm{Re}\, \mathrm{Sp}(a)$.

Let B be a maximal commutative subalgebra of A that contains h and k. By Theorem 15.4

$$\mathrm{Sp}(A, a) = \mathrm{Sp}(B, a) = \{\phi(h) + i\phi(k) : \phi \in \Phi_B\}.$$

Since $h, k \in \mathrm{Her}(A)$,

$$\max \mathrm{Re}\, \mathrm{Sp}(A, a) = \max \mathrm{Sp}(B, h) = \max \mathrm{Sp}(A, h),$$

$$\max \mathrm{Re}\, V(a) = \max V(h).$$

Apply Lemma 3. □

Definition 5. An element k of a complex unital Banach algebra A is *positive* if $V(k) \subset \mathbb{R}^+$. We denote by $\mathrm{pos}(A)$ the set of all positive elements of A. By Lemma 3 we have $k \in \mathrm{pos}(A)$ if and only if $k \in \mathrm{Her}(A)$ and $\mathrm{Sp}(k) \subset \mathbb{R}^+$. Also, $\mathrm{pos}(A)$ is a cone in $\mathrm{Her}(A)$.

Notation. In Lemmas 6, 7, 8, 9 below, A denotes a V-algebra.

Lemma 6. *The mapping* $(h + ik) \to (h - ik)$ *defines a continuous algebra involution on* A.

Proof. Let $h, k \in \mathrm{Her}(A)$, $h + ik = 0$. Then

$$V(h) = -i V(k) \subset \mathbb{R} \cap i\mathbb{R} = \{0\} ,$$

and Theorem 10.14 gives $h = k = 0$. Thus each $a \in A$ has a unique representation of the form $h + ik$ with $h, k \in \mathrm{Her}(A)$. It is now elementary that * is a linear involution on A, where $(h + ik)^* = h - ik$.

Given $h \in \mathrm{Her}(A)$ let $h^2 = p + iq$ where $p, q \in \mathrm{Her}(A)$. Then $h(p + iq) = (p + iq)h$, $hp - ph = i(qh - hq) = c$, say. By Lemma 2, $c, ic \in \mathrm{Her}(A)$, $c = 0$. Therefore $hp = ph$, $h^2 p = ph^2$, $pq = qp$ and Lemma 4 gives $V(h^2) = \mathrm{co} \, \mathrm{Sp}(h^2) \subset \mathbb{R}$, $h^2 \in \mathrm{Her}(A)$. Given $h, k \in \mathrm{Her}(A)$ we now have $h^2, k^2, (h + k)^2 \in \mathrm{Her}(A)$ and so $hk + kh \in \mathrm{Her}(A)$. This, with Lemma 2 and Lemma 12.7, shows that * is an algebra involution on A.

To prove the continuity of *, let $h, k \in \mathrm{Her}(A)$. Then by Theorem 10.17,

$$\|h\| = \max \{|f(h)| : f(1) = 1 = \|f\|\}$$
$$\leqslant \max \{|f(h) + i f(k)| : f(1) = 1 = \|f\|\}$$
$$\leqslant \|h + ik\| ,$$

and similarly $\|k\| \leqslant \|h + ik\|$. Therefore

$$\|(h + ik)^*\| = \|h - ik\| \leqslant \|h\| + \|k\| \leqslant 2 \|h + ik\| . \quad \square$$

Lemma 7. *Let* $k \in \mathrm{pos}(A)$ *and let* C *denote the least closed subalgebra of* A *containing* 1 *and* k. *Then there exists* $u \in C \cap \mathrm{pos}(A)$ *such that* $u^2 = k$.

Proof. We may suppose without loss that $r(k) \leqslant 1$. Let $a = 1 - k$ so that $r(a) \leqslant 1$, and let P denote the set of polynomials in a with co-efficients in $\mathbb{R}^+$. Let $x_0 = 0$ and let

$$x_n = \tfrac{1}{2}(a + x_{n-1}^2) \qquad (n \in \mathbb{N}) .$$

Since $a x_n = x_n a$, an induction argument gives $r(x_n) \leqslant 1$ $(n \in \mathbb{N})$. Also, by induction, $x_n \in P$, and

$$x_{n+1} - x_n = \tfrac{1}{2}(x_n + x_{n-1})(x_n - x_{n-1}) \in P \qquad (n \in \mathbb{N}) .$$

Since $\mathrm{Sp}(C,a) \subset [0,1]$, $\{x_n^\wedge\}$ is an increasing sequence of non-negative functions on Φ_C bounded above by 1 and so pointwise convergent to a function w with $0 \leqslant w \leqslant 1$ on Φ_C. Since $w = \frac{1}{2}(a^\wedge + w^2)$ and $a^\wedge$ is continuous, w is continuous, and so by Dini's theorem, $x_n^\wedge \to w$ uniformly on Φ_C. Therefore $\{x_n\}$ is a Cauchy sequence with respect to r and so, by Theorem 10.17, with respect to $\|\cdot\|$. Therefore $x_n \to v \in \mathrm{Her}(A)$ where $v^\wedge = w$, and $(1-v^\wedge)^2 = 1 - a^\wedge = k^\wedge$. Let $v = 1 - u$. Then $(u^2 - k)^\wedge = 0$, and, since $u^2 - k \in \mathrm{Her}(A)$, $\|u^2 - k\| = r(u^2 - k) = 0$, $u^2 = k$. Finally

$$\mathrm{Sp}(A,u) \subset \mathrm{Sp}(C,u) \subset [0,1],$$

and so $u \in \mathrm{pos}(A)$ by Lemma 3. $\square$

Note that Lemma 7 requires a sharper argument than that employed in Proposition 8.13. There is an alternative proof of Lemma 7 which uses the fact that the power series for $(1-t)^{\frac{1}{2}}$ converges absolutely for $t = 1$.

Lemma 8. *Given* $h \in \mathrm{Her}(A)$ *there exist* $p,q \in \mathrm{pos}(A)$ *such that*

$$h = p - q, \quad pq = qp = 0.$$

Proof. Let C be a maximal commutative subalgebra of A containing h. By Lemma 7 there exists $u \in \mathrm{pos}(A) \cap C$ such that $u^2 = h^2$. Then

$$\mathrm{Sp}(A, u \pm h) = \mathrm{Sp}(C, u \pm h) = \{\phi(u) \pm \phi(h) : \phi \in \Phi_C\}.$$

For $\phi \in \Phi_C$, we have

$$\phi(h)^2 = \phi(h^2) = \phi(u^2) = \phi(u)^2, \quad \phi(h) = \pm \phi(u).$$

Since $u \in \mathrm{pos}(A)$, it follows that $\mathrm{Sp}(A, u \pm h) \subset \mathbb{R}^+$, and so $u \pm h \in \mathrm{pos}(A)$, by Lemma 3. Let $p = \frac{1}{2}(u+h)$, $q = \frac{1}{2}(u-h)$. $\square$

Lemma 9. $a^* a \in \mathrm{pos}(A)$ $(a \in A)$.

Proof. Let $b \in A$ and let $-b^* b \in \mathrm{pos}(A)$. Since

$$\mathrm{Sp}(A, b^* b) \setminus \{0\} = \mathrm{Sp}(A, bb^*) \setminus \{0\},$$

it follows from Lemma 3 that $-bb^* \in \mathrm{pos}(A)$. Let $b = h + ik$ where $h, k \in \mathrm{Her}(A)$. Then $b^* b = 2h^2 + 2k^2 + (-bb^*) \in \mathrm{pos}(A)$, since $\mathrm{pos}(A)$ is closed under addition. Therefore $V(b^* b) = \{0\}$, and so $b^* b = 0$, by Theorem 10.14. Similarly $bb^* = 0$ and hence $h^2 + k^2 = 0$. This gives $V(h^2) = V(k^2) = \{0\}$, $h = k = 0$, $b = 0$.

Given $a \in A$ we have $a^* a \in \mathrm{Her}(A)$ and so by Lemma 8 there exist $p, q \in \mathrm{pos}(A)$ such that $a^* a = p - q$, $pq = qp = 0$. Let $b = aq$. Then $-b^* b = -q a^* a q = q^3 \in \mathrm{pos}(A)$, $b = 0$. This gives in turn, $a^* aq = 0$, $q^2 = 0$, $q = 0$, $a^* a = p \in \mathrm{pos}(A)$. $\square$

We are now ready to prove the Gelfand-Naimark theorem. A key step in the proof is Lemma 9, which depends on Lemmas 7 and 8. In fact, for B^*-algebras Lemma 7 is immediate from the commutative Gelfand-Naimark theorem (Theorem 35.4).

Theorem 10. *Let A be a B^*-algebra. Then there exists a Hilbert space H and an isometric star monomorphism of A into $BL(H)$.*

Proof. We may assume, by Lemma 12.19, that A is unital. Let $D(1)=\{f\in A': f(1)=1=\|f\|\}$. Given $a\in A$, $f\in D(1)$, we have $f(a^*a)\geqslant 0$, by Lemma 9, and so f is positive. Let H_f be the associated Hilbert A-module as in Proposition 37.7, and let H be the l^2 direct sum module

$$H = \sum_{f\in D(1)}^{\oplus} H_f,$$

with associated star representation π. Since

$$|\pi_f(a)| \leqslant (r(a^*a))^{\frac{1}{2}} \leqslant \|a^*a\|^{\frac{1}{2}} = \|a\| \qquad (a\in A),$$

it follows that $|\pi(a)|\leqslant\|a\|$ $(a\in A)$. On the other hand, if u is the coset of 1 in H_f, then $|\pi_f(a)|^2 \geqslant \|\pi_f(a)u\|^2 = f(a^*a)$, and hence

$$|\pi(a)|^2 \geqslant \sup\{f(a^*a): f\in D(1)\}$$
$$= \|a^*a\| = \|a\|^2,$$

by Theorem 10.17. □

Remark. The original Gelfand-Naimark theorem [152] assumed that A is a unital B^*-algebra for which Lemma 9 holds. The proof of the key Lemma 9 was first given by Fukamiya [140] and Kaplansky, and the extension to the non-unital case is due to Yood.

The argument in Theorem 10 would apply to V-algebras except for the very last equation. To complete this step we need some technical results on B^*-algebras. Theorem 12 is due to Russo and Dye [326] and Theorem 13 to Palmer [299]. The proofs we give are due to Harris [184], and are based on the generalized Möbius transformation introduced by Potapov [306].

Notation. In Lemma 11, Theorems 12, 13 below, A denotes a unital B^*-algebra and Γ denotes the unit circle in $\mathbb{C}$. We write

$$U = \{u\in A: uu^* = u^*u = 1\},$$
$$E = \{\exp(ih): h\in \text{Sym}(A)\}.$$

Let $x\in A$, $0<\|x\|<1$. Then $r(x^*)=r(x)<1$, and

$$r(xx^*)=r(x^*x)=\|x\|^2<1.$$

We may therefore define F by

$$F(\lambda)=(1-xx^*)^{-\frac{1}{2}}(\lambda+x)(1+\lambda x^*)^{-1}(1-x^*x)^{\frac{1}{2}} \qquad (|\lambda|<\|x\|^{-1}). \quad (1)$$

Lemma 11. *Given* $x \in A$, $\|x\| < 1$, *the mapping F defined by* (1) *is holomorphic, and* $F(\lambda) \in U$ $(\lambda \in \Gamma)$.

Proof. It is routine that F is holomorphic on the given open disc in $\mathbb{C}$. Let $\lambda \in \Gamma$. The following identities are elementary:

$$(1 + \lambda x^*)^{-1}(\lambda + x) = x + \lambda(1 + \lambda x^*)^{-1}(1 - x^* x), \qquad (2)$$

$$(\lambda + x)(1 + \lambda x^*)^{-1} = x + \lambda(1 - xx^*)(1 + \lambda x^*)^{-1}. \qquad (3)$$

Since $r(x) < 1$, $F(\lambda)$ is invertible, and, by (2), (3),

$$\begin{aligned}
(F(\lambda)^{-1})^* &= (1 - xx^*)^{\frac{1}{2}}(\lambda^* + x^*)^{-1}(1 + \lambda^* x)(1 - x^* x)^{-\frac{1}{2}} \\
&= (1 - xx^*)^{\frac{1}{2}}(1 + \lambda x^*)^{-1}(\lambda + x)(1 - x^* x)^{-\frac{1}{2}} \\
&= (1 - xx^*)^{-\frac{1}{2}} a (1 - x^* x)^{\frac{1}{2}},
\end{aligned}$$

where

$$\begin{aligned}
a &= (1 - xx^*)(1 + \lambda x^*)^{-1}(\lambda + x)(1 - x^* x)^{-1} \\
&= (1 - xx^*)\{x + \lambda(1 + \lambda x^*)^{-1}(1 - x^* x)\}(1 - x^* x)^{-1} \\
&= x + \lambda(1 - xx^*)(1 + \lambda x^*)^{-1} \\
&= (\lambda + x)(1 + \lambda x^*)^{-1}.
\end{aligned}$$

This shows that $F(\lambda)$ is unitary for each $\lambda \in \Gamma$. ☐

Theorem 12. *The closed unit ball of A is the closed convex hull of U.*

Proof. Let $x \in A$, $\|x\| < 1$, and let F be given by (1). Since F is holomorphic we have, by Proposition 6.7,

$$F(0) = \frac{1}{2\pi} \int\limits_0^{2\pi} F(e^{i\theta}) d\theta.$$

Consideration of the power series expansion of $(1 - y)^{\frac{1}{2}}$ for $\|y\| < 1$ gives

$$(1 - xx^*)^{\frac{1}{2}} x = x(1 - x^* x)^{\frac{1}{2}}.$$

Therefore $x = F(0) \in \overline{\text{co}}\, U$, by Lemma 11. ☐

Theorem 13. *The closed unit ball of A is the closed convex hull of E.*

Proof. By Theorem 12, it is enough to show that $U \subset \overline{\text{co}}\, E$. Given $u \in U$, $0 < t < 1$, let $x = tu$. Let $\lambda \in \Gamma$. Since $xx^* = x^* x$, we have

$$\begin{aligned}
(\lambda + F(\lambda))(1 + \lambda x^*) &= \{\lambda + (\lambda + x)(1 + \lambda x^*)^{-1}\}(1 + \lambda x^*) \\
&= \lambda + \lambda^2 x^* + \lambda + x \\
&= 2\lambda\{1 + \tfrac{1}{2}(\lambda x^* + \lambda^* x)\}.
\end{aligned}$$

Since $\|x\| < 1$, it follows that $\lambda + F(\lambda)$ is invertible, $-\lambda \notin \mathrm{Sp}(F(\lambda))$. Therefore $\mathrm{Sp}(F(\lambda))$ is a proper compact subset of Γ and so we may define $h = -i \log F(\lambda)$. Since $F(\lambda) \in U$, the functional calculus gives

$$(\log F(\lambda))^* = \log(F(\lambda))^* = \log F(\lambda)^{-1} = -\log F(\lambda).$$

Therefore $h^* = h$, $F(\lambda) = \exp(ih) \in E$. It follows as in Theorem 12 that $x \in \overline{\mathrm{co}}\, E$. Therefore $U \subset \overline{\mathrm{co}}\, E$. □

Remarks. (1) A more elementary proof of Theorem 12 may be obtained via the standard approximation to the integral using n^{th} roots of unity.

(2) Palmer [300] has sharpened Theorem 13 to the inclusion

$$\{x : \|x\| < 1\} \subset \mathrm{co}\, E.$$

Theorem 14. *Let A be a V-algebra. Then A is a B*-algebra with respect to the involution defined by*

$$(h + ik)^* = h - ik \qquad (h, k \in \mathrm{Her}(A)).$$

Proof. We have established in Lemma 6 that * is an algebra involution on A with $\|a^*\| \leqslant 2\|a\|$ $(a \in A)$. Let H be the Hilbert A-module defined in the proof of Theorem 10. The argument in Theorem 10 gives, in this case

$$|\pi(a)|^2 \leqslant \|a^* a\| \leqslant 2\|a\|^2, \qquad |\pi(a)|^2 \geqslant \|a^* a\| \qquad (a \in A).$$

Since $f(a) \geqslant 0$ $(f \in D(1), a \in \mathrm{pos}(A))$ we have, by Lemma 37.6(ii),

$$|f(a)|^2 \leqslant f(a^* a) \qquad (a \in A),$$

and hence $\max\{|\lambda| : \lambda \in V(a)\} \leqslant \|a^* a\|^{\frac{1}{2}}$. Theorem 10.14 now gives

$$\|a\| \leqslant e|\pi(a)| \qquad (a \in A)$$

so that π is a homeomorphism. It also follows from Theorem 10.17 that $|\pi(h)| = \|h\|$ $(h \in \mathrm{Sym}(A))$.

We have now shown that A is a B*-algebra with norm defined by

$$|a| = |\pi(a)| \qquad (a \in A).$$

Given $a \in A$ with $|a| = 1$, by Lemma 13 there exists a sequence $\{b_n\}$ in $\mathrm{co}\, E$ such that $|b_n - a| \to 0$. Therefore $\|b_n - a\| \to 0$ and so $\|b_n\| \to \|a\|$. Since $\|\exp(ih)\| = 1$ $(h \in \mathrm{Her}(A))$ we have $\|b_n\| \leqslant 1$ $(n \in \mathbb{N})$ and so $\|a\| \leqslant 1$. This proves that

$$\|a\| \leqslant |a| \qquad (a \in A).$$

If for some $x \in A$ we have $\|x\| < |x|$, then

$$\|x^* x\| \leqslant \|x^*\|\, \|x\| < |x^*|\, |x| = |x^* x| = \|x^* x\|.$$

This contradiction completes the proof. □

Remarks. (1) Vidav [398] proved that a *V*-algebra, with an extra condition, is homeomorphic and star isomorphic to a *C**-algebra. Berkson [48] and Glickfeld [160] independently established the isometry. Palmer [299] removed the extra condition and gave a simpler proof of the isometry. For this reason we have called the theorem the "Vidav-Palmer theorem". Burckel [92] gives a simple proof of the commutative case.

(2) The Vidav-Palmer theorem gives a geometrical characterization of *B**-algebras. To be precise, if *A* is a complex unital Banach algebra then the existence of an involution on *A* for which *A* is a *B**-algebra depends only on the Banach space structure of *A*, in fact on the shape of the unit ball near 1.

We give below some simple applications of the Vidav-Palmer theorem; further results are given by Moore [282] and the authors [77].

Theorem 15. *Let A be a unital Banach star algebra such that*

$$\|a^* a\| = \|a^*\| \, \|a\| \qquad (a \in A).$$

Then A is a B*-algebra.

Proof. By the Vidav-Palmer theorem it is sufficient to show that $\operatorname{Sym}(A) \subset \operatorname{Her}(A)$. Let $h \in \operatorname{Sym}(A)$, $\mu = \max \operatorname{Re} V(ih)$, $\lambda = \min \operatorname{Re} V(ih)$. For $t > 0$, Theorem 10.10 gives

$$\|1 + ith\| = 1 + t\mu + o(t), \qquad \|1 - ith\| = 1 - t\lambda + o(t) \qquad (t \to 0).$$

Therefore

$$
\begin{aligned}
1 + t(\mu - \lambda) + o(t) &= \|1 + ith\| \, \|(1 + ith)^*\| \\
&= \|1 + t^2 h^2\| \\
&= 1 + O(t^2) \qquad (t \to 0).
\end{aligned}
$$

Thus $\lambda = \mu$, $V(ih) \subset \lambda + i\mathbb{R}$, $V(h + i\lambda) \subset \mathbb{R}$, and so $h = k - i\lambda$ for some $k \in \operatorname{Her}(A)$. Since $\operatorname{Sp}(h^*) = \operatorname{Sp}(h)^*$ and since $\operatorname{Sp}(k) \subset \mathbb{R}$, it follows that $\lambda = 0$, $h \in \operatorname{Her}(A)$. ☐

Remark. Theorem 15 is due to Glimm and Kadison [166]. Vowden [401] has shown that the result holds without the assumption that *A* be unital.

Theorem 16. *Let A be a unital Banach star algebra such that*
   (i) $\operatorname{Sp}(h) \subset \mathbb{R}$ $(h \in \operatorname{Sym}(A))$,
   (ii) $r(1 + \lambda h) = \|1 + \lambda h\|$ $(h \in \operatorname{Sym}(A), \lambda \in \mathbb{C})$.
Then A is a B*-algebra.

Proof. Let $h \in \operatorname{Sym}(A)$. Conditions (i) and (ii) give

$$V(h) = \operatorname{co} \operatorname{Sp}(h) \subset \mathbb{R}. \qquad ☐$$

Theorem 17. *Let A be a unital Banach star algebra such that* $D(1) \subset \operatorname{Sym}(A')$. *Then A is a B*-algebra.*

Proof. Let $h \in \operatorname{Sym}(A)$. Then

$$V(h) = \{ f(h) : f \in D(1) \} \subset \{ f(h) : f \in \operatorname{Sym}(A') \} \subset \mathbb{R}. \quad \Box$$

Remark. Moore [282] has given a deeper 'dual' characterization of B*-algebras as follows. Let A be a complex unital Banach algebra and let $\operatorname{Her}(A')$ denote the real linear span of $D(1)$. Then A admits an involution with respect to which it is a B*-algebra if and only if $\operatorname{Her}(A') \cap i\operatorname{Her}(A') = \{0\}$.

Theorem 18. (Segal [338], Kaplansky [233]) *Let A be a B*-algebra and let J be a closed bi-ideal of A. Then* $J^* = J$ *and* A/J *is a B*-algebra.*

Proof. Let A be unital. We write $[a]$ for the canonical image of a in A/J. Given $\phi \in D([1])$ let

$$\tilde{\phi}(a) = \phi([a]) \quad (a \in A).$$

Then $\tilde{\phi} \in D(1)$. Given $h \in \operatorname{Sym}(A)$, $\phi([h]) = \tilde{\phi}(h) \in \mathbb{R}$. Since

$$A/J = \{ [h] + i[k] : h, k \in \operatorname{Her}(A) \},$$

the Vidav-Palmer theorem shows that A/J is a B*-algebra with the canonical norm and involution

$$([h] + i[k])^* = [h] - i[k].$$

Then $a \to [a]$ is a star homomorphism, $J^* = J$. The non-unital case is a routine application of Lemma 12.19. $\quad \Box$

Let A be a Banach star algebra with continuous involution. There are natural continuous linear involutions on A' and A'' defined by

$$f^*(a) = \overline{f(a^*)} \quad (a \in A, \ f \in A'),$$
$$F^*(f) = \overline{F(f^*)} \quad (f \in A', \ F \in A'').$$

Theorem 19. *Let A be a B*-algebra. Then the second dual* A'' *with the Arens product and the natural involution* $F \to F^*$ *is a B*-algebra.*

Proof. As in Theorem 18 it is enough to assume that A is unital. Let $F \in \operatorname{Sym}(A'')$, $f \in D(1)$. Then $f^* = f$ and

$$F(f) = F^*(f) = \overline{F(f^*)} = \overline{F(f)}.$$

Thus $F(f) \in \mathbb{R}$ $(f \in D(1))$ and it follows from [77, Corollary 12.3] that $F \in \operatorname{Her}(A'')$. Apply the Vidav-Palmer theorem. $\quad \Box$

§ 39. *B*-Semi-Norms*

Let A be a Banach star algebra.

Definition 1. A *B*-semi-norm* on A is an algebra-semi-norm p on A such that

$$p(a^* a) = (p(a))^2 \quad (a \in A).$$

We denote by $\mathrm{bs}(A)$ the set of all B^*-semi-norms on A, and by $\leqslant$ the pointwise ordering on $\mathrm{bs}(A)$. An *A*-algebra* is a Banach star algebra having a B^*-semi-norm which is a norm (called an *auxiliary norm*).

We show that $\mathrm{bs}(A)$ has a greatest member and is compact in the topology of pointwise convergence. This latter fact makes available for the study of B^*-semi-norms the standard analytic tools. We relate B^*-semi-norms on A and positive linear functionals on A. In particular, each B^*-semi-norm on A has an associated family of positive linear functionals on A which is a weak* compact convex subset of A' and therefore has a rich supply of extreme points. The significance of these extreme points for representation theory is discussed in § 40. B^*-semi-norms were introduced in the context of B^*-algebras by Fell [135] and Effros [131].

Notation. Given $p \in \mathrm{bs}(A)$, we denote by J_p the ideal defined by

$$J_p = \{a \in A : p(a) = 0\},$$

and by $\tilde{p}$ the algebra-norm defined on the algebra A/J_p by

$$\tilde{p}(b) = p(a) \quad (a \in b \in A/J_p).$$

We denote by B_p the completion of A/J_p with respect to the norm $\tilde{p}$ and we denote also by $\tilde{p}$ the usual extension of this norm to B_p.

Lemma 2. *Let* $p \in \mathrm{bs}(A)$. *Then*
(i) $p(a^*) = p(a)$ $(a \in A)$,
(ii) $p(a) \leqslant (r(a^* a))^{\frac{1}{2}}$ $(a \in A)$.

Proof. (i) $(p(a))^2 = p(a^* a) \leqslant p(a^*) p(a)$.
(ii) Apply the argument of Lemma 37.2 to the factor algebra A/J_p. □

Theorem 3. *Each B*-semi-norm on A is continuous.*

Proof. Let $p \in \mathrm{bs}(A)$, let $B = A/\mathrm{rad}(A)$, and let $[a]$ denote the canonical image in B of $a \in A$. By Proposition 25.1 and Lemma 12.10, $\mathrm{rad}(A)$ is a star ideal of A and therefore B is a semi-simple Banach star algebra with the natural involution given by

$$b^* = [a^*] \quad (a \in b \in B).$$

By Theorem 36.2 there is a positive constant κ such that

$$\|b^*\| \leqslant \kappa \|b\| \qquad (b \in B).$$

Given $x \in \mathrm{rad}(A)$ we have $r(x^* x) = 0$ and so, by Lemma 2(ii), $p(x) = 0$. Thus a B^*-semi-norm q is defined on B by

$$q(b) = p(a) \qquad (a \in b \in B).$$

Then

$$(q(b))^2 \leqslant r(b^* b) \leqslant \|b^*\| \, \|b\| \leqslant \kappa \|b\|^2 \qquad (b \in B),$$
$$p(a) = q([a]) \leqslant \kappa^{\frac{1}{2}} \|[a]\| \leqslant \kappa^{\frac{1}{2}} \|a\| \qquad (a \in A). \quad \square$$

Corollary 4. *Let* $p \in \mathrm{bs}(A)$. *Then* A/J_p *is an* A^*-*algebra with* $\tilde{p}$ *as its auxiliary norm, and* B_p *with the norm* $\tilde{p}$ *is a* B^*-*algebra.*

Proof. Since p is continuous and $p(a^*) = p(a)$, J_p is a closed star ideal of A, and so A/J_p with the canonical norm and the involution induced from A is a Banach star algebra. The rest is clear. $\quad \square$

Theorem 5. $\mathrm{bs}(A)$ *is compact in the topology of pointwise convergence, and* m, *defined on* A *by*

$$m(a) = \sup \{p(a) : p \in \mathrm{bs}(A)\},$$

is a B^*-*semi-norm on* A.

Proof. By Lemma 2(ii) and Tychonoff's theorem, $\mathrm{bs}(A)$ is compact in the topology of pointwise convergence, and also $m(a)$ is finite. That $m \in \mathrm{bs}(A)$ is now clear. $\quad \square$

Visibly m is the greatest element of $\mathrm{bs}(A)$ in the pointwise ordering.

Definition 6. A B^*-semi-norm p on A is *neat* if p belongs to every subset Q of $\mathrm{bs}(A)$ such that

$$p(a) = \max \{q(a) : q \in Q\} \qquad (a \in A).$$

We denote by $\mathrm{nbs}(A)$ the set of all neat B^*-semi-norms on A.

The following alternative formulation will also be convenient. We say that an element b of A *peaks* at a B^*-semi-norm p if $q = p$ whenever q is an element of $\mathrm{bs}(A)$ satisfying:

$$q \leqslant p, \qquad q(b) = p(b).$$

Lemma 7. *A* B^*-*semi-norm* p *is neat if and only if there exists an element of* A *that peaks at* p.

Proof. Suppose that b peaks at p, and that for some subset Q of $\mathrm{bs}(A)$,

$$p(a) = \max \{q(a) : q \in Q\} \qquad (a \in A).$$

Then there exists $q \in Q$ with $q(b) = p(b)$. Since also $q \leqslant p$, it follows that $p = q \in Q$, and so p is neat.

Suppose now that no element of A peaks at p. Given $a \in A$, let $N(a)$ denote the set of all $q \in \mathrm{bs}(A)$ such that

$$q \leqslant p, \qquad q(a) = p(a), \qquad q \neq p.$$

Since a does not peak at p, $N(a)$ is non-void. Let $Q = \bigcup \{N(a) : a \in A\}$. Then

$$p(a) = \max \{q(a) : q \in Q\} \qquad (a \in A),$$

but $p \notin Q$, and so p is not neat.   □

Theorem 8. *Let* $p \in \mathrm{bs}(A)$. *Then*

$$p(a) = \max \{q(a) : q \in \mathrm{nbs}(A), q \leqslant p\} \qquad (a \in A).$$

In particular,

$$m(a) = \max \{q(a) : q \in \mathrm{nbs}(A)\} \qquad (a \in A).$$

Proof. Given $a \in A$, let

$$Q = \{q \in \mathrm{bs}(A) : q \leqslant p, q(a) = p(a)\}.$$

It is sufficient to show that $Q \cap \mathrm{nbs}(A) \neq \emptyset$. Q is a pointwise closed, and therefore compact, non-void subset of $\mathrm{bs}(A)$, by Theorem 5; and so there exists a minimal element μ of Q with respect to the pointwise ordering. We have $\mu(a) = p(a)$; and so, whenever $q \leqslant \mu$ and $q(a) = \mu(a)$, we have $q \in Q$, and therefore $q = \mu$ by the minimality of μ. Therefore a peaks at μ, and so μ is neat, as required.   □

Example 9. Let f be a positive linear functional on A, let π_f be the associated star representation, and let

$$p_f(a) = |\pi_f(a)| \qquad (a \in A).$$

Then $p_f \in \mathrm{bs}(A)$.

For the remainder of this section we discuss the construction of positive linear functionals dominated by a given B^*-semi-norm.

Lemma 10. *Let* f *be a positive linear functional on* A, *let* $p \in \mathrm{bs}(A)$, *and let* $|f(a)| \leqslant \kappa p(a)$ $(a \in A)$ *for some positive constant* κ. *Then* $p_f \leqslant p$.

Proof. Since f vanishes on J_p we may define a linear functional $\tilde{f}$ on A/J_p by

$$\tilde{f}(b) = f(a) \qquad (a \in b \in A/J_p).$$

Also, $|\tilde{f}(b)| \leqslant \kappa \tilde{p}(b)$ $(b \in A/J_p)$, and so $\tilde{f}$ has a unique continuous extension to a positive linear functional on the B^*-algebra B_p, which we also denote by $\tilde{f}$. Lemma 37.6 (iii) gives

$$0 \leqslant \tilde{f}(y^* b^* b y) \leqslant r(b^* b) \tilde{f}(y^* y) = \tilde{p}(b^* b) \tilde{f}(y^* y) \qquad (b, y \in B_p).$$

Therefore

$$0 \leqslant f(x^* a^* a x) \leqslant p(a^* a) f(x^* x) \qquad (a, x \in A),$$

which proves that $p_f \leqslant p$. □

Lemma 11. *Let* $p \in \mathrm{bs}(A)$ *and let* $b \in A$. *Then there exists a positive linear functional f on A such that*

(i) $|f(a)| \leqslant p(a)$ $(a \in A)$, (ii) $f(b^* b) = p(b^* b)$, (iii) $f(1) = 1$ *if A has a unit.*

Proof. We note first that it is enough to prove the lemma in the special case when A is a B^*-algebra. For, given a positive linear functional g on B_p such that

$$|g(b)| \leqslant \tilde{p}(b) \quad (b \in B_p), \qquad g([b]^* [b]) = \tilde{p}([b]^* [b]),$$

the linear functional f defined on A by

$$f(a) = g([a]) \qquad (a \in A)$$

has the required properties.

Suppose now that A is a B^*-algebra. We may suppose, by Lemma 12.19, that A is unital. Let $c = b^* b$, and let C be the least closed subalgebra of A containing 1 and c. By Lemma 38.9, $\mathrm{Sp}(c) \subset \mathbb{R}^+$. Therefore $r(c) \in \mathrm{Sp}(C, c)$ and there exists $\phi \in \Phi_C$ with $\phi(c) = r(c) = \|c\| = \|b^* b\|$. We have $\|\phi\| = \phi(1) = 1$, and so, by the Hahn-Banach theorem, ϕ can be extended to a linear functional ψ on A such that $\|\psi\| = \psi(1) = 1$. We have $\psi(b^* b) = \|b^* b\|$, and, as in the proof of Theorem 38.10, ψ is positive. □

Corollary 12. *Let A have a unit. Then*

$$m(a)^2 = \sup \{ f(a^* a): f \text{ a positive linear functional on } A, \ f(1) = 1 \}.$$

Proof. Given $p \in \mathrm{bs}(A)$, $b \in A$, let f be as in Lemma 11. Note that $f(1) = 1$ and hence

$$p_f(b)^2 = \sup \{ f(x^* b^* b x): f(x^* x) = 1 \} \geqslant f(b^* b) = p(b)^2.$$

But $p_f \leqslant p$, by Lemma 10. Apply Theorem 5. □

Notation. Given $p \in \mathrm{bs}(A)$, we denote by $U(p)$ the set of all positive linear functionals f on A such that

$$|f(a)| \leqslant p(a) \qquad (a \in A).$$

Lemma 13. *Let* $p \in \mathrm{bs}(A)$, *and let* $b \in A$. *Then there exists an extreme point f of $U(p)$ such that $f(b^* b) = p(b^* b)$.*

Proof. Let $K = \{ f \in U(p): f(b^* b) = p(b^* b) \}$. $U(p)$ is a weak* compact convex subset of A', and therefore, by Lemma 11, K is non-void, weak* compact and convex. By the Krein-Milman theorem, K has an extreme point f; and it is trivial to check that f is also an extreme point of $U(p)$.   □

We need next a technical result for B^*-algebras which is of general interest.

Lemma 14. *A B^*-algebra has a two-sided approximate identity consisting of self-adjoint elements of norm 1.*

Proof. Let B be a B^*-algebra and let Λ be the family of all finite subsets of B ordered by inclusion. Given $\lambda \in \Lambda$, $\lambda = \{x_1, \ldots, x_n\}$, let $h = x_1^* x_1 + \cdots + x_n^* x_n$, so that $h \in \mathrm{pos}(B)$, by Lemma 38.9. Let $k_\lambda = nh(1+nh)^{-1}$. Then $k_\lambda \in \mathrm{Sym}(B)$, and an application of Theorem 35.5 to the closed algebra generated by h gives

$$\|k_\lambda\| \leqslant 1, \qquad \|(1-k_\lambda)h(1-k_\lambda)\| \leqslant \frac{1}{4n}.$$

Let $v_i = x_i(1-k_\lambda)$. Since

$$\sum_{i=1}^{n} v_i^* v_i = (1-k_\lambda)h(1-k_\lambda),$$

we have, using Theorem 38.10,

$$\|v_i^* v_i\| \leqslant \|(1-k_\lambda)h(1-k_\lambda)\| \leqslant \frac{1}{4n},$$

and so $\|v_i\| \leqslant \frac{1}{2}n^{-\frac{1}{2}}$ $(i = 1, \ldots, n)$.

Given $b \in B$, $\varepsilon > 0$, choose λ_0 to be any finite set of n elements of B such that $b \in \lambda_0$, $n > \varepsilon^{-2}$. Then

$$\|b - bk_\lambda\| < \varepsilon \qquad (\lambda \geqslant \lambda_0),$$

and so $\lim\limits_{\lambda} bk_\lambda = b$ $(b \in B)$. The continuity of the involution gives $\lim\limits_{\lambda} k_\lambda b = b$ $(b \in B)$. Clearly $\lim\limits_{\lambda} \|k_\lambda\| = 1$, and hence $\{\|k_\lambda\|^{-1}k_\lambda\}$ gives the required two-sided approximate identity.   □

Lemma 15. *If $p \neq 0$ and $f \in U(p)$, then*

$$|f(a)|^2 \leqslant f(a^* a) \qquad (a \in A).$$

Proof. If A has a unit, then, by Lemma 37.6(ii), for $a \in A$,

$$|f(a)|^2 \leqslant f(1)f(a^* a) \leqslant p(1)f(a^* a) = f(a^* a).$$

In the general case we may suppose, as in Lemma 11, that A is a B^*-algebra. Let A have approximate identity $\{e_\lambda\}$, where $e_\lambda^* = e_\lambda$, $\|e_\lambda\| = 1$. Then, for $a \in A$,

$$|f(e_\lambda a)|^2 \leqslant f(e_\lambda^* e_\lambda)f(a^* a) \leqslant f(a^* a) \qquad (\lambda \in \Lambda),$$
$$|f(a)|^2 \leqslant f(a^* a).   □$$

Corollary 16. *Each* $f \in U(p)$ *is representable. Moreover* p, f *may be extended to* $\tilde{p}, \tilde{f}$ *on the unitization* B *of* A *such that* $\tilde{p} \in \mathrm{bs}(B)$, $\tilde{f} \in U(\tilde{p})$, $\tilde{f}(1) = 1$.

Proof. The statement and proof of Theorem 37.11. ☐

Theorem 17. *Let* $p \in \mathrm{nbs}(A)$. *Then* $p = p_f$ *for some extreme point* f *of* $U(p)$.

Proof. Given $b \in A$, Lemma 13 gives the existence of g extreme in $U(p)$ such that $g(b^*b) = p(b^*b)$, and, by Lemma 10, $p_g \leqslant p$. Since $g \in U(p)$, Lemma 15 gives $|g(a)|^2 \leqslant g(a^*a)$ $(a \in A)$. In particular

$$(g(b^*b))^2 \leqslant g(b^*bb^*b) \leqslant (p_g(b^*))^2\, g(b^*b),$$

from which

$$p(b^*b) = g(b^*b) \leqslant (p_g(b^*))^2 = (p_g(b))^2 \leqslant (p(b))^2 = p(b^*b).$$

Therefore $p_g(b) = p(b)$, and we have now proved that the set Q of B^*-semi-norms of the form p_f with f extreme in $U(p)$ satisfies

$$p(a) = \max \{q(a) : a \in Q\} \qquad (a \in A).$$

Since p is neat, $p \in Q$. ☐

Corollary 18. *Each neat* B^*-*semi-norm is of the form* p_f *with* f *an extreme point of* $U(m)$.

Proof. Given $p \in \mathrm{nbs}(A)$, we know that $p = p_f$ with f an extreme point of $U(p)$. Since $p \leqslant m$, $f \in U(m)$ and we need only prove that f is also extreme in $U(m)$. Let $f = \alpha f_1 + (1 - \alpha) f_2$ with $0 < \alpha < 1$, $f_1, f_2 \in U(m)$. By Lemma 15 (with $p = m$),

$$|f_1(a)|^2 \leqslant f_1(a^*a) \leqslant \alpha^{-1} f(a^*a) \leqslant \alpha^{-1} p(a^*a)$$

and so $|f_1(a)| \leqslant \alpha^{-\frac{1}{2}} p(a)$ $(a \in A)$. Lemma 10 gives $p_{f_1} \leqslant p$, and so

$$f_1(x^*a^*ax) \leqslant (p(a))^2 f_1(x^*x) \qquad (a, x \in A).$$

Thus,

$$|f_1(a^*a)|^2 \leqslant f_1(a^*aa^*a) \leqslant (p(a^*))^2 f_1(a^*a),$$

from which $f_1(a^*a) \leqslant (p(a))^2$, $|f_1(a)| \leqslant p(a)$ $(a \in A)$. Therefore $f_1 \in U(p)$, and similarly $f_2 \in U(p)$. Since f is extreme in $U(p)$, we now have $f = f_1 = f_2$ ☐

Corollary 19. $m(a) = \max \{(f(a^*a))^{\frac{1}{2}} : f$ *an extreme point of* $U(m)\}$ $= \max \{p_f(a) : f$ *an extreme point of* $U(m)\}$.

Proof. Lemma 13, Theorem 8 and Corollary 18. ☐

Remarks. (1) The connection between extreme positive linear functionals and topologically irreducible star representations is established in § 40.

(2) Effros [131] considers a weaker property than neatness in which the sets Q considered in Definition 6 are restricted to sets with two points.

§ 40. Topologically Irreducible Star Representations

We noted in § 25 the lack of a satisfactory theory for topologically irreducible modules over arbitrary Banach algebras. For Hilbert modules over Banach star algebras there are very satisfactory characterizations of topological irreducibility. The results of this section depend heavily on the special properties of Hilbert spaces and it is not clear how they may be extended to modules over arbitrary Banach algebras. Topologically irreducible modules do not fit well into the general representation theory of Chapter III, but Kadison [224] proved the remarkable result that topologically irreducible Hilbert modules over B^*-algebras are automatically [1]strictly irreducible. For a proof of Kadison's theorem and its implications for B^*-algebras see Dixmier [119] or Sakai [327].

Notation. Throughout this section A denotes a Banach star algebra, H a Hilbert A-module and π the corresponding star representation of A on H. Recall that π is topologically irreducible if and only if $\{0\}$ and H are the only closed submodules of H.

Definition 1. The *commutant* $\operatorname{com}\pi$ of π in $BL(H)$ is defined by
$$\operatorname{com}\pi = \{T \in BL(H) : T\pi(a) = \pi(a)T \quad (a \in A)\}.$$
Clearly $\operatorname{com}\pi$ is a closed star subalgebra of $BL(H)$ containing the identity operator I.

Theorem 2. *The star representation π is topologically irreducible if and only if* $\operatorname{com}\pi = \mathbb{C}I$.

Proof. Let $\operatorname{com}\pi = \mathbb{C}I$, let M be a closed submodule of H, and let P be the orthogonal projection of H onto M. Given $a \in A$, we have $\pi(a)M \subset M$ and hence $P\pi(a)P = \pi(a)P$,
$$P\pi(a) = ((\pi(a))^* P)^* = (P\pi(a^*)P)^* = P\pi(a)P = \pi(a)P.$$
Therefore $P \in \operatorname{com}\pi$, $P = 0$ or I; π is topologically irreducible.

Let π be topologically irreducible. Let S be a self-adjoint element of the unital C^*-algebra $\operatorname{com}\pi$, and let C be the smallest closed star subalgebra of $BL(H)$ containing S and I. By Theorem 35.4 there exists

---

[1] 'Strictly irreducible' is here used to denote 'irreducible' in the sense of Definition 24.3.

an isometric star isomorphism of C onto $C(\Phi_C)$. If Φ_C contains more than one point then by Urysohn's lemma there exist non-zero $T_1, T_2 \in C$ with $T_1 T_2 = 0$. Choose $x \in H$ with $T_2 x \neq 0$. Then, since $C \subset \operatorname{com} \pi$,

$$T_1 H = T_1(\pi(A) T_2 x)^- \subset (T_1 \pi(A) T_2 x)^- = (\pi(A) T_1 T_2 x)^- = \{0\}.$$

This contradiction shows that Φ_C is a singleton and so also is $\operatorname{Sp}(C, S)$. Therefore $\operatorname{Sp}(BL(H), S)$ is a singleton, and since S is self-adjoint, $S = \lambda I$ for some $\lambda \in \mathbb{R}$, $\operatorname{com} \pi = \mathbb{C} I$. ☐

Corollary 3. *Let A be commutative and let π be topologically irreducible. Then π is strictly irreducible.*

Proof. Since A is commutative, we have $\pi(A) \subset \operatorname{com} \pi$, $\pi \in \Phi_A$. ☐

Recall that if p is a B^*-semi-norm on A then $U(p)$ denotes the set of all positive linear functionals f on A such that

$$|f(a)| \leqslant p(a) \qquad (a \in A).$$

Lemma 4. *Let $p \in \operatorname{bs}(A)$ and let $f, g \in U(p)$. The following statements are equivalent:*
(i) *There exist a unit vector $u \in H_f$ and a positive operator $P \in \operatorname{com} \pi_f$ such that $\langle P u, u \rangle_f = 1$ and*

$$f(a) = \langle \pi_f(a) u, u \rangle_f, \qquad g(a) = \langle P \pi_f(a) u, u \rangle_f \quad (a \in A).$$

(ii) *There exist $\phi \in U(p)$ and $0 < \alpha < 1$ such that*

$$f = \alpha g + (1 - \alpha) \phi.$$

Proof. We may suppose, by Corollary 39.16, that A has a unit.
(i) ⇒ (ii). Let condition (i) hold. For $a \in A$ we have

$$f(a^* a) = \langle \pi_f(a^*) \pi_f(a) u, u \rangle_f = \|\pi_f(a) u\|^2,$$

$$g(a^* a) = \langle \pi_f(a^*) P \pi_f(a) u, u \rangle_f \leqslant |P| \, \|\pi_f(a) u\|^2 = |P| \, f(a^* a). \qquad (1)$$

With $a = 1$ in (1) we obtain $|P| \geqslant 1$. Choose α such that $0 < \alpha < |P|^{-1}$, and let $\phi = (1 - \alpha)^{-1} (f - \alpha g)$. Then $\phi(1) = 1$, and by (1), ϕ is positive. We have $\phi(a) = 0$ $(a \in J_p)$, and so we may define $\tilde{\phi}$ on A/J_p by $\tilde{\phi}(b) = \phi(a)$ $(a \in b \in A/J_p)$. Apply Lemma 37.6 (iv) with $A = B_p$, $b = 1$, $f = \tilde{\phi}$ to give

$$|\phi(a)| \leqslant r(b^* b)^{\frac{1}{2}} \leqslant \tilde{p}(b^* b)^{\frac{1}{2}} = p(a), \qquad \phi \in U(p).$$

(ii) ⇒ (i). Let condition (ii) hold, and let u be the coset of 1 in H_f. Then $f(a) = \langle \pi_f(a) u, u \rangle_f$ $(a \in A)$ and $X_f = \pi_f(A) u$ is dense in H_f. Let $a \in A$ and let $\pi_f(a) u = 0$. Then $f(a^* a) = 0$, and since g and ϕ are positive it follows that $g(a^*a)=0$, $g(b^*a)=0$ $(b\in A)$. We may therefore define a conjugate bilinear form ψ on X_f by

$$\psi(\pi_f(a)u, \pi_f(b)u)=g(b^*a) \quad (a,b\in A),$$

and we have

$$|\psi(\pi_f(a)u, \pi_f(b)u)|^2 = |g(b^*a)|^2$$
$$\leqslant g(b^*b)g(a^*a)$$
$$\leqslant \alpha^{-2} f(b^*b) f(a^*a)$$
$$= \alpha^{-2}\|\pi_f(a)u\|^2 \|\pi_f(b)u\|^2.$$

Since X_f is dense in H_f, it follows that ψ extends to a bounded conjugate bilinear form on H_f. Therefore there exists a bounded linear operator P on H_f such that

$$\psi(x,y)=\langle Px,y\rangle_f \quad (x,y\in H_f).$$

For $a,b,c\in A$ we have

$$\langle \pi_f(a)P\pi_f(b)u, \pi_f(c)u\rangle_f = \langle P\pi_f(a)u, \pi_f(a^*c)u\rangle_f$$
$$= \psi(\pi_f(a)u, \pi_f(a^*c)u)_f$$
$$= g(c^*ab)$$
$$= \psi(\pi_f(ab)u, \pi_f(c)u)$$
$$= \langle P\pi_f(a)\pi_f(b)u, \pi_f(c)u\rangle_f,$$

and hence $P\in\operatorname{com}\pi_f$, since $\pi_f(A)u$ is dense in H_f. The operator P is positive since

$$\langle P\pi_f(a)u, \pi_f(a)u\rangle_f = g(a^*a)\geqslant 0 \quad (a\in A).$$

Finally,

$$g(a)=\psi(\pi_f(a)u, u)=\langle P\pi_f(a)u,u\rangle_f \quad (a\in A). \quad \square$$

Theorem 5. *Let p be a B^*-semi-norm on A and let $f\in U(p)$. Then π_f is topologically irreducible if and only if f is an extreme point of $U(p)$.*

Proof. Let π_f be topologically irreducible. Then $\operatorname{com}\pi_f=\mathbb{C}I$ by Theorem 2, and Lemma 4 now shows that f is extreme in $U(p)$.

Conversely let f be an extreme point of $U(p)$. Then, by Lemma 4, the only projections in $\operatorname{com}\pi_f$ are scalar multiples of the identity. Therefore $\operatorname{com}\pi_f=\mathbb{C}I$ and π_f is topologically irreducible. $\square$

Theorem 6. *Let p be a neat B^*-semi-norm on A. Then there exists a topologically irreducible star representation π of A such that*

$$p(a)=|\pi(a)| \quad (a\in A).$$

Proof. Theorem 39.17 and Theorem 5. $\square$

Remark. We do not know whether every topologically irreducible star representation corresponds in this way to a neat B^*-semi-norm. An affirmative answer is given in the commutative case by Bucy and Maltese [91] and Pickford [304], and also for certain classes of B^*-algebras by Pickford [304].

Definition 7.   The *star radical* of A, srad(A), is defined to be the intersection of the kernels of all the topologically irreducible star representations of A on Hilbert spaces. A is *star-semi-simple* if srad$(A)=(0)$.

Proposition 8.   (i) srad(A) *is a closed star bi-ideal of A.*
(ii) rad$(A)\subset$ srad(A).
(iii) $A/$srad(A) *is star-semi-simple.*

Proof. (i) Every star representation is continuous.
(ii) Let $q\in$ rad(A) and let π be a star representation of A. Then $r(q^*q)=0$ and it follows from Lemma 37.2 that $\pi(q)=0$.
(iii) Each topologically irreducible star representation of A is zero on srad(A) and hence induces a topologically irreducible star representation of the Banach star algebra $A/$srad(A).   ☐

Theorem 9.   srad$(A)=\{a:m(a)=0\}$, *where m is the greatest B^*-semi-norm on A.*

Proof. Theorem 39.8 and Theorem 6.   ☐

Corollary 10.   srad(A) *is the intersection of the kernels of all star representations of A on Hilbert spaces.*

Corollary 11.   *A star-semi-simple Banach star algebra is an A^*-algebra with auxiliary norm m.*

Theorem 12.   *Let W denote the set of all elements of A of the form $a_1^* a_1 + \cdots + a_n^* a_n$ with $a_1, \ldots, a_n \in A$. Then*

$$\{a\in A: -a^* a\in W^-\}\subset \text{srad}(A), \tag{2}$$

and, if A has a unit, the opposite inclusion also holds.

Proof. Let $a\in A$, $w\in W$, $p\in$ bs(A). By Lemma 39.11 there exists $f\in U(p)$ such that $f(a^* a)=p(a^* a)$. Therefore

$$p(a^*\ a)= f(a^* a)\leqslant f(a^* a+w)\leqslant p(a^* a+w).$$

Let $-a^* a\in W^-$. Given $\varepsilon>0$ there exists $w\in W$ such that $\|a^* a+w\| <\varepsilon$. Then

$$(m(a))^2 = m(a^* a)\leqslant m(a^* a+w)\leqslant r(a^* a+w)\leqslant \|a^* a+w\| <\varepsilon.$$

Therefore $m(a)=0$, $a\in$ srad(A) and (2) is proved.

Let $q \in \mathrm{srad}(A)$ and let μ denote the sublinear functional defined on A by

$$\mu(a) = \inf\{\|a + w\| : w \in W\}.$$

By the Hahn-Banach theorem, there exists a real linear functional g on A with $g(q^* q) = \mu(q^* q)$ and $g(a) \leqslant \mu(a)$ $(a \in A)$. Define f on A by

$$f(u + iv) = g(u) + ig(v) \qquad (u, v \in \mathrm{Sym}(A)).$$

Then f is a complex linear functional, and is positive since

$$-f(a^* a) = g(-a^* a) \leqslant \mu(-a^* a) = 0 \qquad (a \in A).$$

Then $p_f(q) \leqslant m(q) = 0$, and so

$$f(a^* q^* q a) = 0 \qquad (a \in A). \tag{3}$$

If A has a unit, we take $a = 1$ in (3) and obtain

$$\mu(q^* q) = f(q^* q) = 0,$$

from which $-q^* q \in W^-$. ☐

Remarks. (1) The existence of a unit was needed only so that (3) implies $f(q^* q) = 0$. It is not difficult to give other conditions sufficient for this, for example that A has a bounded right approximate identity and a continuous involution.

(2) It is trivial to verify that the Banach star algebra C of Example 37.16 satisfies $\mathrm{srad}(C) = C$.

Most of the results of this section are due in essence to Naimark [284]. Theorem 12 is due to Kelley and Vaught [242].

§ 41. Hermitian Algebras

Definition 1. Let A be a Banach star algebra. Then A is *Hermitian* (or has *Hermitian involution*) if every self-adjoint element of A has real spectrum.

It is immediate from Lemma 5.2 that A is Hermitian if and only if the unitization of A, $A + \mathbb{C}$, is Hermitian. We shall assume throughout this section that A has a unit.

Notation. Throughout this section A denotes a Hermitian Banach star algebra with unit. Given $h \in \mathrm{Sym}(A)$ we write $h \geqslant 0$ if $\mathrm{Sp}(h) \subset \mathbb{R}^+$. We define a real-valued function s on A by

$$s(a) = (r(a^* a))^{\frac{1}{2}} \qquad (a \in A).$$

We prove the Shirali-Ford theorem [350] which asserts that $x^* x \geqslant 0$ for every $x \in A$. The Shirali-Ford theorem, which had been an outstanding conjecture for some years, extends Lemma 38.9. Theorem 35.3 shows that a commutative Banach star algebra is Hermitian if and only if the spectral radius is a B^*-semi-norm. We prove the Pták theorem [309] that s is a B^*-semi-norm on A. In the opposite direction, a Banach star algebra is Hermitian if s coincides with the greatest B^*-semi-norm m.

Lemma 2. $r(a) \leqslant s(a) = s(a^*)$ $(a \in A)$.

Proof. That $s(a) = s(a^*)$ is immediate from Proposition 5.3. It is sufficient to show that $1 \notin \mathrm{Sp}(a)$ whenever $s(a) < 1$. Let $s(a) < 1$. By Proposition 12.11, there exists $h \in \mathrm{Sym}(A)$ such that $h^2 = 1 - a^* a$. Since h is invertible, we have

$$(1 + a^*)(1 - a) = h^2 + a^* - a$$
$$= h\{1 + h^{-1}(a^* - a)h^{-1}\}h.$$

Since $ih^{-1}(a^* - a)h^{-1}$ is self-adjoint, it has real spectrum. Therefore $1 + h^{-1}(a^* - a)h^{-1}$ is invertible, and $1 - a$ has a left inverse. A similar argument applied to $(1 - a)(1 + a^*)$ shows that $1 - a$ has a right inverse. Thus $1 \notin \mathrm{Sp}(a)$. ☐

Lemma 3. $r(hk) \leqslant r(h)r(k)$ $(h, k \in \mathrm{Sym}(A))$.

Proof. Let $h, k \in \mathrm{Sym}(A)$. By Lemma 2 and Proposition 5.3,

$$r(hk) \leqslant s(hk) = (r(khhk))^{\frac{1}{2}} = (r(h^2 k^2))^{\frac{1}{2}}.$$

It follows by induction that

$$r(hk) \leqslant (r(h^{2^n} k^{2^n}))^{2^{-n}} \leqslant \|h^{2^n}\|^{2^{-n}} \|k^{2^n}\|^{2^{-n}} \quad (n \in \mathbb{N}). \quad ☐$$

Lemma 4. *Let* $h, k \in \mathrm{Sym}(A)$.
(i) If $h \geqslant 0$, $k \geqslant 0$, then $h + k \geqslant 0$.
(ii) $r(h + k) \leqslant r(h) + r(k)$.

Proof. (i) Let $h \geqslant 0, k \geqslant 0$. It is sufficient to show that $1 + h + k \in \mathrm{Inv}(A)$. Since $1 + h, 1 + k \in \mathrm{Inv}(A)$, we may define $u, v \in \mathrm{Sym}(A)$ by

$$u = h(1 + h)^{-1}, \quad v = k(1 + k)^{-1}.$$

The spectral mapping theorem gives $r(u) < 1$, $r(v) < 1$. By Lemma 3, $r(uv) < 1$, and so $1 - uv \in \mathrm{Inv}(A)$. Since

$$1 + h + k = (1 + h)(1 - uv)(1 + k),$$

it follows that $1 + h + k \in \mathrm{Inv}(A)$.
(ii) We have $r(h) \pm h \geqslant 0$, $r(k) \pm k \geqslant 0$, and so by (i),

$$r(h) + r(k) \pm (h + k) \geqslant 0.$$

Therefore $r(h + k) \leqslant r(h) + r(k)$. ☐

We give next the Shirali-Ford theorem [350]. Harris [184] and Pták [310] gave alternative proofs; the proof we give is a simplification of Harris's proof due to A. W. Tullo.

Theorem 5. *Let A be a Hermitian Banach star algebra. Then* $x^* x \geqslant 0$ $(x \in A)$.

Proof. Let $\delta = \sup\{-\lambda : \lambda \in \mathrm{Sp}(a^* a), a \in A, s(a) \leqslant 1\}$. Assume that $\delta > 0$. Then there exist $a \in A$, $\xi \in \mathrm{Sp}(a^* a)$ such that $s(a) < 1$, $-\xi > \frac{1}{4}\delta$. Let $b = 2a(1 + a^* a)^{-1}$. Then

$$1 - b^* b = (1 - a^* a)^2 (1 + a^* a)^{-2},$$

and hence

$$\mathrm{Sp}(b^* b) = \{1 - (\phi(\lambda))^2 : \lambda \in \mathrm{Sp}(a^* a)\},$$

where $\phi(\lambda) = (1 - \lambda)(1 + \lambda)^{-1}$. Thus $\mathrm{Sp}(b^* b) \subset]-\infty, 1[$.
Let $b = h + ik$, where $h, k \in \mathrm{Sym}(A)$. Then

$$b b^* = 2h^2 + 2k^2 - b^* b.$$

By Lemma 4(i), we have $2h^2 + 2k^2 - b^* b + 1 \geqslant 0$, and so $\mathrm{Sp}(b b^*) \subset [-1, \infty[$. By Proposition 5.3 we now have $\mathrm{Sp}(b^* b) \subset [-1, 1]$, $s(b) \leqslant 1$. By definition of δ, we have $-\{1 - (\phi(\xi))^2\} \leqslant \delta$, $\phi(\xi) \leqslant (1 + \delta)^{\frac{1}{2}}$. Since $\phi(\phi(\xi)) = \xi$, and ϕ is decreasing, it follows that $\xi \geqslant \phi\{(1 + \delta)^{\frac{1}{2}}\}$,

$$-\xi \leqslant \frac{(1 + \delta)^{\frac{1}{2}} - 1}{(1 + \delta)^{\frac{1}{2}} + 1} \leqslant \frac{\frac{1}{2}\delta}{2} = \frac{1}{4}\delta.$$

This contradiction shows that $\delta \leqslant 0$, as required. □

Remark. The converse of Theorem 5 is rather more elementary. In fact, let B be a Banach star algebra with unit such that $a^* a \geqslant 0$ $(a \in B)$. Suppose that $h \in \mathrm{Sym}(B)$, $\alpha + i\beta \in \mathrm{Sp}(h)$, $\alpha, \beta \in \mathbb{R}$, $\beta \neq 0$. Let $k = \beta^{-1}(h - \alpha)$. Then $i \in \mathrm{Sp}(k)$, $0 \in \mathrm{Sp}(1 + k^2)$, $1 + k^* k \in \mathrm{Sing}(B)$, and this contradicts the fact that $\mathrm{Sp}(k^* k) \subset \mathbb{R}^+$.

Lemma 6. $r(\frac{1}{2}(a + a^*)) \leqslant s(a)$ $(a \in A)$.

Proof. Let $a = h + ik$, where $h, k \in \mathrm{Sym}(A)$. Then, by Lemma 4(i), $r(h^2 + k^2) - h^2 = \{r(h^2 + k^2) - (h^2 + k^2)\} + k^2 \geqslant 0$, and so

$$\begin{aligned}
(r(h))^2 = r(h^2) \leqslant r(h^2 + k^2) &= \tfrac{1}{2}r(a^* a + a a^*)\\
&\leqslant \tfrac{1}{2}r(a^* a) + \tfrac{1}{2}r(a a^*)\\
&= (s(a))^2. \quad □
\end{aligned}$$

Theorem 7. *s is a B*-semi-norm on A.*

Proof. Let $a,b \in A$. Then, by Lemma 3,

$$(s(ab))^2 = r(b^* a^* ab) = r(a^* abb^*)$$
$$\leqslant r(a^* a) r(bb^*)$$
$$= (s(a))^2 (s(b))^2 .$$

Also, by Lemma 4 (ii),

$$(s(a+b))^2 = r((a^* + b^*)(a+b)) \leqslant r(a^* a) + r(b^* b) + r(a^* b + b^* a)$$
$$= (s(a))^2 + (s(b))^2 + r(a^* b + b^* a) .$$

Lemma 6 gives

$$r(a^* b + b^* a) \leqslant 2s(a^* b) \leqslant 2s(a^*)s(b) = 2s(a)s(b) ,$$

and hence

$$(s(a+b))^2 \leqslant (s(a) + s(b))^2 .$$

Finally,

$$(s(a^* a))^2 = r(a^* aa^* a) = (r(a^* a))^2 = (s(a))^4 . \quad \square$$

Corollary 8. $s = m$, *the greatest B*-semi-norm on A.*

Proof. Theorem 7 and Lemma 39.2 (ii). $\quad \square$

Theorem 9. $\mathrm{rad}(A) = \mathrm{srad}(A) = \{a \in A : s(a) = 0\}$.

Proof. Let $a \in \mathrm{rad}(A)$. Then $a^* a \in \mathrm{rad}(A)$, $s(a) = (r(a^* a))^{\frac{1}{2}} = 0$. Conversely, let $s(a) = 0$. Then, by Lemma 2, for every $b \in A$,

$$r(ba) \leqslant s(ba) \leqslant s(b)s(a) = 0 ,$$

and so $a \in \mathrm{rad}(A)$. Finally, apply Corollary 8 and Theorem 40.9. $\quad \square$

Corollary 10. *A semi-simple Hermitian Banach star algebra is an A*-algebra.*

Proof. Theorem 9 and Corollary 40.11. $\quad \square$

Remark. Gelfand and Naimark [153] give an example of a A^*-algebra which is not Hermitian.

Theorem 11. *Let B be a Banach star algebra with unit such that $s = m$. Then B is Hermitian.*

Proof. Let $a \in B$, let $\alpha = r(a^* a)$, and let $h = \alpha - a^* a$. Let f be a positive linear functional on B with $f(1) = 1$. Then

$$f(h^2) = \alpha^2 - 2\alpha f(a^* a) + f(a^* aa^* a) .$$

By Lemma 37.6, $f(a^* aa^* a) \leqslant r(a^* a)f(a^* a)$, and so

$$f(h^2) \leqslant \alpha^2 - \alpha f(a^* a) \leqslant \alpha^2 .$$

It follows from Corollary 39.12 that

$$(r(h))^2 = r(h^2) = r(h^* h) = m(h)^2 \leqslant \alpha^2,$$

so that $r(\alpha - a^* a) \leqslant \alpha$. It follows that for all $a \in B$,

$$\mathrm{Sp}(a^* a) \cap \{-\lambda : \lambda > 0\} = \emptyset,$$

$1 + a^* a \in \mathrm{Inv}(B)$. If B is not Hermitian there exists $k \in \mathrm{Sym}(B)$ and there exists $\alpha + \beta i \in \mathrm{Sp}(k)$ with $\alpha, \beta \in \mathbb{R}$, $\beta \neq 0$. Then $u = \beta^{-1}(k - \alpha) \in \mathrm{Sym}(B)$, $i \in \mathrm{Sp}(u)$, $-1 \in \mathrm{Sp}(u^2)$ and $1 + u^* u$ is singular. $\quad\square$

We end this section with two rather technical results for Hermitian algebras which are of some interest.

Proposition 12. *Each self-adjoint singular element of A is a joint topological divisor of zero. If the involution on A is continuous, then each singular element of A is either a left or a right topological divisor of zero.*

Proof. Let $h \in \mathrm{Sym}(A) \cap \mathrm{Sing}(A)$. Since $\mathrm{Sp}(h) \subset \mathbb{R}$, $h - n^{-1} i \in \mathrm{Inv}(A)$ for each $n \in \mathbb{N}$. Therefore $h \in \partial \mathrm{Inv}(A)$ and Theorem 2.14 applies.

Let $a \in \mathrm{Sing}(A)$. Then either $a^* a$ or $a a^*$ belongs to $\mathrm{Sing}(A)$. The rest is routine (see Rickart [321, Lemma (1.5.1)]). $\quad\square$

Observe that Proposition 12 may be applied in particular to any unital C^*-algebra. Rickart [321, pp. 278, 279] proves the further interesting fact that if $A = BL(H)$ where H is an infinite-dimensional Hilbert space, then $\mathrm{Sing}(A)$ has interior points.

Proposition 13. *Let A be star-normed. Let $a = x + y$, where $x, y \in A$ and $\|1 - x - y^*\| < 1$. Then there exists $\varepsilon > 0$ such that $r(1 - \varepsilon a) < 1$, and consequently a is invertible in any closed subalgebra of A containing 1 and a.*

Proof. We have

$$\|1 - x^* - y\| = \|1 - x - y^*\| < 1,$$

and so $\|1 - \tfrac{1}{2}(a + a^*)\| < 1$. Let $a = h + ik$, where $h, k \in \mathrm{Sym}(A)$. Then $\|1 - h\| < 1$. Let $\alpha \in \mathbb{R}$, $\alpha \geqslant 0$ and let $t = (1 + \alpha)^{-1}(1 - h)$. Then $\alpha + h = (1 + \alpha)(1 - t)$ and so there exists $u \in \mathrm{Sym}(A) \cap \mathrm{Inv}(A)$ such that $u^2 = \alpha + h$. Let $\lambda = -\alpha + \beta i$ where $\alpha, \beta \in \mathbb{R}$, $\alpha \geqslant 0$. Then

$$\lambda - a = i(\beta - k) - (\alpha + h) = u(iw - 1)u,$$

where $w = u^{-1}(\beta - k)u^{-1}$. Since $w \in \mathrm{Sym}(A)$ and A is Hermitian, $iw - 1 \in \mathrm{Inv}(A)$, and hence $\lambda - a \in \mathrm{Inv}(A)$. This proves that

$$\mathrm{Re}\,\lambda > 0 \quad (\lambda \in \mathrm{Sp}(a)).$$

Since $\mathrm{Sp}(a)$ is compact, there exists $\gamma > 0$ such that

$$|\lambda - \gamma| < \gamma \quad (\lambda \in \mathrm{Sp}(a)).$$

Let $\varepsilon = \gamma^{-1}$. Then $|\lambda - 1| < 1 \; (\lambda \in \mathrm{Sp}(\varepsilon a))$ and so $r(1 - \varepsilon a) < 1$. $\quad \square$

Cohen [104] used a special case of Proposition 13 to give a simple proof of Wermer's theorem [409] that the disc algebra is a maximal closed subalgebra of $C(\Gamma)$, where Γ is the unit circle in $\mathbb{C}$. The first author [68] used Proposition 13 as the starting point for a general theory of maximal subalgebras of star-normed Hermitian algebras; see also Gorin [170].

Chapter VI. Cohomology

§ 42. Tensor Products

Let X, Y, Z be normed linear spaces over the same field $\mathbb{F}$. A mapping $\phi: X \times Y \to Z$ is said to be *bilinear* if
   (i) for each $y \in Y$, the mapping $x \to \phi(x, y)$ is linear,
   (ii) for each $x \in X$, the mapping $y \to \phi(x, y)$ is linear.
When $Z = \mathbb{F}$, such a mapping is called a *bilinear functional* or *bilinear form*. A bilinear mapping $\phi: X \times Y \to Z$ is said to be *bounded* if there exists $M > 0$ such that

$$\|\phi(x, y)\| \leqslant M \|x\| \, \|y\| \qquad (x \in X, \, x \in Y).$$

The norm of ϕ, $\|\phi\|$, is then defined by

$$\|\phi\| = \sup \{ \|\phi(x, y)\| : \|x\| \leqslant 1, \, \|y\| \leqslant 1 \}.$$

When X, Y, Z are Banach spaces, each separately continuous bilinear mapping $\phi: X \times Y \to Z$ is bounded (by the uniform boundedness theorem). The set of all bounded bilinear mappings from $X \times Y$ to Z is denoted by $BL(X, Y; Z)$.

Proposition 1. $BL(X, Y; Z)$ *is a normed space with the usual pointwise operations, and is a Banach space if Z is a Banach space.*

Proof. Routine.   ☐

The above remarks may clearly be extended to the case of n-linear mappings from $X_1 \times X_2 \times \cdots \times X_n$ to Z. The corresponding space of bounded n-linear mappings is denoted by $BL(X_1, \ldots, X_n; Z)$. For the case $X_1 = X_2 = \cdots = X_n = X$, we write this more simply as $BL^n(X; Z)$.

Definition 2. Let X, Y be normed spaces over $\mathbb{F}$ with dual spaces X', Y'. Given $x \in X$, $y \in Y$, let $x \otimes y$ be the element of $BL(X', Y'; \mathbb{F})$ defined by

$$x \otimes y(f, g) = f(x) \, g(y) \qquad (f \in X', \, g \in Y').$$

The *algebraic tensor product* of X and Y, $X \otimes Y$, is defined to be the linear span of $\{ x \otimes y : x \in X, y \in Y \}$ in $BL(X', Y'; \mathbb{F})$.

Let $\tau: X \times Y \rightarrow X \otimes Y$ be the bilinear mapping defined by

$$\tau(x,y) = x \otimes y \qquad (x \in X, y \in Y).$$

We show in Theorem 6 below that, given any bilinear mapping $\phi: X \times Y \rightarrow Z$, there exists a unique linear mapping $\sigma: X \otimes Y \rightarrow Z$ such that $\phi = \sigma \circ \tau$. This is the key property of the algebraic tensor product.

Lemma 3. *Given* $u \in X \otimes Y$, *there exist linearly independent sets* $\{x_i\}, \{y_i\}$ *such that* $u = \sum_{i=1}^{n} x_i \otimes y_i$.

Proof. Choose a representation $u = \sum_{i=1}^{n} x_i \otimes y_i$, where n is as small as possible. Suppose that $y_n = \sum_{i=1}^{n-1} c_i y_i$. Then

$$u = \sum_{i=1}^{n-1} (x_i + c_i x_n) \otimes y_i,$$

and this contradicts the minimality of n. It follows that $\{y_i\}$ is a linearly independent set, and a similar argument applies to $\{x_i\}$. $\Box$

Lemma 4. *Let* $\sum_{i=1}^{n} x_i \otimes y_i = 0$, *where* $\{x_i\}$ *is a linearly independent set. Then* $y_i = 0$ $(i = 1, 2, \ldots, n)$.

Proof. We obtain in succession

$$f\left(\sum_{i=1}^{n} g(y_i) x_i\right) = \sum_{i=1}^{n} f(x_i) g(y_i) = 0 \qquad (f \in X', g \in Y'),$$

$$\sum_{i=1}^{n} g(y_i) x_i = 0 \qquad (g \in Y'),$$

$$g(y_i) = 0 \qquad (g \in Y', i = 1, \ldots, n),$$

$$y_i = 0 \qquad (i = 1, \ldots, n). \quad \Box$$

Lemma 5. *Let* $\{x_1, \ldots, x_m\}, \{y_1, \ldots, y_n\}$ *be linearly independent subsets of* X, Y *respectively. Then* $\{x_i \otimes y_j : i = 1, \ldots, m, j = 1, \ldots, n\}$ *is a linearly independent subset of* $X \otimes Y$.

Proof. Suppose that $\sum_{i=1}^{m} \sum_{j=1}^{n} \alpha_{ij} x_i \otimes y_j = 0$. Using Lemma 4 we obtain in succession

$$\sum_{i=1}^{m} x_i \otimes \left(\sum_{j=1}^{n} \alpha_{ij} y_j\right) = 0,$$

$$\sum_{j=1}^{n} \alpha_{ij} y_j = 0 \qquad (i = 1, \ldots, m),$$

$$\alpha_{ij} = 0 \qquad (i = 1, \ldots, m, j = 1, \ldots, n). \quad \Box$$

Theorem 6. *Given a bilinear mapping* $\phi: X \times Y \to Z$, *there exists a unique linear mapping* $\sigma: X \otimes Y \to Z$ *such that*

$$\sigma(x \otimes y) = \phi(x, y) \quad (x \in X, \, y \in Y).$$

Proof. We show first that, if $\sum_{r=1}^{k} x_r \otimes y_r = 0$, then $\sum_{r=1}^{k} \phi(x_r, y_r) = 0$. To see this, let $\{a_i\}$, $\{b_j\}$ be bases for the linear spans of $\{x_r\}$, $\{y_r\}$ respectively, and let

$$x_r = \sum_i \alpha_{ir} a_i, \qquad y_r = \sum_j \beta_{jr} b_j.$$

We now have

$$\sum_i \sum_j \sum_r \alpha_{ir} \beta_{jr} a_i \otimes b_j = 0,$$

and Lemma 5 gives $\sum_r \alpha_{ir} \beta_{jr} = 0$ for all i and j. Therefore

$$\sum_r \phi(x_r, y_r) = \sum_i \sum_j \sum_r \alpha_{ir} \beta_{jr} \phi(a_i, b_j) = 0.$$

We thus obtain a well defined linear mapping σ of $X \otimes Y$ into Z by taking

$$\sigma\left(\sum_{r=1}^{k} x_r \otimes y_r\right) = \sum_{r=1}^{k} \phi(x_r, y_r). \quad \square$$

Remark. There are alternative approaches to the definition of the algebraic tensor product $X \otimes Y$. For example, it can be represented as a set of finite rank linear mappings from X' to Y by defining

$$\left(\sum_{i=1}^{n} x_i \otimes y_i\right)(f) = \sum_{i=1}^{n} f(x_i) y_i \quad (f \in X').$$

In particular, the algebraic tensor product of a Hilbert space H with itself can be identified with the set of all finite rank bounded linear operators on H.

Definition 7. Given normed spaces X, Y, the linear space $X \otimes Y$ inherits a natural norm from $BL(X', Y'; \mathbb{F})$. This norm is called the *weak tensor norm* and is denoted by w. We thus have

$$w(u) = \sup\left\{\left|\sum_i f(x_i) g(y_i)\right| : \|f\| \leqslant 1, \, \|g\| \leqslant 1\right\},$$

where $u = \sum_i x_i \otimes y_i$. In particular $w(x \otimes y) = \|x\| \, \|y\|$. The completion of $X \otimes Y$ with respect to w, i.e. the closure of $X \otimes Y$ in $BL(X', Y'; \mathbb{F})$ is called the *weak tensor product* of X and Y. The weak tensor product of X and Y will be denoted here by $X \otimes_w Y$. Another notation for it is $X \hat{\otimes} Y$.

Example 8. Let E, F be compact Hausdorff spaces. Then there exists an isometric linear isomorphism of $C(E) \otimes_w C(F)$ onto $C(E \times F)$. By Theorem 6 there exists a unique linear mapping $T : C(E) \otimes C(F) \to C(E \times F)$ such that

$$T(a \otimes b)(s, t) = a(s) b(t) \qquad (s \in E, t \in F, a \in C(E), b \in C(F)).$$

We have

$$\left\| T\left(\sum_i a_i \otimes b_i \right) \right\|_\infty = \sup_{E \times F} \left| \sum_i a_i(s) b_i(t) \right|$$

$$= \sup_F \left\| \sum_i b_i(t) a_i \right\|_\infty$$

$$= \sup_{\|\mu\| \leqslant 1} \sup_F \left| \sum_i b_i(t) \mu(a_i) \right|$$

$$= \sup_{\|\mu\| \leqslant 1} \left\| \sum_i \mu(a_i) b_i \right\|_\infty$$

$$= \sup_{\|\mu\| \leqslant 1} \sup_{\|\nu\| \leqslant 1} \left| \sum_i \mu(a_i) \nu(b_i) \right|$$

$$= w\left(\sum_i a_i \otimes b_i \right).$$

Therefore T extends to a linear isometry of $C(E) \otimes_w C(F)$ into $C(E \times F)$. By the Stone-Weierstrass theorem the range of T is a dense subset of $C(E \times F)$, and therefore the range of the extended mapping is the whole of $C(E \times F)$.

Definition 9. Given normed spaces X, Y, the *projective tensor norm p* on $X \otimes Y$ is defined by

$$p(u) = \inf \left\{ \sum_i \|x_i\| \, \|y_i\| : u = \sum_i x_i \otimes y_i \right\}$$

where the infimum is taken over all (finite) representations of u.

Lemma 10. *The projective tensor norm p is a norm on $X \otimes Y$ and*
(i) $p(u) \geqslant w(u) \quad (u \in X \otimes Y),$
(ii) $p(x \otimes y) = \|x\| \, \|y\| \quad (x \in X, y \in Y).$

Proof. It is elementary that p is a semi-norm on $X \otimes Y$. Given $u = \sum_i x_i \otimes y_i$, and given $f \in X', g \in Y'$, we have

$$|u(f, g)| = \left| \sum_i f(x_i) g(y_i) \right| \leqslant \|f\| \, \|g\| \sum_i \|x_i\| \, \|y_i\|,$$

and hence $w(u) \leqslant p(u)$. Given $x \in X, y \in Y$, we have

$$\|x\| \, \|y\| = w(x \otimes y) \leqslant p(x \otimes y),$$

and $p(x \otimes y) \leqslant \|x\| \, \|y\|$ is clear from the definition of p. ☐

Definition 11. The completion of $(X \otimes Y, p)$ is called the *projective tensor product* of X and Y, and will be denoted here by $X \otimes_p Y$. Another notation for it is $X \hat{\otimes} Y$.

Proposition 12. $X \otimes_p Y$ can be represented as the linear subspace of $BL(X', Y'; \mathbb{F})$ consisting of all elements of the form $u = \sum_{n=1}^{\infty} x_n \otimes y_n$ where $\sum_{n=1}^{\infty} \|x_n\| \|y_n\| < \infty$. Moreover $p(u)$ is the infimum of the sums $\sum_{n=1}^{\infty} \|x_n\| \|y_n\|$ over all such representations of u.

Proof. Routine. $\Box$

We now identify the dual space of $X \otimes_p Y$. Given $F \in (X \otimes_p Y)'$, let ϕ_F be the bilinear mapping on $X \times Y$ defined by

$$\phi_F(x, y) = F(x \otimes y) \qquad (x \in X, \, y \in Y).$$

Proposition 13. The mapping $F \to \phi_F$ is an isometric linear isomorphism of $(X \otimes_p Y)'$ onto $BL(X, Y; \mathbb{F})$.

Proof. Given $F \in (X \otimes_p Y)'$, we clearly have $\phi_F \in BL(X, Y; \mathbb{F})$ and $\|\phi_F\| \leq \|F\|$. Conversely, given $\phi \in BL(X, Y; \mathbb{F})$, by Theorem 6 there exists a unique linear functional F on $X \otimes Y$ such that

$$F(x \otimes y) = \phi(x, y) \qquad (x \in X, \, y \in Y).$$

Then

$$\left| F\left(\sum_i x_i \otimes y_i \right) \right| = \left| \sum_i \phi(x_i, y_i) \right| \leq \|\phi\| \sum_i \|x_i\| \|y_i\|,$$

and so $|F(u)| \leq \|\phi\| p(u)$ $(u \in X \otimes Y)$. Therefore F has a unique extension to an element $\tilde{F}$ of $(X \otimes_p Y)'$, $\|\tilde{F}\| \leq \|\phi\|$, and $\phi = \phi_{\tilde{F}}$. $\Box$

Remark. It is elementary to verify that the mapping $\phi \to T_\phi$, where

$$(T_\phi x)(y) = \phi(x, y) \qquad (x \in X, \, y \in Y)$$

is an isometric isomorphism of $BL(X, Y; \mathbb{F})$ onto $BL(X, Y')$.

Example 14. Let μ, ν be positive σ-finite measures on measure spaces M, N respectively, and let $\mu \times \nu$ be the corresponding product measure on $M \times N$. Then there exists an isometric linear isomorphism of $L^1(\mu) \otimes_p L^1(\nu)$ onto $L^1(\mu \times \nu)$. By Theorem 6 there exists a linear mapping $T : L^1(\mu) \otimes L^1(\nu) \to L^1(\mu \times \nu)$ such that

$$T(a \otimes b)(s, t) = a(s) b(t) \qquad (s \in M, \, t \in N, \, a \in L^1(\mu), \, b \in L^1(\nu)).$$

It is easy to see that $\|Tu\|_1 \leq p(u)$ $(u \in L^1(\mu) \otimes L^1(\nu))$, and so T extends to a norm reducing linear mapping of $L^1(\mu) \otimes_p L^1(\nu)$ into $L^1(\mu \times \nu)$. The rest of the proof involves standard integration theory.

Definition 15. Given normed spaces X, Y, a norm α on $X \otimes Y$ is said to be a *cross-norm* if $\alpha(x \otimes y) = \|x\| \|y\|$ $(x \in X, y \in Y)$.

The projective and weak tensor norms on $X \otimes Y$ are cross-norms. In fact, p is the largest cross-norm on $X \otimes Y$; for if α is a cross-norm on $X \otimes Y$ and $u = \sum_i x_i \otimes y_i$, then

$$\alpha(u) \leqslant \sum_i \alpha(x_i \otimes y_i) = \sum_i \|x_i\| \, \|y_i\|$$

and so $\alpha(u) \leqslant p(u)$.

Proposition 16. *Let A, B be linear subspaces of $BL(X)$, $BL(Y)$ respectively. Then $A \otimes B$ can be embedded in $BL(X \otimes_p Y)$ and the induced norm on $A \otimes B$ is a cross-norm.*

Proof. Given $S \in A$, $T \in B$, by Theorem 6, there exists a unique linear operator $S \,\square\, T$ on $X \otimes Y$ such that

$$S \,\square\, T(x \otimes y) = Sx \otimes Ty \quad (x \in X, y \in Y).$$

Since

$$p\left(\sum_i S x_i \otimes T y_i \right) \leqslant \sum_i \|S x_i\| \, \|T y_i\| \leqslant \|S\| \, \|T\| \sum_i \|x_i\| \, \|y_i\|,$$

$S \,\square\, T$ can be extended to an element of $BL(X \otimes_p Y)$ with $\|S \,\square\, T\| \leqslant \|S\| \, \|T\|$. Moreover

$$\|S \,\square\, T\| \geqslant \sup\{p(S \,\square\, T(x \otimes y)) : p(x \otimes y) = 1\}$$
$$= \sup\{\|S x\| \, \|T y\| : \|x\| = \|y\| = 1\}$$
$$= \|S\| \, \|T\|.$$

By Theorem 6 there is a linear mapping $\sigma : A \otimes B \to BL(X \otimes_p Y)$ such that $\sigma(S \otimes T) = S \,\square\, T$, and it is easily checked that σ is injective. $\square$

Finally we consider tensor products of Banach algebras.

Proposition 17. *Let A, B be normed algebras over $\mathbb{F}$. There exists a unique product on $A \otimes B$ with respect to which $A \otimes B$ is an algebra and*

$$(a \otimes b)(c \otimes d) = ac \otimes bd \quad (a, c \in A, b, d \in B).$$

Proof. Given $a \in A$, $b \in B$, by Theorem 6 there exists a unique linear operator $\lambda(a, b)$ on $A \otimes B$ such that

$$\lambda(a, b)(c \otimes d) = ac \otimes bd \quad (c \in A, d \in B).$$

The mapping $(a, b) \to \lambda(a, b)$ is clearly bilinear and so, by Theorem 6 again, there exists a unique linear mapping σ of $A \otimes B$ into $L(A \otimes B)$ such that

$$\sigma(a \otimes b) = \lambda(a, b) \quad (a \in A, b \in B).$$

The required product on $A \otimes B$ is given by

$$(u, v) \to \sigma(u)(v) \quad (u, v \in A \otimes B).$$

The rest is routine verification. $\square$

Proposition 18. *Let A, B be normed algebras over $\mathbb{F}$. Then the projective tensor norm on $A \otimes B$ is an algebra norm.*

Proof. Let $u, v \in A \otimes B$, $u = \sum_{i=1}^{m} a_i \otimes b_i$, $v = \sum_{j=1}^{n} c_j \otimes d_j$. Then

$$uv = \sum_{i=1}^{m} \sum_{j=1}^{n} a_i c_j \otimes b_i d_j,$$

$$\sum_{i=1}^{m} \sum_{j=1}^{n} \|a_i c_j\| \, \|b_i d_j\| \leqslant \sum_{i=1}^{m} \|a_i\| \, \|b_i\| \sum_{j=1}^{n} \|c_j\| \, \|d_j\|$$

and therefore $p(u, v) \leqslant p(u) p(v)$. $\square$

Now let A, B be Banach algebras over $\mathbb{F}$. By Proposition 18 we may extend the product on $A \otimes B$ to $A \otimes_p B$ so that $A \otimes_p B$ becomes a Banach algebra. The projective tensor product $A \otimes_p B$ has a unit element $1 \otimes 1$ if A and B have unit elements and is unital if A and B are unital. Conversely, for commutative A and B, it can be proved that if $A \otimes_p B$ has a unit element so also have A and B, see Gelbaum [147].

Let G, H be groups. By Example 14, there is an isometric linear isomorphism T of $l^1(G) \otimes_p l^1(H)$ onto $l^1(G \times H)$. It is easily verified in this case that T is an algebra isomorphism. Other examples of projective tensor products of Banach algebras may be found in Bonsall and Duncan [78], and Varopoulos [396]. We shall use the projective tensor product in the next sections to develop the cohomology of Banach algebras. For examples of other algebra cross-norms see Schatten [334] and Sakai [327].

Proposition 19. *Let A, B be commutative complex Banach algebras. Then $\Phi_A \otimes_p B$ is homeomorphic to $\Phi_A \times \Phi_B$.*

Proof. We give the proof for the case when A and B have unit elements; the general case is rather more complicated. Given $\phi \in \Phi_A$, $\psi \in \Phi_B$, by Theorem 6, there exists a unique linear functional $\phi \,\square\, \psi$ on $A \otimes B$ such that

$$\phi \,\square\, \psi(a \otimes b) = \phi(a) \psi(b) \qquad (a \in A, \ b \in B).$$

Clearly $\phi \,\square\, \psi$ extends uniquely to an element of $\Phi_{A \otimes_p B}$. Conversely, given $\chi \in \Phi_{A \otimes_p B}$ let

$$\phi(a) = \chi(a \otimes 1) \qquad (a \in A),$$

$$\psi(b) = \chi(1 \otimes b) \qquad (b \in B).$$

Then $\phi \in \Phi_A$, $\psi \in \Phi_B$ and $\chi = \phi \,\square\, \psi$.

It is straightforward to prove that the mapping $(\phi, \psi) \rightarrow \phi \,\square\, \psi$ is weak* continuous, using the inequality

$$\|\phi \,\square\, \psi\| \leqslant 1 \qquad (\phi \in \Phi_A, \ \psi \in \Phi_B),$$

and the inverse mapping is continuous by compactness. $\square$

An equally straightforward argument applies to the Shilov boundaries to show that $\check{S}(A \otimes_p B)$ is homeomorphic to $\check{S}(A) \times \check{S}(B)$ under the restriction of the above homeomorphism.

Now let A, B be arbitrary complex Banach algebras. If $A \otimes_p B$ is semi-simple, then A and B are semi-simple (see Gelbaum [147]); the converse is false even for commutative algebras as Milne [280] shows using a construction of Enflo [133]. Tomiyama [389] shows that if for each $\varepsilon > 0$ and each compact subset K of A there exists $S \in BL(A)$ of finite rank such that $\|Sa - a\| < \varepsilon$ $(a \in K)$, then $A \otimes_p B$ is semi-simple if both A and B are semi-simple commutative Banach algebras. The result below shows that the difficulty in obtaining $A \otimes_p B$ to be semi-simple when A, B are semi-simple lies in passing to the completion of $A \otimes B$. The completion of a semi-simple normed algebra can be a radical Banach algebra.

Proposition 20. *Let A, B be semi-simple commutative Banach algebras. Then the algebra $A \otimes B$ is semi-simple.*

Proof. Given $\phi \in \Phi_A$, $\psi \in \Phi_B$ we have, as in Theorem 19, a corresponding multiplicative linear functional $\phi \,\square\, \psi$ on $A \otimes B$ such that

$$(\phi \,\square\, \psi)(a \otimes b) = \phi(a)\psi(b) \qquad (a \in A, \; b \in B).$$

Let $u \in A \otimes B$ and let $(\phi \,\square\, \psi)(u) = 0$ $(\phi \in \Phi_A, \; \psi \in \Phi_B)$. By Lemma 3 there exist linearly independent subsets $\{a_i\}, \{b_i\}$ of A, B respectively such that $u = \sum_{i=1}^{n} a_i \otimes b_i$. Since A and B are semi-simple we now have in succession

$$\sum_{i=1}^{n} \phi(a_i)\psi(b_i) = 0 \qquad (\phi \in \Phi_A, \; \psi \in \Phi_B),$$

$$\sum_{i=1}^{n} \phi(a_i)b_i = 0 \qquad (\phi \in \Phi_A),$$

$$\phi(a_i) = 0 \qquad (\phi \in \Phi_A, \; i = 1, \ldots, n),$$

$$u = 0. \quad \square$$

§ 43. Amenable Banach Algebras

A will denote a complex Banach algebra with unit, X a Banach A-bimodule, X' the dual Banach A-bimodule (Example 9.13 (iv)), in which the module multiplications are given by

$$(a f)(x) = f(x a), \qquad (f a)(x) = f(a x) \qquad (a \in A, \; f \in X', \; x \in X).$$

Definition 1. *A bounded X-derivation* is a bounded linear mapping D of A into X such that

$$D(ab) = (Da)b + a(Db) \quad (a, b \in A).$$

The set of all bounded X-derivations is denoted by $Z^1(A, X)$; it is a linear subspace of $BL(A, X)$. Given $x \in X$, let δ_x be the mapping of A into X given by

$$\delta_x(a) = ax - xa \quad (a \in A).$$

Then a simple calculation shows that δ_x belongs to $Z^1(A, X)$. We call δ_x an *inner X-derivation*, and denote by $B^1(A, X)$ the set of all inner X-derivations. $B^1(A, X)$ is a linear subspace of $Z^1(A, X)$, and we denote by $H^1(A, X)$ the difference space of $Z^1(A, X)$ modulo $B^1(A, X)$,

$$H^1(A, X) = Z^1(A, X) - B^1(A, X).$$

$H^1(A, X)$ is called the *first cohomology group* of A with coefficients in X.

Interest in $H^1(A, X)$ has concentrated on the question whether $H^1(A, X) = \{0\}$, or in other words whether every bounded X-derivation is inner. In particular, by the theorem of Singer and Wermer (Theorem 18.16), we know already that $H^1(A, A) = \{0\}$ whenever A is semisimple and commutative.

Definition 2. A is said to be *amenable* if $H^1(A, X') = \{0\}$ for every Banach A-bimodule X.

The reason for the choice of the word 'amenable' in this context is not at all self-evident. It began with a pun perpetrated by Day [112]. A *group G is said to be amenable* if there exists an *invariant mean* on G, i.e. a positive linear functional μ on $l^\infty(G)$ such that $\mu(1) = 1$, and

$$\mu(T_h m) = \mu(m) \quad (m \in l^\infty(G), h \in G),$$

where T_h is the left translation operator on $l^\infty(G)$ defined by

$$(T_h m)(g) = m(h^{-1} g) \quad (m \in l^\infty(G), h \in G).$$

The next proposition provides the pretext for the transference of the term 'amenable' to Banach algebras.

Proposition 3. *Let G be a group. Then the group G is amenable if and only if the group algebra $l^1(G)$ is amenable.*

We base half of the proof of Proposition 3 on the following proposition, in which $f a$ is defined as usual for $a \in A$, $f \in A'$ by

$$(f a)(x) = f(ax) \quad (x \in A). \tag{1}$$

Proposition 4. *Let A be amenable and let σ be a multiplicative linear functional on A. Then there exists $F \in A'' \setminus \{0\}$ such that $F(\sigma) = 1$ and*

$$F(f a) = \sigma(a) F(f) \quad (a \in A, f \in A').$$

Proof. We make A' into a Banach A-bimodule by taking the right module multiplication to be defined as in (1) and the left module multiplication to be defined by

$$a f = \sigma(a) f \quad (a \in A, \ f \in A').$$

We know that $\sigma \in A'$ (Proposition 16.3), and so

$$a \sigma = \sigma a = \sigma(a) \sigma. \tag{2}$$

Therefore $\mathbb{C}\sigma$ is a closed A-submodule of A'. Take $X = A' - \mathbb{C}\sigma$, and let π denote the canonical mapping of A' onto X. Then (see Example 9.13 (iv)), the dual mapping π^* is an A-module monomorphism of X' into A''. Choose $v \in A''$ with $v(\sigma) = 1$, and let δ denote the inner A''-derivation given by

$$\delta a = a v - v a \quad (a \in A).$$

Given $a \in A$, we have $\delta a \in A''$, and, by (2),

$$(\delta a)(\sigma) = (a v)(\sigma) - (v a)(\sigma) = v(\sigma a) - v(a \sigma) = 0.$$

Therefore $\delta a \in \pi^* X'$. Since π^* is a monomorphism, there exists a unique element $D a$ of X' with

$$\pi^* D a = \delta a.$$

Since π^* is a module monomorphism, D is an X'-derivation. The algebra A is assumed to be amenable, and so there exists $\psi \in X'$ with

$$D a = a \psi - \psi a \quad (a \in A).$$

Therefore, for all $a \in A$,

$$a(\pi^* \psi) - (\pi^* \psi)(a) = \pi^*(a \psi - \psi a) = \pi^* D a = \delta a = a v - v a.$$

Let $F = v - \pi^* \psi$. Then $F \in A''$, $F(\sigma) = v(\sigma) - \psi(\pi \sigma) = 1$, and $a F = F a$ $(a \in A)$, i.e.

$$F(f a) = F(a f) = \sigma(a) F(f) \quad (a \in A, \ f \in A'). \quad \square$$

Proof of Proposition 3. Take $A = l^1(G)$. We recall that the product in A is convolution, i.e.

$$(a b)(g) = \sum_{h \in G} a(h) b(h^{-1} g) \quad (a, b \in A, \ h \in G).$$

Given $g \in G$, let **g** denote the element of A defined by

$$\mathbf{g}(h) = \begin{cases} 1 & (h = g), \\ 0 & (h \neq g). \end{cases}$$

Then $(a\mathbf{g})(h)=a(hg^{-1})$, $(\mathbf{g}a)(h)=a(g^{-1}h)$. It follows that $\mathbf{e}$ is a unit element for A, where e is the identity of G, and that the mapping $g\to\mathbf{g}$ is a monomorphism of the group G into the group $\mathrm{Inv}(A)$. Let σ be defined on A by

$$\sigma(a)=\sum_{g\in G} a(g)\,.$$

Then σ is a multiplicative linear functional on A.

Suppose first that A is amenable. Then, by Proposition 4, there exists $F\in A''\backslash\{0\}$ with $F(\sigma)=1$ and $F(fa)=\sigma(a)F(f)$ $(a\in A, f\in A')$. Given $m\in l^{\infty}(G)$, let m' denote the linear functional on A given by

$$m'(a)=\sum_{g\in G} a(g)m(g)\,.$$

Then $m\to m'$ is a linear isometry of $l^{\infty}(G)$ onto A'. Also for $g,h\in G$,

$$(T_{g^{-1}}m)(h)=m(gh)\,,$$

and so

$$
\begin{aligned}
(T_{g^{-1}}m)'(a) &= \sum_{u\in G} a(u)m(gu) \\
&= \sum_{h\in G} a(g^{-1}h)m(h) \\
&= \sum_{h\in G} (\mathbf{g}a)(h)m(h) \\
&= m'(\mathbf{g}a) \qquad (a\in A)\,.
\end{aligned}
$$

This proves that $(T_{g^{-1}}m)'=m'\mathbf{g}$. Let ϕ be defined on $l^{\infty}(G)$ by $\phi(m)=F(m')$. Then $\phi\in(l^{\infty}(G))'$, $\phi(1)=F(\sigma)=1$, and $\phi(T_g m)=\phi(m)$ $(g\in G, m\in l^{\infty}(G))$.

The functional ϕ need not be positive, but the existence of an invariant mean on $l^{\infty}(G)$ is now easily established. Let $l^{\infty}_{+}(G)=\{m\in l^{\infty}(G): m\geqslant0\}$; and for $x\in l^{\infty}_{+}(G)$, take

$$\theta(x)=\sup\{\mathrm{Re}\,\phi(y): y\in l^{\infty}(G),\ |y|\leqslant x\}\,.$$

Then θ is positive homogeneous and additive on $l^{\infty}_{+}(G)$, $\theta(1)\neq0$, and the extension of $(\theta(1))^{-1}\theta$ to $l^{\infty}(G)$ is the required invariant mean.

Suppose on the other hand that the group G is amenable with invariant mean μ. Let X be a Banach A-bimodule, let D be an X'-derivation, and let κ be a positive constant with

$$\|af\|\leqslant\kappa\|a\|\|f\|,\qquad \|fb\|\leqslant\kappa\|f\|\|b\|\qquad (a,b\in A,\ f\in X')\,.$$

Given $x\in X$, let $\mathbf{x}$ be the complex valued function defined on G by

$$\mathbf{x}(g)=(\mathbf{g}\,D\mathbf{g}^{-1})(x)\qquad (g\in G)\,.$$

Since $D\mathbf{g}^{-1}\in X'$, we have

$$\|\mathbf{g}\,D\mathbf{g}^{-1}\|\leqslant\kappa\|\mathbf{g}\|\|D\mathbf{g}^{-1}\|\leqslant\kappa\|D\|\,;$$

and so $\mathbf{x} \in l^\infty(G)$ and $\|\mathbf{x}\|_\infty \leqslant \kappa \|D\| \|x\|$. Let f be defined on X by

$$f(x) = \mu(\mathbf{x}) \quad (x \in X).$$

Then $f \in X'$, and we show that $D = \delta_f$.

Each element a of A can be written in the form $a = \sum_{g \in G} a(g)\mathbf{g}$, with the series convergent in norm. Therefore it is enough to prove that

$$(D\mathbf{g})(x) = f(x\mathbf{g} - \mathbf{g}x) \quad (g \in G, \ x \in X). \tag{3}$$

Fix $x \in X$ and $g \in G$, and let $z = x\mathbf{g} - \mathbf{g}x$. For $h \in G$, we have

$$\mathbf{z}(h) = (\mathbf{h}D\mathbf{h}^{-1})(z) = (\mathbf{h}D\mathbf{h}^{-1})(x\mathbf{g}) - (\mathbf{h}D\mathbf{h}^{-1})(\mathbf{g}x).$$

Since $D(\mathbf{h}^{-1}\mathbf{g}) = (D\mathbf{h}^{-1})\mathbf{g} + \mathbf{h}^{-1}D\mathbf{g}$, we have in turn

$$-(\mathbf{h}D\mathbf{h}^{-1})\mathbf{g} = \mathbf{h}\mathbf{h}^{-1}D\mathbf{g} - \mathbf{h}D(\mathbf{h}^{-1}\mathbf{g})$$
$$= D\mathbf{g} - \mathbf{g}(\mathbf{g}^{-1}\mathbf{h})D(\mathbf{g}^{-1}\mathbf{h})^{-1},$$
$$-(\mathbf{h}D\mathbf{h}^{-1})(\mathbf{g}x) = (D\mathbf{g})(x) - \mathbf{g}[(\mathbf{g}^{-1}\mathbf{h})D(\mathbf{g}^{-1}\mathbf{h})^{-1}](x)$$
$$= (D\mathbf{g})(x) - [(\mathbf{g}^{-1}\mathbf{h})D(\mathbf{g}^{-1}\mathbf{h})^{-1}](x\mathbf{g}).$$

Taking $y = x\mathbf{g}$, we now have

$$\mathbf{z}(h) = (\mathbf{h}D\mathbf{h}^{-1})(y) - [(\mathbf{g}^{-1}\mathbf{h})D(\mathbf{g}^{-1}\mathbf{h})^{-1}](y) + (D\mathbf{g})(x)$$
$$= \mathbf{y}(h) - \mathbf{y}(g^{-1}h) + (D\mathbf{g})(x) \quad (h \in G),$$

i.e.

$$\mathbf{z} = \mathbf{y} - T_g\mathbf{y} + (D\mathbf{g})(x)\mathbf{1}.$$

By the invariance of μ, we have $\mu(\mathbf{y}) = \mu(T_g\mathbf{y})$; and so

$$f(z) = \mu(\mathbf{z}) = (D\mathbf{g})(x)\mu(\mathbf{1}) = (D\mathbf{g})(x),$$

which proves (3). □

The class of amenable groups includes all finite groups and all Abelian groups (Proposition 5). The simplest example of a group which is not amenable is provided by the free group on two symbols. For a detailed discussion of amenable groups see Hewitt and Ross [194].

Proposition 5. *If G is Abelian, then G is amenable.*

Proof. We use the notation from the proof of Proposition 3. Let $A = l^1(G)$, $\sigma(a) = \sum_{g \in G} a(g)$ $(a \in A)$. We have

$$\sigma(a\mathbf{g}) = \sigma(\mathbf{g}a) = \sigma(\mathbf{g})\sigma(a) = \sigma(a) \quad (a \in A, \ g \in G),$$

and so

$$\mathbf{g}\sigma = \sigma\mathbf{g} = \sigma \quad (g \in G). \tag{4}$$

Identify A' with $l^\infty(G)$, A'_+ with $l^\infty_+(G)$, and let

$$A''_+ = \{F \in A'' : F(\phi) \geqslant 0 \ (\phi \in A'_+)\},$$
$$K = \{F \in A''_+ : \|F\| = F(\sigma) = 1\}.$$

Then K is non-void, convex, weak* compact. By (4),

$$F(\mathbf{g}\,\sigma) = F(\sigma) = 1 \qquad (F \in K, \ g \in G).$$

So $K\mathbf{g} \subset K$ $(g \in G)$. By the Markoff-Kakutani fixed point theorem, there exists $F \in K$ with

$$F\mathbf{g} = F \qquad (g \in G). \quad \square$$

The following technical lemma shows that in the definition of amenability it is sufficient to consider unit linked Banach A-bimodules (see Definition 9.11).

Lemma 6. *Let* $H^1(A, X') = \{0\}$ *for every unit linked Banach A-bimodule X. Then A is amenable.*

Proof. For this proof only, let e be the unit of A, let X be any Banach A-bimodule and let $D \in Z^1(A, X')$. We have

$$X' = Y_1 \oplus Y_2 \oplus Y_3 \oplus Y_4$$

where $Y_1 = e X' e$, $Y_2 = (1-e) X' e$, $Y_3 = e X'(1-e)$, $Y_4 = (1-e) X'(1-e)$. For $j = 1, 2, 3, 4$, let $\pi_j : X' \to Y_j$ be the associated projection and let $D_j = \pi_j \circ D$. Then $D_j \in Z^1(A, Y_j)$. Since Y_1 is unit linked and isometrically isomorphic to $(e X e)'$, we have $D_1 = \delta_{y_1}$ for some $y_1 \in Y_1$. Let $y_2 = -D_2 e$. Since $Da = (De)a + eDa$ and $aD_2 e = a(1-e)(De)e = 0$, we have, for $a \in A$,

$$D_2 a = (1-e)(De) a e = (D_2 e) a = \delta_{y_2} a.$$

A similar argument applies to D_3 and D_4. It follows that $D = \delta_f$ for some $f \in X'$. $\quad \square$

The projective tensor product $A \otimes_p A$ is a Banach A-bimodule with module multiplications determined by

$$(a \otimes b) c = a \otimes b c, \qquad c(a \otimes b) = c a \otimes b \qquad (a, b, c \in A).$$

Our next goal is a characterization of the amenability of A in terms of the A-bimodule $A \otimes_p A$. Let $\pi : A \otimes_p A \to A$ be the bounded linear mapping determined by

$$\pi(a \otimes b) = a b \qquad (a, b \in A).$$

Definition 7. An *approximate diagonal* for A is a bounded net $\{m_\alpha\}$ in $A \otimes_p A$ such that, for $a \in A$,

$$\lim_\alpha (m_\alpha a - a m_\alpha) = 0, \qquad \lim_\alpha \pi(m_\alpha) a = a.$$

Since we are assuming that A has a unit element 1, the second condition is equivalent to the condition

$$\lim_\alpha \pi(m_\alpha) = 1.$$

A *virtual diagonal* for A is an element M of $(A \otimes_p A)''$ such that, for $a \in A$,

$$a M = M a, \qquad (\pi^{**} M) a = a.$$

Lemma 8. *A has an approximate diagonal $\{m_\alpha\}$ if and only if it has a virtual diagonal M. $\{m_\alpha\}$ can be chosen with $\lim \phi(m_\alpha) = M(\phi)$ ($\phi \in A'$).*

Proof. Let $\{m_\alpha\}$ be an approximate diagonal for A and let M be a weak* cluster point of the canonical image of $\{m_\alpha\}$ in $(A \otimes_p A)''$. Evidently M is a virtual diagonal.

Conversely, let M be a virtual diagonal for A. Then there exists a bounded net $\{\mu_\beta\}$ in $A \otimes_p A$ such that the canonical image of $\{\mu_\beta\}$ converges to M in the weak* topology on $(A \otimes_p A)''$. It follows that, for $a \in A$,

$$\lim_\beta (a \mu_\beta - \mu_\beta a) = 0, \qquad \lim_\beta \pi(\mu_\beta) a = a,$$

with convergence in the weak topology on A. The method of Proposition 11.4 gives an approximate diagonal. □

Theorem 9. *A is amenable if and only if it has a virtual diagonal.*

Proof. Let A have a virtual diagonal M, and let $\{m_\alpha\}$ be an associated approximate diagonal as in Lemma 8. Let X be a unit linked Banach A-bimodule and let $D \in Z^1(A, X')$. Given $x \in X$, let μ_x be the continuous linear functional on $A \otimes_p A$ determined by

$$\mu_x(a \otimes b) = (a D b)(x) \qquad (a, b \in A),$$

and let $f(x) = M(\mu_x)$ ($x \in X$). Then $f \in X'$. We show that $D = \delta_f$. Given $a \in A$, $x \in X$, we have, for $b, c \in A$,

$$\mu_{xa - ax}(b \otimes c) = (\mu_x a - a \mu_x)(b \otimes c) + (b c D a)(x),$$

and so, for $m \in A \otimes_p A$,

$$\mu_{xa - ax}(m) = (\mu_x a - a \mu_x)(m) + (\pi(m) D a)(x).$$

Therefore

$$
\begin{aligned}
(af - fa)(x) &= f(xa - ax) \\
&= M(\mu_{xa-ax}) \\
&= M(\mu_x a - a\mu_x) + \lim_\alpha (\pi(m_\alpha)Da)(x) \\
&= (aM - Ma)(\mu_x) + (1\,Da)(x) \\
&= (Da)(x).
\end{aligned}
$$

Hence $D = \delta_f$, and A is amenable by Lemma 6.

Conversely, let A be amenable, and let u be the canonical image of $1 \otimes 1$ in $(A \otimes_p A)''$. Then $\pi^{**}(\delta_u(a))$ is the canonical image of $\pi(a \otimes 1 - 1 \otimes a)$ which is 0, and so $\delta_u \in Z^1(A, \ker \pi^{**})$. Let $X = (A \otimes_p A)'/(\pi^*(A))^-$. Then $\ker \pi^{**}$ is isometrically isomorphic to X', and, since A is amenable, $\delta_u = \delta_v$ for some $v \in \ker \pi^{**}$. Let $M = u - v$, and let $a \in A$. We have $aM = Ma$ since $\delta_u = \delta_v$, and also $(\pi^{**}M)a = (\pi^{**}v)a = a$. Thus M is a virtual diagonal. $\square$

The above converse argument is closely related to the corresponding part of Proposition 3. Indeed, Proposition 3 can be deduced from Theorem 9 through the identification

$$
l^1(G) \otimes_p l^1(G) = l^1(G \times G).
$$

Theorem 9 can also be used to show that the Banach algebra $l^1(\mathbb{N}, \alpha)$ of Example 1.23 (with $S = \mathbb{N}$) is never amenable. On the other hand, if $A = l^1(\mathbb{N}, \alpha)$ with α as in Example 18.26, then $H^1(A, A) = \{0\}$.

We now give a much simpler example of a Banach algebra which is not amenable.

Example 10. Let A be the disc algebra. Then A is not amenable; for $\mathbb{C}$ is a Banach A-bimodule under the module multiplications

$$
\lambda f = f\lambda = \lambda f(0) \quad (\lambda \in \mathbb{C}, \ f \in A).
$$

Let $Df = f'(0)$ $(f \in A)$. Then $D \in Z^1(A, \mathbb{C})$. But $B^1(A, \mathbb{C}) = \{0\}$ and clearly $\mathbb{C}$ is a dual bimodule, so that A is not amenable.

The next result is a standard tool for establishing the amenability of Banach algebras; it implies that A is amenable if $\mathrm{Inv}(A)$ contains a large bounded amenable subgroup. The exact technique involved is illustrated in Theorem 12, 13.

Proposition 11. *Let A be amenable and let $\phi: A \to B$ be a continuous homomorphism of A onto a dense subalgebra of a Banach algebra B. Then B is amenable.*

Proof. Let X be a Banach B-bimodule and let $D \in Z^1(B, X')$. X is a Banach A-bimodule under the module multiplications

$$ax = \phi(a)x, \qquad xa = x\phi(a) \qquad (x \in X, \ a \in A),$$

and $D \circ \phi \in Z^1(A, X')$. Since A is amenable, there exists $f \in X'$ such that $D \circ \phi = \delta_f$. Therefore

$$Db = bf - fb \qquad (b \in \phi(A)),$$

and so $D = \delta_f$, by continuity.   □

Theorem 12. *Let E be a compact Hausdorff space. Then $C(E)$ is amenable.*

Proof. Let $G = \{\exp(ih) : h \in C_\mathbb{R}(E)\}$. G is amenable by Proposition 5, and so $l^1(G)$ is amenable by Proposition 3. Let $\phi : l^1(G) \to C(E)$ be the norm decreasing homomorphism defined by

$$\phi\left(\sum a_n \exp(ih_n)\right)(t) = \sum a_n \exp(ih_n(t)) \qquad (t \in E),$$

and let $B_0 = \phi(l^1(G))$. Clearly $1 \in B_0$, B_0 is self-adjoint, and B_0 separates the points of E. Apply the Stone-Weierstrass theorem and Proposition 11.   □

The next theorem uses the following elementary properties of amenable groups. A group G is amenable if every finitely generated subgroup is amenable. A group G is amenable if it has a normal subgroup H such that H and G/H are amenable.

Theorem 13. *Let H be a separable Hilbert space and let $A = KL(H) + \mathbb{C} I$. Then A is amenable.*

Proof. Let $\{e_n\}$ be an orthonormal basis for H. Let P be the group of all permutations π of $\mathbb{N}$ such that $\pi(n) = n$ except on a finite set, and let Q be the multiplicative group of all mappings $\chi : \mathbb{N} \to \{-1, 1\}$ such that $\chi(n) = 1$ except on a finite set. Given $\pi \in P$, $\chi \in Q$, let T be the bounded linear operator on H determined by

$$T e_n = \chi(n) e_{\pi(n)} \qquad (n \in \mathbb{N}).$$

Then $T \in A$ since $T - I$ has finite rank. Let G be the subgroup of $\text{Inv}(A)$ consisting of all such operators T, and let J be the subgroup of G consisting of those T for which π is the identity. Then J is an Abelian normal subgroup of G, and G/J is isomorphic to P. Since every finitely generated subgroup of P is finite, P is amenable. Therefore $G/J, J$ are amenable, G is amenable, $l^1(G)$ is amenable. Let C be the linear span of G. Since $G \subset S(A)$, it is now sufficient, by Proposition 11, to show that C is dense in A. For this it is enough to show that the rank one operators $e_j \otimes e_i$

are in C for each $i, j \in \mathbb{N}$. Let π be the permutation which transposes i and j, and let

$$T_1 e_n = e_{\pi(n)}, \quad T_2 e_n = (1 - 2\delta_{in}) e_{\pi(n)} \quad (n \in \mathbb{N}).$$

Then $e_j \otimes e_i = \frac{1}{2}(T_1 - T_2)$. $\quad\square$

It is clear that the conclusion of Theorem 13 holds if H is replaced by any separable Banach space with a Schauder basis $\{e_n\}$ such that

$$\|\sum \zeta_n \chi(n) e_{\pi(n)}\| = \|\sum \zeta_n e_n\| \quad (\pi \in P, \ \chi \in Q).$$

Given an amenable Banach algebra A, it is natural to ask which Banach A-bimodules X, other than dual modules, satisfy $H^1(A, X) = \{0\}$. Proposition 14 shows that this holds whenever X is *commutative*, i.e. $ax = xa$ $(x \in X, a \in A)$. On the other hand, if A is a semi-simple commutative Banach algebra such that $H^1(A, X) = \{0\}$ for every Banach A-bimodule X, then it is immediate from Proposition 16 that A is finite dimensional.

Proposition 14. *Let A be amenable and let X be a commutative Banach A-bimodule. Then $H^1(A, X) = \{0\}$.*

Proof. Let $D \in Z^1(A, X)$ and let τ be the canonical mapping of X into X''. Then $\tau \circ D \in Z^1(A, X'')$. Since A is amenable and X'' is a commutative Banach A-bimodule, $\tau \circ D = 0$, $D = 0$, $H^1(A, X) = \{0\}$. $\quad\square$

Corollary 15. *Let A be commutative and amenable. Then $H^1(A, A) = \{0\}$.*

Proposition 16. *Let A be commutative and let Φ_A be infinite. Then there exists a Banach A-bimodule X such that $H^1(A, X) \neq \{0\}$.*

Proof. Let $E = \Phi_A$ and let $X = \{f \in C(E \times E): f(t, t) = 0 \ (t \in E)\}$. Then X is a Banach A-bimodule under the module multiplications

$$(af)(s, t) = a^\wedge(s) f(s, t),$$

$$(fa)(s, t) = a^\wedge(t) f(s, t),$$

where $s, t \in E$, $f \in X$, $a \in A$. Let $D: A \to X$ be defined by

$$(Da)(s, t) = a^\wedge(s) - a^\wedge(t).$$

Then $D \in Z^1(A, X)$. Suppose that $D = \delta_g$ for some $g \in X$. Then

$$a^\wedge(s)(1 - g(s, t)) = a^\wedge(t)(1 - g(s, t))$$

and so $a^\wedge(s) = a^\wedge(t)$ whenever $g(s, t) \neq 1$. Therefore $g(s, t) = 1$ when $s \neq t$, and so $1 - g$ is the characteristic function of the diagonal of $E \times E$. This shows that each point of E is isolated, and hence E is finite since it is compact. $\quad\square$

The results of this section are mainly due to Johnson [217, 218] where a fuller treatment is given.

§ 44. Cohomology of Banach Algebras

The Hochschild cohomology theory of associative linear algebras has a natural analogue in the setting of Banach algebras. We give here a brief introduction to the cohomology of Banach algebras, see Johnson [217] for a fuller account.

Notation. A denotes a complex Banach algebra with or without unit, and X denotes a Banach A-bimodule. Recall that $BL^n(A; X)$ denotes the Banach space of all bounded n-linear mappings of $A_1 \times A_2 \times \cdots \times A_n$ to X where $A_j = A$ for each j. Let $C_n = BL^n(A; X)$ $(n \in \mathbb{N})$ and let $C_0 = X$.

Definition 1. For $n \in \mathbb{N}$ we define linear mappings $\delta^n : C_{n-1} \to C_n$ as follows:

$$(\delta^1 x)(a) = ax - xa \qquad (x \in X, \, a \in A).$$

For $n \geq 2$, $T \in C_{n-1}$, $a_1, \ldots, a_n \in A$,

$$(\delta^n T)(a_1, \ldots, a_n) = a_1 T(a_2, \ldots, a_n) + \sum_{j=1}^{n-1} (-1)^j T(a_1, \ldots, a_j a_{j+1}, \ldots, a_n)$$

$$+ (-1)^n T(a_1, \ldots, a_{n-1}) a_n.$$

Note in particular that

$$(\delta^2 T)(a_1, a_2) = a_1 T a_2 - T(a_1 a_2) + (T a_1) a_2,$$

so that $\ker \delta^2$ consists of all bounded X-derivations. We now extend Definition 43.1 by taking, for all $n \in \mathbb{N}$, $Z^n(A, X)$ to be $\ker \delta^{n+1}$ and $B^n(A, X)$ to be the range of δ^n.

Lemma 2. For $n \in \mathbb{N}$, $\delta^{n+1} \circ \delta^n = 0$.

Proof. The proof is elementary, but we give only the case $n = 2$. Let $a_1, a_2, a_3 \in A$, $T \in C_1$. Then

$\delta^3 (\delta^2 T)(a_1, a_2, a_3)$
$= a_1 (\delta^2 T)(a_2, a_3) - \delta^2 T(a_1 a_2, a_3) + \delta^2 T(a_1, a_2 a_3) + (\delta^2 T(a_1, a_2)) a_3$
$= a_1 \{a_2 T a_3 - T(a_2 a_3) + (T a_2) a_3\} - \{a_1 a_2 T a_3 - T(a_1 a_2 a_3) + (T(a_1 a_2)) a_3\}$
$\quad + \{a_1 T(a_2 a_3) - T(a_1 a_2 a_3) + (T a_1) a_2 a_3\}$
$\quad - \{a_1 T a_2 - T(a_1 a_2) + (T a_1) a_2\} a_3$
$= 0.$ □

It follows from Lemma 2 that $B^n(A, X)$ is a linear subspace of $Z^n(A, X)$.

Definition 3. The *nth cohomology group* of A with coefficients in X, $H^n(A, X)$, is the difference space of $Z^n(A, X)$ modulo $B^n(A, X)$, i.e.

$$H^n(A, X) = Z^n(A, X) - B^n(A, X).$$

The mappings δ^n are rather complicated and the significance of the higher cohomology groups is not yet fully understood, but it is always possible to express $H^n(A, X)$ as the first cohomology group of A with coefficients in another Banach A-bimodule as we now show.

Let σ^n be the canonical mapping of C_{n+k} to $BL^n(A; C_k)$ defined by

$$((\sigma^n T)(a_1, \ldots, a_n))(a_{n+1}, \ldots, a_{n+k}) = T(a_1, \ldots, a_{n+k}).$$

It is routine to verify that σ^n is a linear isometry of C_{n+k} onto $BL^n(A; C_k)$. The Banach space C_k becomes a Banach A-bimodule under the module multiplications

$$(a\,T)(a_1, \ldots, a_k) = a\,T(a_1, \ldots, a_k),$$
$$(T\,a)(a_1, \ldots, a_k) = T(a\,a_1, \ldots, a_k)$$
$$+ \sum_{j=1}^{k-1} (-1)^j T(a, a_1, \ldots, a_j a_{j+1}, \ldots, a_k)$$
$$+ (-1)^k (T(a, a_1, \ldots, a_{k-1})) a_k.$$

Let Δ^n be the linear mappings corresponding to δ^n when the A-bimodule X is replaced by the A-bimodule C_k.

Lemma 4. For $n \in \mathbb{N}$, $\Delta^n \circ \sigma^{n-1} = \sigma^n \circ \delta^{n+k}$.

Proof. Elementary. $\square$

Proposition 5. The linear spaces $H^{n+k}(A, X)$ and $H^n(A, C_k)$ are isomorphic.

Proof. Apply Lemma 4. $\square$

Proposition 6. Let A be amenable. Then $H^n(A, X') = \{0\}$ for every Banach A-bimodule X and for every $n \in \mathbb{N}$.

Proof. Let Y be the $(n+1)$-fold projective tensor product

$$Y = A \otimes_p A \otimes_p \cdots \otimes_p A \otimes_p X.$$

Then Y becomes a Banach A-bimodule under the module multiplications determined by

$$(a_1 \otimes \cdots \otimes a_n \otimes x) a = a_1 \otimes \cdots \otimes a_n \otimes x a,$$
$$a(a_1 \otimes \cdots \otimes a_n \otimes x) = (a\,a_1) \otimes \cdots \otimes a_n \otimes x$$
$$+ \sum_{j=1}^{n-1} (-1)^j a \otimes a_1 \otimes \cdots \otimes a_j a_{j+1} \otimes \cdots \otimes a_n \otimes x$$
$$+ (-1)^n a \otimes a_1 \otimes \cdots \otimes a_{n-1} \otimes a_n x.$$

Proposition 42.13 and the remark following generalize to give an iso-metric A-bimodule isomorphism of Y' onto $BL^n(A,X')$. By Proposition 5, we have isomorphisms

$$H^{n+1}(A, X') \approx H^1(A, BL^n(A, X')) \approx H^1(A, Y') = \{0\} . \quad \square$$

We show next how the Wedderburn property for Banach algebras is related to the vanishing of a certain second cohomology group.

Definition 7. A has the *Wedderburn property* if there exists a closed subalgebra C of A such that $A = C \oplus \mathrm{rad}(A)$, in the linear space sense. The first Wedderburn theorem asserts that A has this property if it is finite dimensional.

We consider the following illustrative case:

$$R = \mathrm{rad}(A), \quad R^2 = \{0\}, \quad A = X \oplus R \text{ for some closed subspace of } A . \quad (1)$$

Let $B = A/R$ and let π be the canonical homomorphism of A onto B. Then $\pi|_X$ is a bounded linear isomorphism of X onto B, and so, by Banach's isomorphism theorem, there exists a bounded linear mapping ρ of B onto X such that $\pi(\rho(b)) = b$ $(b \in B)$.

Let

$$br = \rho(b)r, \quad rb = r\rho(b) \quad (r \in R, \ b \in B) . \quad (2)$$

Lemma 8. *R is a Banach B-bimodule with module multiplications given by (2).*

Proof. For $b_1, b_2 \in B$, we have

$$\pi(\rho(b_1)\rho(b_2) - \rho(b_1 b_2)) = \pi\rho(b_1)\pi\rho(b_2) - \pi\rho(b_1 b_2)$$
$$= b_1 b_2 - b_1 b_2 = 0,$$

and hence

$$\rho(b_1)\rho(b_2) - \rho(b_1 b_2) \in R \quad (b_1, b_2 \in B) . \quad (3)$$

For $r \in R$,

$$b_1(b_2 r) - (b_1 b_2)r = (\rho(b_1)\rho(b_2) - \rho(b_1 b_2))r = 0,$$

by (3) and the fact that $R^2 = \{0\}$. Similarly, $(rb_1)b_2 = r(b_1 b_2)$, and the rest is clear. $\square$

Theorem 9. *Let R satisfy condition (1) and let $H^2(B, R) = \{0\}$. Then A has the Wedderburn property.*

Proof. Let $T: B \times B \to R$ be defined by

$$T(b_1, b_2) = \rho(b_1)\rho(b_2) - \rho(b_1 b_2) . \quad (4)$$

It follows from (1), (2) and (3) that $T \in Z^2(B, R)$, and so there exists $\xi \in BL(B, R)$ such that $T = \delta^2 \xi$. Let $\phi = \rho - \xi$. Then for $b_1, b_2 \in B$,

$$\phi(b_1)\phi(b_2) - \phi(b_1 b_2) = (\rho(b_1) - \xi(b_1))(\rho(b_2) - \xi(b_2)) - \rho(b_1 b_2) + \xi(b_1 b_2)$$
$$= \rho(b_1)\rho(b_2) - \rho(b_1 b_2) - \rho(b_1)\xi(b_2)$$
$$+ \xi(b_1 b_2) - \xi(b_1)\rho(b_2)$$
$$= T(b_1, b_2) - b_1\xi(b_2) + \xi(b_1 b_2) - \xi(b_1)b_2$$
$$= (T - \delta^2 \xi)(b_1, b_2)$$
$$= 0.$$

Thus ϕ is a continuous algebra homomorphism of B into A. Since

$$\pi\phi(b) = \pi\rho(b) - \pi\xi(b) = \pi\rho(b) = b \qquad (b \in B),$$

$\phi(B)$ is a closed subalgebra of A. Since $\pi|_{\phi(B)}$ is one-one, $\phi(B) \cap R = \{0\}$. Given $a \in A$, we have $a - \phi\pi(a) \in R$ since

$$\pi(a - \phi\pi(a)) = \pi(a) - \pi(a) = 0,$$

and hence $a = \phi\pi(a) + (a - \phi\pi(a)) \in \phi(B) + R$. Therefore A has the Wedderburn property. $\square$

The above proof did not use the full force of the condition $H^2(B, R) = \{0\}$. Suppose conversely that A has the Wedderburn property and that $A = C \oplus R$ where C is a closed subalgebra of A. Let ρ be as above (with C in place of X), and let ϕ be the bicontinuous algebra isomorphism of B onto C, given by the Wedderburn property, such that $\phi\pi(c) = c$ $(c \in C)$. Let T be the member of $Z^2(B, R)$ defined by (4) and let $\xi = \rho - \phi$. Then $\pi\xi = 0$, so that $\xi \in BL(B, R)$. By reversing the corresponding argument in the proof of Theorem 9 we obtain $T = \delta^2 \xi$.

Feldman [446] gives an example of a Banach algebra A such that $\dim \mathrm{rad}(A) = 1$ and A does not have the Wedderburn property.

A useful introduction to the cohomology of operator algebras is included in Kadison and Ringrose [225].

Chapter VII. Miscellany

§ 45. Quasi-Algebraic Elements and Capacity

A will denote a complex Banach algebra with unit. As usual, a complex polynomial in one variable is said to be *monic* if the coefficient of the term of highest degree is 1. We denote by P_n the set of all complex monic polynomials of degree n.

An element a of A is *algebraic* if there exists a monic polynomial p with $p(a)=0$. It is easy to see that an element is algebraic if and only if it belongs to a finite dimensional subalgebra. Moreover, if $a_1, ..., a_n$ are mutually commuting algebraic elements, there exists a finite dimensional subalgebra containing the set $\{a_1, ..., a_n\}$, and therefore every polynomial in $a_1, ..., a_n$ is algebraic. On the other hand the sum and product of two non-commuting algebraic elements can fail to be algebraic. The spectrum of an algebraic element is finite; and, although the finiteness of $\mathrm{Sp}(a)$ does not imply that a is algebraic, it does imply the existence of a monic polynomial p for which $r(p(a))=0$.

The concept of quasi-algebraic element which we consider in this section is due to Halmos [183]. Roughly speaking an element a is quasi-algebraic if some monic polynomial is small øn a, the precise definition is in terms of the related concept of capacity, as follows.

Definition 1. Given $a \in A$,

$$\mathrm{cap}_n(a) = \inf\{\|p(a)\| : p \in P_n\},$$

$$\mathrm{cap}(a) = \inf\left\{(\mathrm{cap}_n(a))^{\frac{1}{n}} : n \in \mathbb{N}\right\},$$

and a is *quasi-algebraic* if $\mathrm{cap}(a)=0$. $\mathrm{cap}(a)$ is called the *capacity* of a.

We shall see that $\mathrm{cap}(a)$ coincides with the capacity, in the classical sense, of $\mathrm{Sp}(a)$.

Lemma 2. *Let* $a \in A$. *Then*
(i) *for each* $n \in \mathbb{N}$, *there exists* $t_n \in P_n$ *such that* $\mathrm{cap}_n(a) = \|t_n(a)\|$;
(ii) $\mathrm{cap}(a) = \lim_{n \to \infty} (\mathrm{cap}_n(a))^{\frac{1}{n}}$;
(iii) $\mathrm{cap}(a) \leqslant r(a)$.

Proof. (i) $\mathrm{cap}_n(a)$ is the infimum of the distances of a^n from the points of the linear span of the set $\{1, a, a^2, \ldots, a^{n-1}\}$, and so the infimum is attained.

(ii) Given $p \in P_n$, $q \in P_m$, we have $pq \in P_{n+m}$ and

$$\|(p\,q)(a)\| \leqslant \|p(a)\|\,\|q(a)\| .$$

Therefore

$$\mathrm{cap}_{n+m}(a) \leqslant \mathrm{cap}_n(a)\,\mathrm{cap}_m(a) ,$$

and the proof is completed as for the limit formula for the spectral radius (Proposition 2.8).

(iii) The polynomial z^n belongs to P_n. □

Proposition 3. $\mathrm{cap}(a)$ *is invariant under change to an equivalent algebra-norm.*

Proof. Lemma 2 (ii). □

Since the spectral radius is an algebra semi-norm on a commutative subalgebra we can define a concept analogous to $\mathrm{cap}(a)$ by using the spectral radius in place of the norm. Somewhat surprisingly we obtain the same value of $\mathrm{cap}(a)$ as we now show.

Definition 4. Given $a \in A$, let

$$\mathrm{Spcap}_n(a) = \inf\{r(p(a)): p \in P_n\} ,$$

$$\mathrm{Spcap}(a) = \inf\left\{(\mathrm{Spcap}_n(a))^{\frac{1}{n}}: n \in \mathbb{N}\right\}.$$

Theorem 5. *For all* $a \in A$, $\mathrm{Spcap}(a) = \mathrm{cap}(a)$.

Proof. It is evident that $\mathrm{Spcap}(a) \leqslant \mathrm{cap}(a)$. Suppose that $\mathrm{Spcap}(a) < \delta$. Then there exists $n \in \mathbb{N}$ with $\mathrm{Spcap}_n(a) < \delta^n$. Therefore there exists $p \in P_n$ such that

$$r(p(a)) < \delta^n .$$

We choose an equivalent algebra-norm on A for which $\|p(a)\| < \delta^n$ (Corollary 4.2), and then have $\mathrm{cap}_n(a) < \delta^n$, $\mathrm{cap}(a) < \delta$, by Proposition 3. □

Definition 6. Given a compact subset E of $\mathbb{C}$, the *capacity* of E, $\mathrm{Cap}(E)$ is defined by

$$\mathrm{Cap}_n(E) = \inf\{\|p\|_\infty : p \in P_n\} ,$$

$$\mathrm{Cap}(E) = \inf\left\{(\mathrm{Cap}_n(E))^{\frac{1}{n}}: n \in \mathbb{N}\right\},$$

where $\|\cdot\|_\infty$ denotes the uniform norm on $C(E)$.

Corollary 7. *For all* $a \in A$, $\mathrm{cap}(a) = \mathrm{Cap}(\mathrm{Sp}(a))$.

Proof. For all $p \in P_n$,
$$r(p(a)) = \max\{|p(z)| : z \in \mathrm{Sp}(a)\},$$
and so $\mathrm{Spcap}(a) = \mathrm{Cap}(\mathrm{Sp}(a))$. Apply Theorem 5. ☐

Theorem 8. *If* $\mathrm{Sp}(a)$ *is countable, then a is quasi-algebraic.*

Proof. It is known (see Tsuji [479]) that every countable compact subset of $\mathbb{C}$ has zero capacity. Apply Corollary 7. ☐

For further results on quasi-algebraic elements see Halmos [183]. It is there proved that every polynomial in a quasi-algebraic element is quasi-algebraic, and Theorem 8 is applied to prove that every bounded linear operator with countable spectrum on a separable Hilbert space is 'quasi-triangular'.

§ 46. Nilpotents and Quasi-Nilpotents

Let A be a complex Banach algebra. Recall that an element $x \in A$ is *nilpotent* if $x^n = 0$ for some $n \in \mathbb{N}$, and *quasi-nilpotent* if $r(x) = 0$.

Definition 1. Let $E \subset A$. We say that E is *nil* if each element of E is nilpotent, and E is *nilpotent* if, for some $n \in \mathbb{N}$, $E^n = \{0\}$, i.e.
$$x_1 x_2 \ldots x_n = 0 \qquad (x_1, x_2, \ldots, x_n \in E).$$

We need the following special case of the Nagata-Higman theorem; see Jacobson [455, Appendix C] for a proof.

Lemma 2. *Let $n \in \mathbb{N}$ and let $x^n = 0$ $(x \in A)$. Then $A^m = \{0\}$, where* $m = 2^n - 1$.

Theorem 3. (Grabiner [173]). *Every nil Banach algebra is nilpotent.*

Proof. Let A be nil, and, for $n \in \mathbb{N}$, let $A_n = \{x \in A : x^n = 0\}$. Then A_n is closed and $A = \bigcup \{A_n : n \in \mathbb{N}\}$. By the Baire category theorem, there exist $m \in \mathbb{N}$, $b \in A$, $r > 0$ such that
$$\{a : \|a - b\| < r\} \subset A_m.$$
Given $x \in A$ there exists $\delta > 0$ such that $b + \lambda x \in A_m$ $(|\lambda| < \delta)$; and so $(b + \lambda x)^m = 0$ $(|\lambda| < \delta)$, $x^m = 0$. Apply Lemma 2. ☐

Theorem 4. (Dixon [120]). *Every nil bi-ideal of A is contained in a sum of nilpotent bi-ideals.*

Proof. Let I be a non-zero nil bi-ideal of A. Since we are not assuming that I is closed, it is necessary to refine the argument of Theorem 3.

Let $y \in I \setminus \{0\}$, and let $A_n = \{x \in A : (xy)^n = 0\}$. Then A_n is closed and $A = \bigcup \{A_n : n \in \mathbb{N}\}$. It follows as in Theorem 3 that there exists $m \in \mathbb{N}$ such that $(xy)^m = 0$ $(x \in A)$. By Lemma 2, $(Ay)^k = \{0\}$ where $k = 2^m - 1$. Let J_y be the bi-ideal generated by y, and let $z \in J_y$. Then z is of the form

$$z = \lambda y + ay + yb + \sum_{j=1}^{r} c_j y d_j,$$

where $\lambda \in \mathbb{C}$, $a, b, c_j, d_j \in A$. Clearly $z^{2k} = 0$, and so I is contained in the sum of the nilpotent bi-ideals J_y $(y \in I)$. $\square$

Corollary 5. *If A is semi-prime, then $\{0\}$ is the only nil bi-ideal of A.*

Remark. Dixon [120] deduces from Theorem 4 that the nil, Levitzki, and Baer lower radicals of A all coincide with the sum of the nilpotent bi-ideals.

Let A be semi-simple. If A is commutative, then, by Corollary 17.7, 0 is the only quasi-nilpotent element of A. When A is not commutative it is natural to ask if A must contain a quasi-nilpotent or even a nilpotent element other than 0. Recall from Corollary 15.7 that A is commutative if $\inf\{r(x) : \|x\| = 1\} > 0$.

Kaplansky proved that every non-commutative B*-algebra contains a non-zero nilpotent element (see Dixmier [119, p. 58]). Theorem 31.6 shows that every non-commutative primitive Banach algebra with minimal one-sided ideals contains a non-zero nilpotent element. Duncan and Tullo [129] give some related results and also the following example.

Example 6. Let F be the free algebra on two symbols u, v, and let $\{w_n\}$ be the standard enumeration of the words given by

$$u, v, u^2, uv, vu, v^2, u^3, u^2 v, \dots .$$

Let A be the algebra of all infinite series $x = \sum \alpha_n w_n$, where $\|x\| = \sum |\alpha_n| < \infty$. Clearly A is a non-commutative Banach algebra. Let $x \in B$, $x \neq 0$, and let α_p be the first non-zero coefficient in the series $\sum \alpha_n w_n$. Then the coefficient of w_p^k in x^k is precisely α_p^k and hence

$$\|x^k\| \geq |\alpha_p^k|, \qquad r(x) \geq |\alpha_p| > 0 .$$

Thus 0 is the only quasi-nilpotent in A.

For the remainder of this section A will denote a radical Banach algebra. Thus every element of A is quasi-nilpotent. Example 18.25 gives a commutative radical Banach algebra which is an integral domain. The following example due to Dixon [120] has the additional property of possessing a bounded approximate identity.

Example 7. Let A be the space of complex measurable functions f on $\mathbb{R}^+$ such that

$$\|f\| = \int_0^\infty \exp(-t\,e^t)|f(t)|\,dt < \infty,$$

and let

$$(f*g)(t) = \int_0^t f(s)g(t-s)\,ds.$$

Then A is a commutative radical Banach algebra. A is an integral domain, by the Titchmarsh convolution theorem [383], and any one-sided Dirac sequence gives a bounded approximate identity.

A modification of the construction in Example 6 gives another interesting radical algebra.

Example 8. Let $\{w_n\}$ be as in Example 6 and let λ_n be the length of the word w_n. Let A be the algebra of all infinite series $x = \sum \alpha_n w_n$ where

$$\|x\| = \sum \frac{|\alpha_n|}{\lambda_n!} < \infty.$$

Then A is a non-commutative radical Banach algebra in which 0 is the only divisor of zero. Note that A can be regarded as an algebra of compact operators on the Banach space l^1.

Grabiner [174, 175] gives some interesting properties of quasi-nilpotent operators; he shows in particular that if A is radical and $x \in A$ is not nilpotent, then the sequence $\{A x^n\}$ is strictly decreasing.

We do not know if there exist radical Banach algebras that are simple or topologically simple.

§ 47. Positiveness of the Spectrum

Throughout this section, A denotes a complex Banach algebra with unit, and $\mathbb{R}^+$ denotes the set of all non-negative real numbers. We consider two kinds of positiveness of the spectrum:

(i) $\mathrm{Sp}(a)$ has non-void intersection with $\mathbb{R}^+$,
(ii) $\mathrm{Sp}(a)$ is contained in $\mathbb{R}^+$.

We discuss conditions on subsets Q of A which imply that the spectrum of every element of Q is positive in one of these senses.

Definition 1. A *wedge* is a non-void subset W of a real or complex linear space such that

$$x, y \in W, \ \alpha \in \mathbb{R}^+ \Rightarrow x+y, \alpha x \in W.$$

A *semi-algebra* in A is a wedge W in A such that

$$x, y \in W \Rightarrow xy \in W.$$

A *local semi-algebra* (l.s.a.) in A is a subset Q of A such that $Q \cap C$ is a semi-algebra for each commutative subalgebra C of A. Equivalently, Q is a local semi-algebra in A if Q is a non-void subset of A and

$$x, y \in Q, \quad \alpha \in \mathbb{R}^+, \quad xy = yx \Rightarrow x+y, xy, \alpha x \in Q.$$

An *elemental semi-algebra* (e.s.a.) in A is a non-void subset Q of A such that

$$a \in Q, n \in \mathbb{N}, \alpha_1, \ldots, \alpha_n \in \mathbb{R}^+ \Rightarrow \alpha_1 a + \alpha_2 a^2 + \cdots + \alpha_n a^n \in Q.$$

It is evident that every semi-algebra is a local semi-algebra and that every local semi-algebra is an elemental semi-algebra. It will appear from the following examples that neither of these implications can be reversed.

Example 2. Let E be a non-void compact Hausdorff space, $B = C(E)$,

$$W = \{f \in B : f(E) \subset \mathbb{R}^+\}, \quad Q = \{f \in B : f(E) \cap \mathbb{R}^+ \neq \emptyset\}.$$

Then W is a semi-algebra, and Q is an e.s.a. which is not a wedge if E has more than one point. Since B is commutative Q is therefore not an l.s.a. We note also that W has the property:

$$f \in W \Rightarrow (1+f)^{-1} \in W.$$

Example 3. Let W be a wedge in a complex Banach space X. Then $\{T \in BL(X) : TW \subset W\}$ is a semi-algebra in the Banach algebra $BL(X)$.

Example 4. Let H be a complex Hilbert space, and let P denote the set of all positive operators on H, i.e. all operators $T \in BL(H)$ such that

$$(Tx, x) \in \mathbb{R}^+ \quad (x \in H).$$

Then P is a wedge and a local semi-algebra. In fact, given $S, T \in P$, we have $ST \in P$ if and only if $ST = TS$. We note that P also satisfies the condition

$$T \in P \Rightarrow (I+T)^{-1} \in P.$$

Definition 5. The *spectral e.s.a.* and the *spectral l.s.a.* are the subsets $\mathrm{spesa}(A)$, $\mathrm{splsa}(A)$ defined by

$$\mathrm{spesa}(A) = \{a \in A : \mathrm{Sp}(a) \cap \mathbb{R}^+ \neq \emptyset\},$$

$$\mathrm{splsa}(A) = \{a \in A : \mathrm{Sp}(a) \subset \mathbb{R}^+\}.$$

Proposition 6. spesa(A) *is a closed elemental semi-algebra in A and*
(i) $a \in \text{spesa}(A) \cap \text{Inv}(A)$, $r(a^{-1}) < 1 \Rightarrow a - 1 \in \text{spesa}(A)$,
(ii) $a \in \text{spesa}(A)$, $r(a) < 1 \Rightarrow a(1-a)^{-1} \in \text{spesa}(A)$,
(iii) $a \in \text{spesa}(A) \Rightarrow \|1 + a\| \geqslant r(1 + a) \geqslant 1$.

Proof. If $a \in A \setminus \text{spesa}(A)$, then $\text{Sp}(a) \subset \mathbb{C} \setminus \mathbb{R}^+$, and Proposition 5.17 gives $\text{Sp}(b) \subset \mathbb{C} \setminus \mathbb{R}^+$ for all b sufficiently near a. Thus spesa(A) is closed, and the rest is clear. ☐

Proposition 7. *Let P be a subset of* $\text{Inv}(A)$ *such that*
(i) $a \in P$, $\alpha > 0 \Rightarrow \alpha a \in P$,
(ii) $a \in P$, $r(a^{-1}) < 1 \Rightarrow a - 1 \in P$,
(iii) $-1 \notin P^-$.
Then $P \subset \text{spesa}(A)$.

Proof. It follows easily from (i) and (ii) that P satisfies:
(ii)′ $a \in P$, $0 < \lambda < (r(a^{-1}))^{-1} \Rightarrow a - \lambda \in P$.
Let $a \in P$, and let $\mu = \sup\{\lambda \in \mathbb{R}^+ : a - \lambda \in P\}$. By (ii)′, $\mu \geqslant (r(a^{-1}))^{-1} > 0$. Also $\mu < \infty$, for if there exists a sequence $\{\lambda_n\}$ with $\lambda_n > 0$, $\lim_{n \to \infty} \lambda_n = \infty$ and $a - \lambda_n \in P$, then, by (i),

$$-1 = \lim_{n \to \infty} (\lambda_n^{-1} a - 1) \in P^-,$$

which contradicts (iii). We prove that $\mu \in \text{Sp}(a)$. Suppose on the contrary that $a - \mu \in \text{Inv}(A)$. There exists a sequence $\{\lambda_n\}$ with $0 < \lambda_n \leqslant \mu$, $a - \lambda_n \in P$, and $\lim_{n \to \infty} \lambda_n = \mu$. By continuity of the inverse mapping on $\text{Inv}(A)$, we have $\lim_{n \to \infty} (a - \lambda_n)^{-1} = (a - \mu)^{-1}$, and since, on a commutative subalgebra, the spectral radius is a semi-norm dominated by the norm,

$$\lim_{n \to \infty} r((a - \lambda_n)^{-1}) = r((a - \mu)^{-1}) > 0.$$

Choose t with $0 < t < (r(a - \mu)^{-1})^{-1}$. Then, for sufficiently large n,

$$0 < t < (r(a - \lambda_n)^{-1})^{-1},$$

and (ii)′ gives $a - \lambda_n - t \in P$. But, for sufficiently large n, we have $\lambda_n + t > \mu$, contradicting the definition of μ. ☐

Theorem 8. *Let Q be a subset of A such that*
(i) $a \in Q$, $\alpha \in \mathbb{R}^+ \Rightarrow \alpha a \in Q$,
(ii) $a \in Q$, $r(a) < 1 \Rightarrow a(1-a)^{-1} \in Q$,
(iii) $-1 \notin Q^-$.
Then $Q \subset \text{spesa}(A)$.

Proof. Let $P = \{a \in \text{Inv}(A) : a^{-1} \in Q\}$. For singular elements a of Q we have $0 \in \text{Sp}(a)$, and so it is enough to prove that $Q \cap \text{Inv}(A) \subset \text{spesa}(A)$.

For $a \in \mathrm{Inv}(A)$ we have $\mathrm{Sp}(a) = (\mathrm{Sp}(a^{-1}))^{-1}$, and so it is enough to prove that $P \subset \mathrm{spesa}(A)$. Given $a \in P$ with $r(a^{-1}) < 1$, we have $a^{-1} \in Q$, and therefore

$$(a-1)^{-1} = a^{-1}(1-a^{-1})^{-1} \in Q .$$

Thus $a - 1 \in P$, and P satisfies condition (ii) of Proposition 7. P obviously satisfies conditions (i) and (iii) of that proposition, and so $P \subset \mathrm{spesa}(A)$. ☐

Corollary 9. *Let Q be an elemental semi-algebra in A such that $-1 \notin Q^-$. Then $Q \subset \mathrm{spesa}(A)$.*

Proof. We may assume that Q is closed. If $a \in Q$ with $r(a) < 1$, then

$$a(1-a)^{-1} = \lim_{n \to \infty} (a + a^2 + \cdots + a^n) \in Q . \qquad ☐$$

Remarks. Theorem 8 is due to Yood [430], Proposition 7 and the present proof of Theorem 8 to Bonsall and Thompson [82]. Other conditions implying that $Q \subset \mathrm{spesa}(A)$ are given in [82]; for example it is proved that condition (ii) in Theorem 8 can be replaced by the purely algebraic condition:

$$a \in Q \;\Rightarrow\; a + a^2 \in Q .$$

On the other hand, it cannot be replaced by:

$$a \in Q \;\Rightarrow\; a^2 \in Q ;$$

for the set $\{\rho e^{i\theta} : \rho \in \mathbb{R}^+, \theta \in \{0, 2\pi/3, 4\pi/3\}\}$ is a multiplicative semi-group in $\mathbb{C}$ satisfying conditions (i) and (iii) of Theorem 8.

Definition 10. A subset Q of A is said to be of *type* 0 if

$$a \in Q \;\Rightarrow\; (1+a)^{-1} \in Q .$$

Proposition 11. $\mathrm{splsa}(A)$ *is a type 0 local semi-algebra in A, and the intersection of $\mathrm{splsa}(A)$ with each closed commutative subalgebra of A is closed.*

Proof. The final statement uses Proposition 5.18, and the rest is clear. ☐

Proposition 12. *Let Q be a type 0 subset of A such that*

$$a \in Q, \; \alpha \in \mathbb{R}^+ \;\Rightarrow\; \alpha a, \; \alpha + a, \; a^2 \in Q .$$

Then $Q \subset \mathrm{splsa}(A)$.

Proof. Let $a \in Q$, and suppose that $r e^{i\theta} \in \mathrm{Sp}(a)$ with $r > 0$ and $\theta \in [-\pi, \pi] \setminus \{0\}$. Since Q is of type 0, $1 + r^{-1} a \in \mathrm{Inv}(A)$, and so either $0 < \theta < \pi$ or $-\pi < \theta < 0$. If $0 < \theta < \pi$, choose $n \in \mathbb{N}$ such that $\pi/2 \leqslant 2^n \theta < \pi$, and take $b=a^{2^n}$. Then $b\in Q$ and $\mathrm{Sp}(b)$ contains a point $-\gamma+i\delta$ with $\gamma\geqslant 0$ and $\delta>0$. We have $(\gamma+b)^2\in Q$ and $-\delta^2\in\mathrm{Sp}((\gamma+b)^2)$. Thus

$$\delta^2+(\gamma+b)^2\notin\mathrm{Inv}(A),$$

contradicting the type 0 condition. A similar argument applies if $-\pi<\theta<0$. $\square$

Corollary 13. *Every type 0 elemental semi-algebra in A is contained in* $\mathrm{splsa}(A)$.

Proof. Clear. $\square$

Remark. In Proposition 12 and its corollary, we have not used the full force of the type 0 condition but only:

$$a\in Q \Rightarrow 1+a\in\mathrm{Inv}(A).$$

Theorem 14. *Let Q be a type 0 elemental semi-algebra in A that has closed intersection with each closed commutative subalgebra of A. Then:*
(i) $a\in Q,\ r(a)\leqslant 1 \Rightarrow 1-a\in Q$;
(ii) *if $a+b\in Q$ whenever a,b are commuting elements of Q, then Q is a local semi-algebra.*

Proof. (i) Let $a\in Q,\ r(a)<1$. Then $(1-a)^{-1}=1+b$ with $b=\sum_{k=1}^{\infty}a^k\in Q$. Therefore $1-a=(1+b)^{-1}\in Q$, by the type 0 condition. If $a\in Q$ with $r(a)=1$, we take $\alpha_n=1-\dfrac{1}{n}$. Then $1-\alpha_n a\in Q$ and so $1-a\in Q$.

(ii) We note first that

$$a\in Q \Rightarrow a(1+a)^{-1}\in Q. \tag{1}$$

For, let $a\in Q$ and $b=(1+a)^{-1}$. By Corollary 13, $\mathrm{Sp}(a)\subset\mathbb{R}^+$, and so $r(b)\leqslant 1$. Also $b\in Q$, and therefore, by (i), $1-b\in Q$, i.e. $a(1+a)^{-1}\in Q$; Q satisfies (1).

Let log denote the branch of the logarithm that is holomorphic on $\mathbb{C}\setminus\mathbb{R}_-$ and satisfies $\log 1=0$. We prove that

$$a\in Q \Rightarrow \log(1+a)\in Q. \tag{2}$$

Let $a\in Q$, and let $c=a(1+a)^{-1}$. By (1), we have $c\in Q$, and, since $\mathrm{Sp}(a)\subset\mathbb{R}^+$, $r(c)<1$. Therefore

$$\log(1-c)^{-1}=c+\tfrac{1}{2}c^2+\tfrac{1}{3}c^3+\cdots\in Q.$$

Since $1-c=(1+a)^{-1}$, this proves that $\log(1+a)\in Q$.

Suppose now that $a+b\in Q$ whenever a,b are commuting elements of Q. Let $a,b\in Q$ with $ab=ba$ and let $\varepsilon>0$. By (2)

$$\log(\varepsilon+a)-\log\varepsilon,\quad \log(\varepsilon+b)-\log\varepsilon\in Q.$$

Thus there exists $q_\varepsilon \in Q$ such that

$$\log((\varepsilon + a)(\varepsilon + b)) = \log \varepsilon^2 + q_\varepsilon,$$

$$(\varepsilon + a)(\varepsilon + b) = \varepsilon^2 \exp(q_\varepsilon) \in Q.$$

Finally, let $\varepsilon \to 0$. ☐

Remark. The type 0 condition is a particular case of the *type n* condition

$$a \in Q \implies a^n (1 + a)^{-1} \in Q.$$

Semi-algebras and wedges of types 0, 1, 2 in $C(E)$ have been considered by Bonsall [64, 65, 66], Pryce [307], and Brown [88, 89].

§ 48. Type 0 Semi-Algebras

Throughout this section A denotes a complex Banach algebra with unit, and Q a closed type 0 semi-algebra in A, i.e. Q is a non-void closed subset of A such that

$$x, y \in Q, \ \alpha \in \mathbb{R}^+ \implies x + y, \alpha x, x y, (1 + x)^{-1} \in Q.$$

We denote by R the real subalgebra of A given by $R = Q - Q = \{x - y : x, y \in Q\}$, and by $\leqslant$ the relation of quasi-order on R given by the wedge Q, i.e. $x \leqslant y$ means $y - x \in Q$.

The axioms satisfied by Q are much stronger than is visible at first sight. We have remarked already in Example 47.4 that the type 0 local semi-algebra P of positive operators on a Hilbert space has the property that the product of two of its elements belongs to P only if they commute. We prove that Q is nearly commutative and that R can be represented by an algebra of real functions in which Q appears as the non-negative functions. We end the section with an account of an interesting construction due to Barbeau [37] for maximal subsemi-algebras of Q.

Lemma 1. $a \in Q, \alpha \geqslant r(a) \implies \alpha - a \in Q.$

Proof. Let $a \in Q$ and $\alpha \geqslant r(a)$. If $\alpha > 0$, then $r\left(\frac{1}{\alpha} a\right) \leqslant 1$, and Theorem 47.14 (i) gives $1 - \frac{1}{\alpha} a \in Q$, $\alpha - a \in Q$. Let $\alpha = 0$. Then $r(a) = 0$ and so, for all $n \in \mathbb{N}$, $r(na) = 0 < 1$, $1 - na \in Q$, $\frac{1}{n} - a \in Q$. Since Q is closed, it follows that $-a \in Q$. ☐

Theorem 2. R *is a closed real subalgebra of* A *and*
(i) $Q = R \cap \mathrm{splsa}(A),$
(ii) $\mathrm{Sp}(a) \subset \mathbb{R} \ (a \in R),$
(iii) $a \in R, r(a) \leqslant 1 \Leftrightarrow -1 \leqslant a \leqslant 1,$
(iv) $a, b \in Q \implies ab - ba \in Q \cap (-Q) \ and \ r(ab - ba) = 0.$

Proof. (i) Let $a = x - y$ with $x, y \in Q$, let $\mathrm{Sp}(a) \subset \mathbb{R}^+$, and let $\alpha = r(y)$. By Lemma 1,

$$a + \alpha = x + (\alpha - y) \in Q .$$

Since $\mathrm{Sp}(a) \subset \mathbb{R}^+$, $r(a + \alpha) = r(a) + \alpha$. Therefore, by Lemma 1 again,

$$r(a) - a = r(a + \alpha) - (a + \alpha) \in Q .$$

Since $\mathrm{Sp}(a) \subset \mathbb{R}^+$, $r(r(a) - a) \leqslant r(a)$, and so a third application of Lemma 1 gives

$$a = r(a) - (r(a) - a) \in Q .$$

We have now proved that $R \cap \mathrm{splsa}(A) \subset Q$, and the opposite inclusion is given by Corollary 47.13.

(ii) Let $a \in R$. Then $a = x - y$ with $x, y \in Q$, and, as in the proof of (i), $a + r(y) \in Q$. Since $Q \subset \mathrm{splsa}(A)$, this gives $\mathrm{Sp}(a) \subset \mathbb{R}$.

(iii) Let $a \in R$, and $r(a) \leqslant 1$. By (ii), $\mathrm{Sp}(a) \subset \mathbb{R}$, and so $\mathrm{Sp}(1 \pm a) \subset \mathbb{R}^+$, $1 \pm a \in R \cap \mathrm{splsa}(A) = Q$, $-1 \leqslant a \leqslant 1$. Conversely, if $1 \pm b \in Q$, then $\mathrm{Sp}(1 \pm b) \subset \mathbb{R}^+$, $\mathrm{Sp}(b) \subset [-1, 1]$, $r(b) \leqslant 1$.

(iv) Define p on R by

$$p(x) = \sup \mathrm{Sp}(x) \qquad (x \in R) .$$

We prove that p is a sublinear functional on R. Given $x, y \in R$, it follows from (ii) and (i) that $p(x) - x \in Q$ and $p(y) - y \in Q$. Therefore

$$p(x) + p(y) - (x + y) \in Q .$$

Since $Q \subset \mathrm{splsa}(A)$, it follows that

$$p(x + y) \leqslant p(x) + p(y) ,$$

and the positive homogeneity of p is obvious.

Let $a \in R$ and let K denote the set of all real linear functionals f on R such that $f(a) = p(a)$ and $f(x) \leqslant p(x)$ $(x \in R)$. Since p is sublinear, K is a non-void weak* compact convex set. We have $p(1) = -p(-1) = 1$ and $p(-x) \leqslant 0$ $(x \in Q)$. Therefore

$$f(1) = 1 \qquad (f \in K), \tag{1}$$

$$f(x) \geqslant 0 \qquad (f \in K, x \in Q) . \tag{2}$$

Since $p(x) - x \in Q$, we have

$$p(x) y - y x \in Q \qquad (x \in R, y \in Q), \tag{3}$$

and therefore, by (2)

$$f(y x) \leqslant p(x) f(y) \qquad (x \in R, y \in Q, f \in K) . \tag{4}$$

By the Krein-Milman theorem there exists an extreme point f of K. We prove that f is a multiplicative linear functional. Since $R = Q - Q$ it is enough to prove that

$$f(ux) = f(u) f(x) \qquad (u \in Q, \ x \in R). \tag{5}$$

Since $f(1) = 1$, (5) holds if it holds with u replaced by $1 + u$. Thus it is enough to prove (5) with $u \in Q$ and $f(u) > 0$. Then, by normalization it is enough to prove (5) with $0 < u < 1$, and $0 \neq f(u) \neq 1$. Given such $u \in Q$, define f_1, f_2 on R by

$$f_1(x) = f(ux)/f(u), \qquad f_2(x) = f((1-u)x)/f(1-u) \qquad (x \in R).$$

By (4), $f_1(x) \leqslant p(x)$, and similarly $f_2(x) \leqslant p(x)$. We have $0 < f(u) < 1$ and

$$f = f(u) f_1 + (1 - f(u)) f_2 .$$

Thus

$$p(a) = f(a) = f(u) f_1(a) + (1 - f(u)) f_2(a) \leqslant p(a),$$

and $f_1(a) = f_2(a) = p(a)$, $f_1, f_2 \in K$. Since f is an extreme point of K it follows that $f_1 = f$ and (5) is proved. We have proved that given $a \in R$, there exists a multiplicative linear functional f with $f(a) = p(a)$.

Suppose now that $a, b \in Q$. Then $ab - ba \in R$ and, by what we have just proved, there exists a multiplicative linear functional ϕ on R with

$$\phi(ab - ba) = p(ab - ba).$$

Therefore $p(ab - ba) = 0$, and similarly $p(ba - ab) = 0$. Therefore $r(ab - ba) = 0$, $ab - ba \in \mathrm{splsa}(A) \cap R = Q$. Similarly $ba - ab \in Q$. Finally, to prove that R is closed, let $x_n \in R$ with $\lim_{n \to \infty} x_n = x \in A$. Then there exists $m \in \mathbb{N}$ such that

$$\|x_p - x_q\| \leqslant 1 \qquad (p, q \geqslant m).$$

It follows from (i) that

$$1 + x_p - x_m \in Q \qquad (p \geqslant m),$$

and so $1 + x - x_m \in Q$, $x \in R$. $\square$

Inspection of the proof of Theorem 2 shows that the type 0 condition is needed only to give $Q \subset \mathrm{splsa}(A)$ and Lemma 1. However, the extra generality obtained by starting from these as axioms would be illusory, as the next proposition shows.

Proposition 3. *Let P be a closed semi-algebra in A such that*
  (i) $a \in P \Rightarrow 1 + a \in \mathrm{Inv}(A)$,
  (ii) $a \in P \Rightarrow \alpha - a \in P$ *for some* $\alpha > \max(1, r(a))$.
Then P is of type 0.

Proof. By the remark following Corollary 47.13, $P \subset \text{splsa}(A)$. Let $a \in P$. Then $\text{Sp}(a) \subset \mathbb{R}^+$ and there exists $\alpha > \max(1, r(a))$ such that $\alpha - a \in P$. We have

$$(\alpha + a)^{-1} = (\alpha - a)(\alpha^2 - a^2)^{-1} = (\alpha - a)\left(\frac{1}{\alpha^2} + \frac{1}{\alpha^4} a^2 + \cdots\right) \in P,$$

and $0 < (\alpha - 1) r((\alpha + a)^{-1}) < 1$; and so

$$(1 + a)^{-1} = (\alpha + a)^{-1}[1 - (\alpha - 1)(\alpha + a)^{-1}]^{-1} \in P. \quad \square$$

Theorem 4. *Let Ψ_R denote the set of non-zero homomorphisms of R into $\mathbb{R}$. Then Ψ_R is a non-void weak* compact subset of the (real) dual space R' of R, the mapping $a \to a^\wedge$ given by*

$$a^\wedge(\phi) = \phi(a) \qquad (a \in R, \ \phi \in \Psi_R),$$

is a homomorphism of R into $C_\mathbb{R}(\Psi_R)$ and, for $a \in R$,
   (i) $\max \text{Sp}(a) = \max a^\wedge(\Psi_R)$,
   (ii) $a^\wedge \geqslant 0$ *if and only if $a \in Q$.*
Moreover the following statements are equivalent:
   (iii) $Q \cap (-Q) = \{0\}$,
   (iv) $a \in Q, \ r(a) = 0 \Rightarrow a = 0$,
   (v) *the mapping $a \to a^\wedge$ is an isomorphism.*

Proof. We have seen in the proof of Theorem 2 that, for each $a \in R$, there exists $\phi \in \Psi_R$ with

$$\phi(a) = \max \text{Sp}(a).$$

Since $a^\wedge(\Psi_R) \subset \text{Sp}(a)$, (i) is now clear. We have proved that Ψ_R is non-void, and it is now routine to verify that Ψ_R is a weak* compact subset of R' and that the mapping $a \to a^\wedge$ is a homomorphism of R into $C_\mathbb{R}(\Psi_R)$.

(ii) If $a \in Q$, we have $\max \text{Sp}(-a) \leqslant 0$, $-a^\wedge(f) \leqslant 0$ $(f \in \Psi_R)$, and so $a^\wedge \geqslant 0$. Conversely, if $a \in R$ and $a^\wedge \geqslant 0$, we have $\max \text{Sp}(-a) \leqslant 0$, $a \in \text{splsa}(A) \cap R = Q$.

(iii) $\Rightarrow$ (iv). Assume (iii) and let $a \in Q$ with $r(a) = 0$. Then $-a \in \text{splsa}(A) \cap R = Q$, $a \in Q \cap (-Q) = \{0\}$.

(iv) $\Rightarrow$ (v). Assume (iv) and let $a \in R$ with $a^\wedge = 0$. Then $\max \text{Sp}(a) = \max \text{Sp}(-a) = 0$, $r(a) = 0$. Then $a \in \text{splsa}(A) \cap R = Q$, $a = 0$.

(v) $\Rightarrow$ (iii). Assume (v) and let $a \in Q \cap (-Q)$. Then $a^\wedge \geqslant 0$ and $-a^\wedge = (-a)^\wedge \geqslant 0$. Thus $a^\wedge = 0$, $a = 0$. $\quad \square$

Corollary 5. *Suppose that there exists a positive constant κ such that*

$$\|a\| \leqslant \kappa r(a) \qquad (a \in Q).$$

Then the mapping $a \to a^\wedge$ defined in Theorem 4 is a bicontinuous isomorphism of R onto $C_\mathbb{R}(\Psi_R)$.

Proof. Let $a \in R$. Then $r(a) - a \in Q$, and so

$$\|a\| \leqslant \|a - r(a)\| + \|r(a)\|$$
$$= \|r(a) - a\| + r(a)\|1\|$$
$$\leqslant \kappa r(r(a) - a) + \kappa r(a) \leqslant 3\kappa r(a).$$

Thus r is an algebra-norm on R equivalent to the given algebra-norm. Therefore R is complete with respect to r, and the rest is routine. □

Example 6. Let $P = \mathbb{R}^+ + \mathrm{rad}(A)$. Then P is a closed type 0 semi-algebra, and $P \cap (-P) = \mathrm{rad}(A)$. To see that P is type 0, let $\alpha \in \mathbb{R}^+$, $a \in \mathrm{rad}(A)$. Then $1 + \alpha + a \in \mathrm{Inv}(A)$ and

$$(1 + \alpha + a)^{-1} = \frac{1}{1 + \alpha} - \frac{a}{(1 + \alpha)^2} + \frac{a^2}{(1 + \alpha)^3} - \cdots \in P.$$

To see that P is closed, let $\alpha_n \in \mathbb{R}^+$, $a_n \in \mathrm{rad}(A)$, $\lim_{n \to \infty} (\alpha_n + a_n) = b$. Then the sequence $\{\alpha_n\}$ is bounded, for otherwise $1 \in \mathrm{rad}(A)$. That $b \in P$ follows easily.

Example 7. Let A be commutative and let $P = \mathrm{splsa}(A)$. Then P is a closed type 0 semi-algebra, and $P \cap (-P) = \mathrm{rad}(A)$.

Notation. For the rest of this section we assume that A is commutative, Q as before is a closed type 0 semi-algebra in A, $R = Q - Q$, R', Ψ_R are as in Theorem 4. We denote by R'^+ the set of all $\sigma \in R'$ such that

$$\sigma(a) \geqslant 0 \quad (a \in Q).$$

Given $\phi \in R'$, we denote by γ_ϕ the 'geometric mean' on $Q \cap \mathrm{Inv}(A)$ defined by

$$\gamma_\phi(a) = \exp(\phi(\log a)) \quad (a \in Q \cap \mathrm{Inv}(A)).$$

Here log denotes the branch of the logarithm that is holomorphic on $\mathbb{C} \setminus \mathbb{R}$ and satisfies $\log 1 = 0$. That $\phi(\log a)$ is well defined follows from the next lemma.

Lemma 8. $\exp(R) = Q \cap \mathrm{Inv}(A)$, $\log(Q \cap \mathrm{Inv}(A)) = R$.

Proof. By Theorem 2, R is a closed real subalgebra of A, and so $\exp(R) \subset R$. Also, for all $x \in R$, we have $\mathrm{Sp}(x) \subset \mathbb{R}$, and so $\exp(x) \in \mathrm{splsa}(A)$. Therefore, by Theorem 2 again, $\exp(x) \in Q \cap \mathrm{Inv}(A)$. On the other hand, given $a \in Q \cap \mathrm{Inv}(A)$, we have $a - \varepsilon \varepsilon \in R \cap \mathrm{splsa}(A) = Q$ for some $\varepsilon > 0$. Thus, as in the proof of Theorem 47.14,

$$\log a = \log(1 + \varepsilon^{-1}(a - \varepsilon)) + \log \varepsilon \varepsilon \in R. □$$

Lemma 9. *Let* $\phi \in R'$. *Then* γ_ϕ *is continuous on* $Q \cap \mathrm{Inv}(A)$, $\gamma_\phi(1) = 1$, *and, for all* $a, b \in Q \cap \mathrm{Inv}(A)$,

$$\gamma_\phi(a) > 0, \qquad \gamma_\phi(ab) = \gamma_\phi(a)\gamma_\phi(b).$$

Proof. Clear. ☐

Proposition 10. *Let* $\phi \in R'$, $\psi \in \Psi_R$, *and let*

$$G(\phi, \psi) = \{a \in Q \cap \mathrm{Inv}(A): \psi(a) \leqslant \gamma_\phi(a)\}.$$

Then either $G(\phi, \psi) = Q \cap \mathrm{Inv}(A)$, *or* $G(\phi, \psi)$ *is a maximal element of the set of (multiplicative) semi-groups properly contained in* $Q \cap \mathrm{Inv}(A)$.

Proof. Since $Q \subset \mathrm{splsa}(A)$, we have $\psi(a) > 0$ $(a \in Q \cap \mathrm{Inv}(A))$. Therefore $G(\phi, \psi)$ is a semi-group. Let $u \in (Q \cap \mathrm{Inv}(A)) \backslash G(\phi, \psi)$. Then $\psi(u) > \gamma_\phi(u)$, and so $\psi(u^{-1}) < \gamma_\phi(u^{-1})$. By continuity of ψ and γ_ϕ, there exists an open neighbourhood V of u^{-1} in $Q \cap \mathrm{Inv}(A)$ with $\psi(v) < \gamma_\phi(v)$ $(v \in V)$ and so with $V \subset G(\phi, \psi)$. Let S be a semi-group containing u and $G(\phi, \psi)$. Then uV is a neighbourhood of 1 and is contained in S. We prove that $Q \cap \mathrm{Inv}(A) \subset S$.

Given $a \in Q \cap \mathrm{Inv}(A)$, we have $a^{\frac{1}{n}} = \exp\left(\frac{1}{n}\log a\right) \in Q \cap \mathrm{Inv}(A)$, and $\lim_{n \to \infty} a^{\frac{1}{n}} = 1$. Therefore there exists $n \in \mathbb{N}$ with $a^{\frac{1}{n}} \in uV \subset S$. But then $a = \left(a^{\frac{1}{n}}\right)^n \in S$. ☐

Lemma 11. *Let* $\sigma \in R'^+$ *with* $\sigma(1) = 1$. *Then, for all* $a, b \in Q \cap \mathrm{Inv}(A)$,
(i) $0 < \gamma_\sigma(a) \leqslant \sigma(a) \leqslant r(a)$,
(ii) $\gamma_\sigma(a+b) \geqslant \gamma_\sigma(a) + \gamma_\sigma(b)$,
(iii) $b - a \in Q \cap \mathrm{Inv}(A) \Rightarrow \gamma_\sigma(a) < \gamma_\sigma(b)$.

Proof. (i) Let $a \in Q \cap \mathrm{Inv}(A)$. For all $\varepsilon > 0$, $r(a) + \varepsilon - a \in Q$, and so $r(a) + \varepsilon - \sigma(a) \geqslant 0$, and therefore $\sigma(a) \leqslant r(a)$. Also there exists $\delta > 0$ such that $a - \delta \in Q$, and so

$$\sigma(a) \geqslant \delta > 0.$$

Let $x = a - 1 - \log a$. Then $x \in R$ and

$$\psi(x) = \psi(a) - 1 - \psi(\log a) = \psi(a) - 1 - \log(\psi(a)) \geqslant 0 \qquad (\psi \in \Phi_A).$$

Therefore $x \in R \cap \mathrm{splsa}(A) = Q$, and so

$$\sigma(a) - 1 - \sigma(\log a) \geqslant 0. \tag{6}$$

Since $\sigma(a) > 0$, we may replace a by $(\sigma(a))^{-1} a$ in (6) and obtain

$$\sigma(\log a) = \sigma\left(\log((\sigma(a))^{-1} a)\right) + \log \sigma(a)$$
$$\leqslant \sigma((\sigma(a))^{-1} a) - 1 + \log \sigma(a) = \log \sigma(a).$$

(ii) Let $a, b \in Q \cap \mathrm{Inv}(A)$. Then $a(a+b)^{-1}$, $b(a+b)^{-1} \in Q \cap \mathrm{Inv}(A)$, and (i) gives

$$\gamma_\sigma(a(a+b)^{-1}) + \gamma_\sigma(b(a+b)^{-1}) \leqslant \sigma(a(a+b)^{-1}) + \sigma(b(a+b)^{-1}) = \sigma(1) = 1 .$$

The multiplicative property of γ_σ now gives (ii).

(iii) Let $a, b, b - a \in Q \cap \mathrm{Inv}(A)$. By (ii),

$$\gamma_\sigma(b) = \gamma_\sigma((b-a)+a) \geqslant \gamma_\sigma(b-a) + \gamma_\sigma(a) > \gamma_\sigma(a) . \quad \square$$

Notation. Given $\sigma \in R'^{+}$ with $\sigma(1) = 1$, we extend γ_σ to the whole of Q by taking

$$\gamma_\sigma(a) = \lim_{\lambda \to 0+} \gamma_\sigma(\lambda + a) \qquad (a \in Q) .$$

If $a \in Q$ and $0 < \lambda < \mu$, then $\lambda + a$, $\mu + a$, $(\mu + a) - (\lambda + a) \in Q \cap \mathrm{Inv}(A)$, and so, by Lemma 11, $0 < \gamma_\sigma(\lambda + a) < \gamma_\sigma(\mu + a)$. Thus $\lim_{\lambda \to 0+} \gamma_\sigma(\lambda + a)$ exists in $\mathbb{R}^{+}$. Also the two definitions coincide on $Q \cap \mathrm{Inv}(A)$, by the continuity there of γ_σ as originally defined.

Lemma 12. *Let* $\sigma \in R'^{+}$ *with* $\sigma(1) = 1$. *Then, for all* $a, b \in Q$,
   (i) $0 \leqslant \gamma_\sigma(a) \leqslant \sigma(a) \leqslant r(a)$,
   (ii) $\gamma_\sigma(a+b) \geqslant \gamma_\sigma(a) + \gamma_\sigma(b)$,
   (iii) $b - a \in Q \Rightarrow \gamma_\sigma(a) \leqslant \gamma_\sigma(b)$,
   (iv) $\gamma_\sigma(ab) = \gamma_\sigma(a) \gamma_\sigma(b)$,
   (v) $\gamma_\sigma(\alpha a) = \alpha \gamma_\sigma(a)$ $(\alpha \in \mathbb{R}^{+})$.

Proof. (i) By Lemma 11, we have

$$\gamma_\sigma(\lambda + a) \leqslant \sigma(\lambda + a) = \lambda + \sigma(a) \qquad (a \in Q, \ \lambda > 0) .$$

Likewise (ii), (iii), (iv) follow easily from Lemma 11. Finally, (v) follows from (iv) since for $\alpha \in \mathbb{R}^{+}$, we have $\alpha \in Q$ and $\gamma_\sigma(\alpha) = \alpha$. $\quad \square$

Theorem 13. *Let* $\sigma \in R'^{+}$ *with* $\sigma(1) = 1$, *let* $\psi \in \Psi_R$, *and let*

$$H(\sigma, \psi) = \{a \in Q : \psi(a) \leqslant \gamma_\sigma(a)\} .$$

Then either $H(\sigma, \psi) = Q$ *or* $H(\sigma, \psi)$ *is a closed semi-algebra which is maximal in the set of all closed semi-algebras properly contained in* Q.

Proof. That $H(\sigma, \psi)$ is a semi-algebra is clear from Lemma 12. To prove that $H(\sigma, \psi)$ is closed, let $a_n \in H(\sigma, \psi)$, $a \in Q$, $\|a - a_n\| < \varepsilon_n$, $\lim_{n \to \infty} \varepsilon_n = 0$. Then $r(a - a_n) < \varepsilon_n$, and so $\varepsilon_n + a - a_n \in Q$ and $\psi(a - a_n) < \varepsilon_n$. Thus

$$\psi(a) < \psi(a_n) + \varepsilon_n \leqslant \gamma_\sigma(a_n) + \varepsilon_n \leqslant \gamma_\sigma(\varepsilon_n + a) + \varepsilon_n ,$$

and $\lim_{n \to \infty} \gamma_\sigma(\varepsilon_n + a) + \varepsilon_n = \gamma_\sigma(a)$. Therefore $\psi(a) \leqslant \gamma_\sigma(a)$, $a \in H(\sigma, \psi)$.

In the notation of Proposition 10, we have

$$G(\sigma, \psi) = H(\sigma, \psi) \cap \mathrm{Inv}(A) .$$

Let S be a closed semi-algebra with

$$H(\sigma, \psi) \subset S \subset Q.$$

Then $G(\sigma, \psi) \subset S \cap \text{Inv}(A)$, and, by Proposition 10, either $S \cap \text{Inv}(A)$ $= G(\sigma, \psi)$ or $S \cap \text{Inv}(A) = Q \cap \text{Inv}(A)$. In the first case $S \cap \text{Inv}(A)$ $= H(\sigma, \psi) \cap \text{Inv}(A)$. Given $a \in S$, $\varepsilon > 0$, we have $a + \varepsilon e \in S \cap \text{Inv}(A) \subset H(\sigma, \psi)$, and, since $H(\sigma, \psi)$ is closed, $a \in H(\sigma, \psi)$. Thus in this case $S = H(\sigma, \psi)$. In the second case, given $a \in Q$, $\varepsilon > 0$, we have $a + \varepsilon e \in Q \cap \text{Inv}(A) \subset S$, and, since S is closed, $a \in S$. Thus $S = Q$. ☐

Theorems 2 and 4 for commutative algebras are essentially due to Brown [89], and Theorem 13 together with further results on maximal subsemi-algebras to Barbeau [37]. The application of the Krein-Milman theorem in the proof of Theorem 2 is similar to its use in the proof of Stone's theorem on ordered algebras (see Kadison [456, Lemma 3.2]).

§ 49. Locally Compact Semi-Algebras

For Banach algebras, local compactness is an uninteresting axiom in that it forces the algebra to have finite dimension. For semi-algebras this does not happen, and there are important examples of locally compact semi-algebras with infinite dimension. At the same time, local compactness has significant structural consequences for semi-algebras.

Let A denote a Banach algebra (real or complex), let B be a semi-algebra in A, and, given $r > 0$, let

$$B_r = \{x \in B: \|x\| \leqslant r\}.$$

Definition 1. B is a *locally compact semi-algebra* if B_1 is a compact set (in the norm topology).

It is easily verified that B is a locally compact semi-algebra if and only if it is a closed subset of A and is a locally compact space in the induced topology. For, given $b \in B$ and $r > \|b\|$, B_r is a compact relative neighbourhood of b; and the rest is clear.

Examples 2. (i) Every closed semi-algebra in a finite dimensional normed algebra is a locally compact semi-algebra.

(ii) Let T be a compact linear operator on a complex Banach space X, and let $0 \neq r(T) \in \text{Sp}(T)$. Then the least closed semi-algebra in $BL(X)$ containing T is a locally compact semi-algebra, and in general has infinite dimension. For the proof, see Bonsall and Tomiuk [83], where it is remarked that the compactness of the operator T is only used to establish that $r(T)$ is an isolated point of the spectrum with a spectral projection of finite rank. Thus the statement holds for Riesz operators

T and indeed whenever $r(T)$ is a pole of the resolvent of T (see Kaashoek and West [222, 223]).

(iii) For an example of an infinite dimensional locally compact semi-algebra of non-negative continuous functions, see Bonsall [67].

Notation. B will denote a locally compact semi-algebra with

$$\{0\} \neq B \subset A .$$

Various notations and concepts are used relative to B. Given a subset E of B,

$$S(E) = \{x \in E : \|x\| = 1\}, \quad \mathrm{ran}(E) = \{x \in B : Ex = \{0\}\},$$

$$\mathrm{lan}(E) = \{x \in B : xE = \{0\}\} .$$

A *right ideal* of B is a subsemi-algebra J of B such that $JB \subset J$, and similarly for left ideals and bi-ideals. B is *semi-prime* if $\{0\}$ is the only bi-ideal J of B with $J^2 = \{0\}$, *prime* if, for all non-zero bi-ideals J of B,

$$\mathrm{ran}(J) = \mathrm{lan}(J) = \{0\} .$$

A *minimal closed right ideal* is a minimal element of the class of non-zero closed right ideals of B, and similarly for ideals of other kinds.

It is easily seen that each non-zero closed right ideal J of B contains a minimal closed right ideal, for $S(J)$ is a non-void compact set.

Lemma 3. *Let E be a closed subset of B such that $\mathbb{R}^+ E \subset E$, let $a \in B$, and let $\mathrm{ran}(a) \cap E = \{0\}$. Then aE is closed.*

Proof. Assume that $E \neq \{0\}$, and let $y = \lim_{n \to \infty} a x_n$ with $x_n \in E$ $(n \in \mathbb{N})$. For each $n \in \mathbb{N}$ there exists $s_n \in S(E)$ with $x_n = \|x_n\| s_n$. There exists a subsequence $\{s_{n_k}\}$ of $\{s_n\}$ that converges to $s \in S(E)$. Then $\lim_{k \to \infty} a s_{n_k} = as \neq 0$, since $\mathrm{ran}(a) \cap E = \{0\}$. Since

$$\lim_{k \to \infty} \|x_{n_k}\| a s_{n_k} = \lim_{k \to \infty} a x_{n_k} = y,$$

it follows that the sequence $\{\|x_{n_k}\|\}$ is bounded, and therefore has a cluster point $\lambda \in \mathbb{R}^+$. Then $y = \lambda as = a(\lambda s) \in aE$. $\quad\square$

Theorem 4. *Let M be a minimal closed right ideal of B with $M^2 \neq \{0\}$. Then M contains an idempotent e such that $M = eB$.*

Proof. Since $M^2 \neq \{0\}$, there exists $a \in M$ with $aM \neq \{0\}$, and so

$$\mathrm{ran}(a) \cap M \neq M .$$

By minimality of M, it follows that $\mathrm{ran}(a) \cap M = \{0\}$, and so, by Lemma 3, aM is a closed right ideal. We have $\{0\} \neq aM \subset M$, and therefore $aM = M$, and there exists $e \in M$ with $ae = a$. Since $a \neq 0$, we have $r(e) \geq 1$; we prove that $r(e) = 1$. Let

$$\kappa = \inf\{\|ax\| : x \in S(M)\} .$$

Since $S(M)$ is compact, this infimum is attained. Since $\operatorname{ran}(a) \cap M = \{0\}$, it follows that $\kappa > 0$. Since $ae^n = a$, we have $\|a\| = \|ae^n\| \geqslant \kappa \|e^n\|$. Therefore $\|e^n\| \leqslant \kappa^{-1} \|a\|$ ($n \in \mathbb{N}$), $r(e) \leqslant 1$; $r(e) = 1$.

For $\lambda > 1$, we now have $\sum_{k=1}^{\infty} \lambda^{-k} e^k = b_\lambda \in M$, and

$$\lambda b_\lambda - e b_\lambda = e, \tag{1}$$

from which $b_\lambda \neq 0$. Let $\{\lambda_n\}$ be a sequence with $\lambda_n > 1$ and $\lim_{n \to \infty} \lambda_n = 1$. By (1), for each n, there exists $m_n \in S(M)$ with

$$\lambda_n m_n - e m_n \in M .$$

By compactness of $S(M)$, it follows that there exists $m \in S(M)$ such that

$$m - e m \in M . \tag{2}$$

Let $J = \{x \in M : x - ex \in M\}$. Then J is a closed right ideal, and by (2), $\{0\} \neq J \subset M$. Therefore $J = M$, i.e.

$$x - ex \in M \quad (x \in M) .$$

But $a(x - ex) = (a - ae)x = 0$, and so

$$x - ex \in \operatorname{ran}(a) \cap M = \{0\} \quad (x \in M) .$$

Thus $e = e^2$ and $M = eB$. $\quad\Box$

Theorem 4 provides the basis for the construction of a socle when B is semi-prime, the details, which are analogous to those for semi-prime Banach algebras with minimal ideals, are given in [67]. The next theorem however has no analogue for algebras.

Theorem 5. *Let B semi-prime. Then the set of minimal closed bi-ideals of B is finite and non-void.*

Proof. Let $\mathfrak{M}$ denote the set of all minimal closed bi-ideals of B. Since B is a non-zero closed bi-ideal and is a locally compact semi-algebra, $\mathfrak{M}$ is non-void. We have

$$\bigcap \{\operatorname{ran}(M) : M \in \mathfrak{M}\} = \{0\} , \tag{3}$$

for otherwise this intersection, which is a closed bi-ideal, contains a minimal closed bi-ideal M_0. But then $M_0^2 = \{0\}$, which is impossible, since B is semi-prime.

If $\mathfrak{M}$ is infinite, we select a sequence $\{M_n\}$ of distinct members of $\mathfrak{M}$ and select $m_n \in S(M_n)$. Since $S(B)$ is compact, we may assume that the sequences have been chosen so that $\lim_{n \to \infty} m_n = a \in S(B)$. Given $M \in \mathfrak{M}$, we have $M M_n \subset M \cap M_n = \{0\}$ whenever $M_n \neq M$. Therefore $M m_n = \{0\}$ for all sufficiently large n, and so $M a = \{0\}$. Since this holds for all $M \in \mathfrak{M}$, we have contradicted (3). $\quad\Box$

Suppose now that B is semi-prime, commutative, and strict (i.e. $B \cap (-B) = \{0\}$). Then B has a finite non-void set of minimal closed ideals each of which is of the form $\mathbb{R}^+ e$ with e an idempotent, and there exists a positive constant κ such that

$$\|a\| \leqslant \kappa r(a) \qquad (a \in B).$$

When B is prime, there is exactly one minimal closed ideal $\mathbb{R}^+ e$ with e an idempotent, and the set K, given by

$$K = \{a \in B : r(a) = 1\},$$

satisfies $K = \{a \in B : ae = e\}$. Moreover K is a multiplicative semi-group and a compact convex base for the cone B. For proofs of these results see Bonsall [67] and Bonsall and Tomiuk [83]. See also Kaashoek [220, 221].

§ 50. Q-Algebras

Throughout this section A denotes a complex commutative Banach algebra with unit. By a uniform algebra we mean here a closed subalgebra of $C(X)$ for some compact Hausdorff space X.

Definition 1. A is a Q-*algebra* if there exist a uniform algebra B and a closed ideal J of B such that A is bicontinuously isomorphic to B/J. When A is semi-simple the isomorphism is automatically bicontinuous, by Theorem 17.8. If there is an isometric isomorphism of A onto B/J we say that A is an $I Q$-*algebra*. The concept of Q-algebra is due to N. Th. Varopoulos.

Example 2. Let A be the algebra generated by 1 and x where $x^2 = 0$. Then A is a Q-algebra. To see this take B to be the disc algebra and $J = \{f \in B : f(0) = f'(0) = 0\}$.

Proposition 3. *Let A be a Q-algebra, let C be a closed subalgebra of A, and let K be a closed ideal of A. Then C and A/K are Q-algebras.*

Proof. Routine. ☐

Proposition 4 characterizes semi-simple Q-algebras as those Banach algebras that are isomorphic to restriction algebras of uniform algebras.

Proposition 4. *Let B be a uniform algebra.*
(i) *Let E be a subset of Φ_B closed in the hull-kernel topology and let $J = \ker E$. Then B/J is semi-simple.*
(ii) *Let J be a closed ideal of B such that B/J is semi-simple. Then $J = \ker E$ for some subset E of Φ_B closed in the hull-kernel topology.*
In both cases B/J is isomorphic to $B|_E$.

Proof. Proposition 23.5 gives (i), and (ii) is routine.   □

The following characterization of Q-algebras in terms of polynomial inequalities is due to I. G. Craw (see Davie [109]).

Proposition 5. *A is a Q-algebra if and only if there exist* $\delta, M > 0$ *such that, for all* n,

$$x_j \in A, \quad \|x_j\| \leqslant \delta \quad (j = 1, \ldots, n) \;\Rightarrow\; \|P(x_1, \ldots, x_n)\| \leqslant M$$

whenever P is a polynomial in n variables satisfying

$$|P(z_1, \ldots, z_n)| \leqslant 1 \quad (|z_j| \leqslant 1, \; j = 1, \ldots, n).$$

Proof. Let $C = B/J$ where J is a closed ideal in a uniform algebra B and let π be the canonical homomorphism of B onto C. Let $x_j \in C$, $\|x_j\| \leqslant 1$ $(j = 1, \ldots, n)$. Given $\varepsilon > 0$ there exist $y_j \in x_j$ with $\|y_j\| \leqslant 1 + \varepsilon$. Then

$$\|P(x_1, \ldots, x_n)\| \leqslant \|P(y_1, \ldots, y_n)\|_\infty \leqslant \sup\{|P(z_1, \ldots, z_n)| : |z_j| \leqslant 1 + \varepsilon\}.$$

This gives the necessity of the condition.

Conversely, let the stated condition hold, let $\Lambda = \{x \in A : \|x\| \leqslant \delta\}$, let $\Delta = \{z \in \mathbb{C} : |z| \leqslant 1\}$ and let E be the Cartesian product Δ^Λ with the product topology. Then E is a compact Hausdorff space. For $x \in \Lambda$, let η_x be the evaluation mapping defined by

$$\eta_x(f) = f(x) \quad (f \in E).$$

Then $\eta_x \in C(E)$, and $\|\eta_x\|_\infty \leqslant 1$. Let B_0 be the subalgebra of $C(E)$ generated by $\{\eta_x : x \in \Lambda\}$, and let B be the closure of B_0. Let $\pi : B_0 \to A$ be the homomorphism defined by

$$\pi(P(\eta_{x_1}, \ldots, \eta_{x_n})) = P(x_1, \ldots, x_n),$$

whenever $(x_1, \ldots, x_n)$ is an ordered n-tuple of distinct elements of Λ. Then π is continuous by the stated condition, and so extends to a continuous homomorphism of B onto A. Then A is isomorphic to $B/\ker \pi$.   □

Remark. It is easy to see that A is an IQ-algebra if and only if the above condition holds with $\delta = M = 1$.

Corollary 6. $l^1(\mathbb{Z})$ *is not a Q-algebra.*

Proof. There is a constant $M > 0$ such that, for each $n \in \mathbb{N}$, there exists a polynomial P,

$$P(z) = \alpha_0 + \alpha_1 z + \cdots + \alpha_n z^n,$$

satisfying $|P(z)| \leqslant 1 \, (|z| \leqslant 1), |\alpha_0| + \cdots + |\alpha_n| \geqslant M n^{\frac{1}{2}}$ (see Zygmund [488]).   □

Theorem 7, the main result of this section, is due to B. Cole (see Wermer [411]).

Theorem 7. *Let A be an IQ-algebra. Then there exists a Hilbert space H and an isometric monomorphism of A into $BL(H)$.*

Proof. We may suppose that $A = B/J$ where J is a closed ideal of B, B a closed unital subalgebra of $C(X)$ for some compact Hausdorff space X. Given a probability measure μ on X, let $H^2(\mu)$ be the closure of B in $L^2(\mu)$, and let K be the closure of J in $H^2(\mu)$. Let H_μ be the orthogonal complement of K in $H^2(\mu)$, and let P_μ be the corresponding orthogonal projection of $H^2(\mu)$ onto H_μ. For $f \in B$, $h \in H_\mu$, let

$$(T_\mu f)(h) = P_\mu(f h) .$$

Then T_μ is a norm-decreasing linear mapping of B into $BL(H_\mu)$. Given $f, g \in B$, $h \in H_\mu$, we have $gh - P_\mu gh \in K$ and so $f(gh - P_\mu gh) \in K$. Therefore

$$(T_\mu f g)(h) = P_\mu(f g h) = P_\mu(f P_\mu g h) = (T_\mu f)(T_\mu g)(h) ,$$

and T_μ is a homomorphism satisfying $T_\mu J = \{0\}$. Let π be the canonical homomorphism of B onto A, and let

$$S_\mu a = T_\mu f \qquad (\pi f = a) .$$

Then S_μ is a norm-decreasing homomorphism of A into $BL(H_\mu)$. To complete the proof it is sufficient to show that for each $a \in A$ there exists a probability measure μ on X such that $\|S_\mu a\| = \|a\|$, for then we take $H = \sum^\oplus H_\mu$ the sum being over all probability measures μ.

Let $a \in A$, $\|a\| = 1$, and let $f \in B$ where $\pi f = a$. Then $d(f, J) = 1$, and so, by the Hahn-Banach theorem and the Riesz representation theorem, there exists a Borel measure v on X such that $\|v\| = v(f) = 1$, $v(g) = 0$ $(g \in J)$. Let $\{g_n\}$ be a sequence in J such that $\lim_{n \to \infty} \|f - g_n\|_\infty = 1$, and let g be a weak* cluster point of $\{g_n\}$ in $L^\infty(|v|)$. Thus $\|f - g\|_\infty \leqslant 1$, and

$$\int (f - g) d v = \int f d v = 1 .$$

Therefore $(f - g) v = |v|$, and $|f - g| = 1$ almost everywhere. Let $\mu = |v|$. Then

$$\int (f - g)^* h d \mu = \int h d v = 0 \qquad (h \in J),$$

and so

$$(S_\mu a)(1) = (T_\mu f)(1) = P_\mu f = f - g .$$

But

$$\| f - g \|^2 = \int (f - g)(f - g)^* d \mu = \int (f - g) d v = 1 ,$$

so that $\|S_\mu a\| = 1$.   □

We do not know whether every commutative closed subalgebra of $BL(H)$ is an IQ-algebra. Bernard [440] has obtained a non-commutative extension of Theorem 7 as follows. Let B be a closed subalgebra of $BL(H)$ and let J be a closed bi-ideal of B. Then there exist a Hilbert space K and an isometric isomorphism of B/J into $BL(K)$.

Davie [109] shows that examples of Q-algebras include $C^1[0, 1]$, and l^p $(1 \leqslant p < \infty)$ with product defined coordinatewise, while $BV[0, 1]$, with product defined pointwise, is not a Q-algebra.

Bibliography

The following bibliography is not intended to be comprehensive. We have made a selection of items that relate to the general theory of Banach algebras and have largely ignored the areas of function algebras, C^* and W^* algebras, harmonic analysis, numerical ranges, and general topological algebras.

1. Ackermans, S. T. M.: On the principal extension of complex sets in a Banach algebra. Indag. Math. **29**, 146—150 (1967).
2. Ackermans, S. T. M.: A case of strong spectral continuity. Indag. Math. **30**, 455—459 (1968).
3. Ackermans, S. T. M.: On the existence of inverses in a Banach algebra. J. Math. Anal. Appl. **27**, 208—209 (1969).
4. Akemann, C. A., Rosenfeld, M.: Maximal one-sided ideals in operator algebras. Amer. J. Math. **94**, 723—728 (1972).
5. Alexander, F. E.: On complemented and annihilator algebras. Glasgow Math. J. **10**, 38—45 (1969).
6. Alexander, F. E.: Representation theorems for complemented algebras. Trans. Amer. Math. Soc. **148**, 385—398 (1970).
7. Alexander, F. E., Tomiuk, B. J.: Complemented B^*-algebras. Trans. Amer. Math. Soc. **137**, 459—480 (1969).
8. Alexander, J. C.: Compact Banach algebras. Proc. London Math. Soc. (3) **18**, 1—18 (1968).
9. Alexander, J. C.: On Riesz operators. Proc. Edinburgh Math. Soc. (2) **16**, 227—232 (1969).
10. Allan, G. R.: A note on B^*-algebras. Proc. Cambridge Philos. Soc. **61**, 29—32 (1965).
11. Allan, G. R.: A form of local characterization of Gelfand transforms. J. London Math. Soc. **43**, 623—625 (1968).
12. Allan, G. R.: An extension of the Shilov-Arens-Calderon theorem. J. London Math. Soc. **44**, 595—601 (1968).
13. Allan, G. R.: On lifting analytic relations in commutative Banach algebras. J. Functional Analysis **5**, 37—43 (1970).
14. Allan, G. R.: Some aspects of the theory of commutative Banach algebras and holomorphic functions of several complex variables. Bull. London Math. Soc. **3**, 1—17 (1971).
15. Allan, G. R.: Embedding the algebra of formal power series in a Banach algebra. Proc. London Math. Soc. (3) **25**, 329—340 (1972).
16. Altman, M.: Factorisation dans les algèbres de Banach. C. R. Acad. Sci. Paris Sér. A–B **272**, 1388—1389 (1971).
17. Altman, M.: Contracteurs dans les algèbres de Banach. C. R. Acad. Sci. Paris Sér. A–B **274**, 399—400 (1972).
18. Ambrose, W.: Structure theorems for a special class of Banach algebras. Trans. Amer. Math. Soc. **57**, 364—386 (1945).

19. Arens, R.: On a theorem of Gelfand and Naimark. Proc. Nat. Acad. Sci. U.S.A. **32**, 237—239 (1946).
20. Arens, R.: Representation of *-algebras. Duke Math. J. **14**, 269—282 (1947).
21. Arens, R.: Approximation in, and representation of, certain Banach algebras. Amer. J. Math. **71**, 763—790 (1949).
22. Arens, R.: Operations induced in function classes. Monatsh. Math. **55**, 1—19 (1951).
23. Arens, R.: The adjoint of a bilinear operation. Proc. Amer. Math. Soc. **2**, 839—848 (1951).
24. Arens, R.: Inverse producing extensions of normed algebras. Trans. Amer. Math. Soc. **88**, 536—548 (1958).
25. Arens, R.: Extensions of Banach algebras. Pacific J. Math. **10**, 1—16 (1960).
26. Arens, R.: The group of invertible elements of a commutative Banach algebra. Studia Math. **21**, 21—23 (1963).
27. Arens, R.: Ideals in Banach algebra extensions. Studia Math. **31**, 29—34 (1968).
28. Arens, R., Calderón, A. P.: Analytic functions of several Banach algebra elements. Ann. of Math. (2) **62**, 204—216 (1955).
29. Asimov, L. A., Ellis, A. J.: On Hermitian functionals on unital Banach algebras. Bull. London Math. Soc. **4**, 333—336 (1973).
30. Bachelis, G. F.: Homomorphisms of annihilator Banach algebras. Pacific J. Math. **25**, 229—247 (1968).
31. Bachelis, G. F.: Homomorphisms of annihilator Banach algebras II. Pacific J. Math. **30**, 283—291 (1969).
32. Bade, W. G., Curtis, P. C.: Homomorphisms of commutative Banach algebras. Amer. J. Math. **82**, 589—608 (1960).
33. Bade, W. G., Curtis, P. C.: The Wedderburn decomposition of commutative Banach algebras. Amer. J. Math. **82**, 851—866 (1960).
34. Bade, W. G., Curtis, P. C.: Embedding theorems for commutative Banach algebras. Pacific J. Math. **18**, 391—409 (1966).
35. Baker, J. W., Pym, J. S.: A remark on continuous bilinear mappings. Proc. Edinburgh Math. Soc. (2) **17**, 245—248 (1971).
36. Banaschewski, B.: Analytic discs in the maximal ideal space of a Banach algebra. Bull. Acad. Polon. Sci. Sér. Sci. Math. Astronom. Phys. **14**, 137—144 (1966).
37. Barbeau, E. J.: The principal semi-algebra in a Banach algebra. Trans. Amer. Math. Soc. **120**, 1—16 (1965).
38. Barnes, B. A.: Modular annihilator algebras. Canad. J. Math. **18**, 566—578 (1966).
39. Barnes, B. A.: Algebras with the spectral expansion property. Illinois J. Math. **11**, 284—290 (1967).
40. Barnes, B. A.: A generalized Fredholm theory for certain maps in the regular representations of an algebra. Canad. J. Math. **20**, 495—504 (1968).
41. Barnes, B. A.: On the existence of minimal ideals in a Banach algebra. Trans. Amer. Math. Soc. **133**, 511—517 (1968).
42. Barnes, B. A.: Subalgebras of modular annihilator algebras. Proc. Cambridge Philos. Soc. **66**, 5—12 (1969).
43. Barnes, B. A.: Examples of modular annihilator algebras. Rocky Mountain J. Math. **1**, 657—665 (1971).
44. Barnes, B. A.: Irreducible algebras of operators which contain a minimal idempotent. Proc. Amer. Math. Soc. **30**, 337—342 (1971).
45. Barnes, B. A.: Density theorems for algebra of operators and annihilator Banach algebras. Michigan Math. J. **19**, 149—155 (1972).
46. Barnes, B. A.: Strictly irreducible *-representations of Banach *-algebras. Trans. Amer. Math. Soc. **170**, 459—469 (1972).

47. Bergman, G. M.: A ring primitive on the left but not on the right. Proc. Amer. Math. Soc. **15**, 473—475 (1964).
48. Berkson, E.: Some characterizations of C^*-algebras. Illinois J. Math. **10**, 1—8 (1966).
49. Berkson, E., Porta, H.: Representations of $B(X)$. J. Functional Analysis **3**, 1—34 (1969).
50. Bernard, A.: Algèbre quotients d'algèbres uniformes. C. R. Acad. Sci. Paris Sér. A–B **272**, 1101 (1971).
51. Beurling, A.: Sur les integrales de Fourier absolument convergentes. Congrès des Math. Scand., Helsingfors 1938.
52. Birtel, F. T.: Banach algebras of multipliers. Duke Math. J. **28**, 203—211 (1961).
53. Birtel, F. T.: Isomorphisms of isometric multipliers. Proc. Amer. Math. Soc. **13**, 204—210 (1962).
54. Birtel, F. T.: On a commutative extension of a normed algebra. Proc. Amer. Math. Soc. **13**, 815—822 (1962).
55. Birtel, F. T. (editor): Function algebras. Glenview: Scott-Foresman 1966.
56. Bishop, E. R.: Generalized Lipschitz algebras. Canad. Math. Bull. **12**, 1—19 (1969).
57. Blum, E. K.: The fundamental group of the principal component of a commutative Banach algebra. Proc. Amer. Math. Soc. **4**, 397—400 (1953).
58. Blum, E. K.: A theory of analytic functions in Banach algebras. Trans. Amer. Math. Soc. **78**, 343—370 (1955).
59. Boel, J.: Sur certaines extensions d'une algèbre de Banach abelienne a élément unité. Acad. Roy. Belg. Bull. Cl. Sci. (5) **49**, 470—489 (1963).
60. Bohnenblust, H. F., Karlin, S.: Geometrical properties of the unit sphere of Banach algebras. Ann. of Math. (2) **62**, 217—229 (1955).
61. Bollobás, B.: Extremal algebras and the theory of numerical ranges (to appear).
62. Bollobás, B.: Adjoining inverses to commutative Banach algebras (to appear).
63. Bonsall, F. F.: A minimal property of the norm in some Banach algebras. J. London Math. Soc. **29**, 156—164 (1954).
64. Bonsall, F. F.: Semi-algebras of continuous functions. Proc. London Math. Soc. **10**, 122—140 (1960).
65. Bonsall, F. F.: Semi-algebras of continuous functions. Proc. Int. Symp. on Linear Spaces, Jerusalem 1960, 101—114.
66. Bonsall, F. F.: Type 2 semi-algebras of continuous functions. Proc. London Math. Soc. **12**, 133—143 (1962).
67. Bonsall, F. F.: Locally compact semi-algebras. Proc. London Math. Soc. (3) **13**, 51—70 (1963).
68. Bonsall, F. F.: Maximal subalgebras of Banach *-algebras. J. London Math. Soc. **40**, 540—550 (1965).
69. Bonsall, F. F.: Compact operators from an algebraic standpoint. Glasgow Math. J. **8**, 41—49 (1967).
70. Bonsall, F. F.: Operators that act compactly on an algebra of operators. Bull. London Math. Soc. **1**, 163—170 (1969).
71. Bonsall, F. F.: Stability theorems for cones and wedges of continuous functions. J. Functional Analysis **4**, 135—145 (1969).
72. Bonsall, F. F.: The numerical range of an element of a normed algebra. Glasgow Math. J. **10**, 68—72 (1969).
73. Bonsall, F. F.: A survey of Banach algebra theory. Bull. London Math. Soc. **2**, 257—274 (1970).
74. Bonsall, F. F., Crabb, M. J.: The spectral radius of a Hermitian element of a Banach algebra. Bull. London Math. Soc. **2**, 178—180 (1970).
75. Bonsall, F. F., Duncan, J.: Dual representations of Banach algebras. Acta Math. **117**, 79—102 (1967).

76. Bonsall, F. F., Duncan, J.: Dually irreducible representations of Banach algebras. Quart. J. Math. Oxford Ser. (2) **19**, 97—111 (1968).
77. Bonsall, F. F., Duncan, J.: Numerical ranges of operators on normed spaces and of elements of normed algebras. London Math. Soc. Lecture Note Series 2, Cambridge 1971.
78. Bonsall, F. F., Duncan, J.: Numerical ranges II. London Math. Soc. Lecture Note Series 10, Cambridge 1973.
79. Bonsall, F. F., Goldie, A. W.: Algebras which represent their linear functionals. Proc. Cambridge Philos. Soc. **49**, 1—14 (1953).
80. Bonsall, F. F., Goldie, A. W.: Annihilator algebras. Proc. London Math. Soc. (3) **4**, 154—167 (1954).
81. Bonsall, F. F., Stirling, D. S. G.: Square roots in Banach *-algebras. Glasgow Math. J. **13**, 74 (1972).
82. Bonsall, F. F., Thompson, A. C.: Banach algebra elements whose spectra intersect $\mathbb{R}^+$ (to appear in Proc. Edinburgh Math. Soc.).
83. Bonsall, F. F., Tomiuk, B. J.: The semi-algebra generated by a compact linear operator. Proc. Edinburgh Math. Soc. (3) **14**, 177—196 (1965).
84. Bourbaki, N.: Élements de mathématique. Fasc. XXXII. Theorie spectrales. Paris: Hermann 1967.
85. Browder, A.: Introduction to function algebras. New York: Benjamin 1969.
86. Browder, A.: On Bernstein's inequality and the norm of Hermitian operators. Amer. Math. Monthly **78**, 871—873 (1971).
87. Brown, D. T.: A class of Banach algebras with a unique norm topology. Proc. Amer. Math. Soc. **17**, 1429—1434 (1966).
88. Brown, G.: Relatively type 0 semi-algebras. Quart. J. Math. Oxford Ser. (2) **18**, 289—291 (1967).
89. Brown, G.: Type 0 semi-algebras in Banach algebras. J. London Math. Soc. **43**, 482—486 (1968).
90. Brown, G.: Stability of wedges and semi-algebras. Proc. Cambridge Philos. Soc. **64**, 365—376 (1968).
91. Bucy, R. S., Maltese, G.: A representation theorem for positive functionals on involutive algebras. Math. Ann. **162**, 364—367 (1966).
92. Burckel, R. B.: A simpler proof of the commutative Glickfeld-Berkson theorem. J. London Math. Soc. (2) **2**, 403—404 (1970).
93. Calkin, J. W.: Two-sided ideals and congruences in the ring of bounded operators in Hilbert space. Ann. of Math. **42**, 839—873 (1941).
94. Chernoff, P. R.: Elements of a normed algebra whose 2^nth powers lie close to the identity. Proc. Amer. Math. Soc. **23**, 386—387 (1969).
95. Ching, W. M., Wong, J. S. W.: Multipliers and H^*-algebras. Pacific J. Math. **22**, 387—395 (1967).
96. Civin, P.: Extensions of homomorphisms. Pacific J. Math. **11**, 1223—1233 (1961).
97. Civin, P., White, C. C.: Maximal closed preprimes in Banach algebras. Trans. Amer. Math. Soc. **147**, 241—260 (1970).
98. Civin, P., Yood, B.: Regular Banach algebras with a countable space of maximal regular ideals. Proc. Amer. Math. Soc. **7**, 1005—1010 (1956).
99. Civin, P., Yood, B.: Involutions on Banach algebras. Pacific J. Math. **9**, 415—436 (1959).
100. Civin, P., Yood, B.: The second conjugate space of a Banach algebra as an algebra. Pacific J. Math. **11**, 847—870 (1961).
101. Civin, P., Yood, B.: Lie and Jordan structures in Banach algebras. Pacific J. Math. **15**, 775—797 (1965).

278 Bibliography

102. Cleveland, S.B.: Homomorphisms of non-commutative *-algebras. Pacific J. Math. **13**, 1097—1109 (1963).
103. Cohen, P. J.: Factorization in group algebras. Duke Math. J. **26**, 199—205 (1959).
104. Cohen, P. J.: A note on constructive methods in Banach algebras. Proc. Amer. Math. Soc. **12**, 159—163 (1961).
105. Crownover, R. M.: Principal ideals which are maximal ideals in Banach algebras. Studia Math. **33**, 299—304 (1969).
106. Crownover, R. M.: One-dimensional point derivation spaces in Banach algebras. Studia Math. **35**, 249—259 (1970).
107. Curtis, P. C.: Order and commutativity in Banach algebras. Proc. Amer. Math. Soc. **9**, 643—646 (1958).
108. Curtis, P. C.: Derivations of commutative Banach algebras. Bull. Amer. Math. Soc. **67**, 271—273 (1961).
109. Davie, A.M.: Quotient algebras of uniform algebras. J. London Math. Soc. (2) **7**, 31—40 (1973).
110. Davie, A.M.: A counterexample on dual Banach algebras. Bull. London Math. Soc. **5**, 79—80 (1973).
111. Davie, A. M.: The approximation problem for Banach spaces (to appear).
112. Day, M. M.: Amenable semigroups. Illinois J. Math. **1**, 509—544 (1957).
113. De Leeuw, K., Katznelson, Y.: Functions that operate on non-selfadjoint algebras. J. Analyse Math. **11**, 207—219 (1963).
114. Detraz, J.: Sous-algèbre de codemension finie d'une algèbre de Banach. C. R. Acad. Sci. Paris Sér. A–B **266**, 117—119 (1968).
115. Dickson, L. E.: Algebras and their arithmetics. New York: Dover 1960.
116. Ditkin, V. A.: On the structure of ideals in certain normed rings. Uch. Zap. Mosk. Univ. Mat. **30**, 83—130 (1939).
117. Dixmier, J.: Sur les automorphismes des algèbres de Banach. C.R. Acad. Sci. Paris Sér. A–B **264**, 729—731 (1967).
118. Dixmier, J.: Les algèbres d'opérateurs dans l'espace Hilbertien. Paris: Gauthier-Villars 1969.
119. Dixmier, J.: Les C*-algebres et leurs representations. Paris: Gauthier-Villars 1968.
120. Dixon, P.G.: Semiprime Banach algebras. J. London Math. Soc. (2) **6**, 676—678 (1973).
121. Dixon, P.G.: Approximate identities in Banach algebras. Proc. London Math. Soc. (3) **26**, 485—496 (1973).
122. Dixon, P. G.: Locally finite Banach algebras (preprint).
123. Domar, Y.: On the ideal structure of certain Banach algebras. Math. Scand. **14**, 197—212 (1964).
124. Donoghue, W. F.: The Banach algebra l^1 with an application to linear transformations. Duke Math. J. **23**, 533—537 (1956).
125. Donoghue, W.F.: The lattice of invariant subspaces of a completely continuous quasi-nilpotent transformation. Pacific J. Math. **7**, 1031—1035 (1957).
126. Duncan, J.: The continuity of the involution on Banach *-algebras. J. London Math. Soc. **41**, 701—706 (1966).
127. Duncan, J.: B^{*} modular annihilator algebras. Proc. Edinburgh Math. Soc. (2) **15**, 89—102 (1966).
128. Duncan, J.: Compact operators and Banach algebras. Proc. Cambridge Philos. Soc. **69**, 79—85 (1971).
129. Duncan, J., Tullo, A. W.: Finite dimensionality, nilpotents and quasinilpotents in Banach algebras (to appear in Proc. Edinburgh Math. Soc.).
130. Dunford, N., Schwartz, J. T.: Linear operators. Part I. New York: Interscience 1958.

131. Effros, E. G.: A decomposition theory for representations of C^*-algebras. Trans. Amer. Math. Soc. **107**, 83—106 (1963).

132. Elliott, G. A.: A weakening of the axioms for a C^*-algebra. Math. Ann. **189**, 257—260 (1970).

133. Enflo, P.: A counterexample to the approximation problem. Acta Math. **130**, 309—317 (1973).

134. Erdos, J. A.: On certain elements of C^*-algebras. Illinois J. Math. **15**, 682—693 (1971).

135. Fell, J. M. G.: The structure of algebras of operator fields. Acta Math. **106**, 233—280 (1961).

136. Fell, J. M. G.: The dual spaces of Banach algebras. Trans. Amer. Math. Soc. **114**, 227—250 (1965).

137. Ford, J. W. M.: A square root lemma for Banach (*)-algebras. J. London Math. Soc. **42**, 521—522 (1967).

138. Freundlich, M.: Completely continuous elements of a normed ring. Duke Math. J. **16**, 273—283 (1949).

139. Fukamiya, M.: On B^*-algebras. Proc. Japan Acad. **27**, 321—327 (1951).

140. Fukamiya, M.: On a theorem of Gelfand and Naimark and the B^*-algebra. Kumamoto J. Sci. Ser. A **1**, 17—22 (1952).

141. Gamelin, T. W.: Uniform algebras. New Jersey: Prentice-Hall 1969.

142. Gardner, L. T.: An invariance theorem for representations of Banach algebras. Proc. Amer. Math. Soc. **16**, 983—986 (1965).

143. Gardner, L. T.: Square roots in Banach algebras. Proc. Amer. Math. Soc. **17**, 132—134 (1966).

144. Garnir, H. G., De Wilde, M., Schmets, J.: Constructive proof of the existence of multiplicative functionals in commutative separable Banach algebras. Bull. Amer. Math. Soc. **73**, 564—566 (1967).

145. Gelbaum, B. R.: Tensor products of Banach algebras. Canad. J. Math. **11**, 297—310 (1959).

146. Gelbaum, B. R.: Note on tensor products of Banach algebras. Proc. Amer. Math. Soc. **12**, 750—757 (1961).

147. Gelbaum, B. R.: Tensor products and related questions. Trans. Amer. Math. Soc. **103**, 525—548 (1962).

148. Gelbaum, B. R.: Tensor products over Banach algebras. Trans. Amer. Math. Soc. **118**, 131—149 (1965).

149. Gelbaum, B. R.: Tensor products of Banach algebras II. Proc. Amer. Math. Soc. **25**, 470—474 (1970).

150. Gelfand, I. M.: Normierte Ringe. Mat. Sb. **9**, 3—21 (1941).

151. Gelfand, I. M.: Ideale und primäre Ideale in normierten Ringen. Mat. Sb. **9**, 41—48 (1941).

152. Gelfand, I. M., Naimark, M. A.: On the embedding of normed rings into the ring of operators in Hilbert space. Mat. Sb. **12**, 197—213 (1943).

153. Gelfand, I. M., Naimark, M. A.: Normed rings with their involution and their representations. Izv. Akad. Nauk SSSR **12**, 445—480 (1948).

154. Gelfand, I. M., Raikov, D. A., Shilov, G. E.: Commutative normed rings. New York: Chelsea 1964.

155. Gelfand, I. M., Shilov, G. E.: Über verschiedene Methoden der Einführung der Topologie in die Menge der maximalen Ideale eines normierten Ringes. Mat. Sb. **9**, 25—39 (1941).

156. Gil de Lamadrid, J.: Uniform cross-norms and tensor products of Banach algebras. Bull. Amer. Math. Soc. **69**, 797—803 (1967).

157. Gil de Lamadrid, J.: Topological modules: Banach algebras, tensor products, algebras of kernels. Trans. Amer. Math. Soc. **126**, 361—419 (1967).

158. Ginsberg, J. I., Newman, D. J.: Generators of certain radical algebras. J. Approximation Theory **3**, 229—235 (1970).

159. Gleason, A. M.: A characterization of maximal ideals. J. Analyse Math. **19**, 171—172 (1967).

160. Glickfeld, B. W.: A metric characterization of $C(X)$ and its generalizations to C^*-algebras. Illinois J. Math. **10**, 547—566 (1966).

161. Glickfeld, B. W.: The Riemann sphere of a commutative Banach algebra. Trans. Amer. Math. Soc. **134**, 1—28 (1968).

162. Glickfeld, B. W.: The theory of analytic functions in commutative Banach algebras with involution. Ann. Mat. Pura Appl. **(4) 86**, 61—77 (1970).

163. Glickfeld, B. W.: Meromorphic functions of elements of a commutative Banach algebra. Trans. Amer. Math. Soc. **151**, 293—307 (1970).

164. Glickfeld, B. W.: On the inverse function theorem in commutative Banach algebras. Illinois J. Math. **15**, 212—221 (1971).

165. Glicksberg, I.: Banach algebras with scattered structure spaces. Trans. Amer. Math. Soc. **98**, 518—526 (1961).

166. Glimm, J. G., Kadison, R. V.: Unitary operators in C^*-algebras. Pacific J. Math. **10**, 547—556 (1960).

167. Godement, R.: Extension à une groupe abelien quelconque des théorèmes tauberiens de N. Wiener et d'un théorème de A. Beurling. C. R. Acad. Sci. Paris Sér. A–B **223**, 16—18 (1946).

168. Gohberg, I. C.: Criteria for one-sided reversibility of elements in normed rings and their applications. Dokl. Acad. Nauk SSSR **145**, 971—974 (1962).

169. Gohberg, I. C.: The factorization problem in normed rings, functions of isometric and symmetric operators, and singular integral equations. Uspehi Mat. Nauk **19**, 71—124 (1964).

170. Gorin, E. A.: Maximal subalgebras of commutative Banach algebras with involution. Mat. Zametki **1**, 173—178 (1967).

171. Gorin, E. A.: Commutative Banach algebras generated by the group of unitary elements. Funkcional. Anal. i Priložen **1**, 86—87 (1967).

172. Gorin, E. A., Lin, Ja. V.: A condition on the radical of a Banach algebra guaranteeing strong decomposability. Mat. Zametki **2**, 589—592 (1967).

173. Grabiner, S.: The nilpotency of Banach nil algebras. Proc. Amer. Math. Soc. **21**, 510 (1969).

174. Grabiner, S.: Ranges of quasi-nilpotent operators. Illinois J. Math. **15**, 150—152 (1971).

175. Grabiner, S.: A formal power series operational calculus for quasi-nilpotent operators. Duke Math. J. **38**, 641—658 (1971).

176. Graham, C. C.: The weak density of the non-invertible elements of a commutative Banach algebra. Bull. Austral. Math. Soc. **4**, 179—182 (1971).

177. Grasselli, J.: Selbstadjungierte Elemente der Banach-Algebra ohne Einheit. Publ. Dept. Math. (Ljubljana) **1**, 5—21 (1964).

178. Grothendieck, A.: Produits tensoriels topologiques et espaces nucléaires. Mem. Amer. Math. Soc. **16** (1955).

179. Guichardet, A.: Sur les caracterès des algèbres de Banach à involution. C. R. Acad. Sci. Paris Sér. A–B **252**, 2800—2802 (1961).

180. Guichardet, A.: Sur l'homologie et la cohomologie des algèbres de Banach. C. R. Acad. Sci. Paris Sér. A–B **262**, 38—41 (1966).

181. Guichardet, A.: Special topics in topological algebras. New York–London–Paris: Gordon and Breach 1968.
182. Gunning, R. C., Rossi, H.: Analytic functions of several complex variables. New Jersey: Prentice-Hall 1965.
183. Halmos, P.R.: Capacity in Banach algebras. Indiana Math. J. **20**, 855—863 (1971).
184. Harris, L. A.: Banach algebras with involution and Möbius transformations. J. Functional Analysis **11**, 1—16 (1972).
185. Helemskii, A. Ja.: A description of annihilator commutative Banach algebras. Soviet Math. Dokl. **5**, 902—904 (1964).
186. Helemskii, A. Ja.: Commutative normed rings with a finite dimensional radical. Vestnik Moskov. Univ. Ser. I Math. Mech. **6**, 7—16 (1964).
187. Helemskii, A. Ja.: Annihilator extensions of commutative Banach algebras. Izv. Akad. Nauk SSSR **29**, 945—956 (1965).
188. Helemskii, A. Ja.: On a certain analytic condition for the radical of a commutative Banach algebra and associated decomposability problems. Soviet Math. Dokl. **7**, 430—432 (1966).
189. Helemskii, A. Ja.: The homological dimension of normed modules over Banach algebras. Mat. Sb. **81 (123)**, 430—444 (1970).
190. Helgason, S.: The derived algebra of a Banach algebra. Proc. Nat. Acad. Sci. U.S.A. **40**, 994—995 (1954).
191. Helgason, S.: Multipliers of Banach algebras. Ann. of Math. (2) **64**, 240—254 (1956).
192. Hennefeld, J.: A note on the Arens product. Pacific J. Math. **26**, 115—119 (1968).
193. Herman, R. H.: On the uniqueness of the ideals of compact and strictly singular operators. Studia Math. **29**, 161—165 (1968).
194. Hewitt, E., Ross, K.A.: Abstract harmonic analysis. Vol. I. Berlin–Göttingen–Heidelberg: Springer 1963.
195. Hewitt, E., Ross, K.A.: Abstract harmonic analysis. Vol. II. Berlin–Heidelberg–New York: Springer 1970.
196. Hille, E.: On roots and logarithms of elements of a complex Banach algebra. Math. Ann. **136**, 46—57 (1958).
197. Hille, E., Phillips, R. S.: Functional analysis and semigroups. Amer. Math. Soc. Coll. 31, Providence 1957.
198. Hirschfeld, R. A., Rolewicz, S.: A class of non-commutative Banach algebras without divisors of zero. Bull. Acad. Polon. Sci. Sér. Sci. Math. Astronom. Phys. **17**, 751—753 (1969).
199. Hirschfeld, R. A., Żelazko, W.: On spectral norm Banach algebras. Bull. Acad. Polon. Sci. Sér. Sci. Math. Astronom. Phys. **16**, 195—199 (1968).
200. Holmes, R. B.: A formula for the spectral radius of an operator. Amer. Math. Monthly **75**, 163—166 (1968).
201. Hörmander, L.: An introduction to complex analysis in several variables. New York: Van Nostrand 1966.
202. Inglestam, L.: A vertex property for Banach algebras with identity. Math. Scand. **11**, 22—32 (1962).
203. Inglestam, L.: Real Banach algebras. Ark. Mat. **5**, 239—270 (1964).
204. Inglestam, L.: Symmetry in real Banach algebras. Math. Scand. **18**, 53—68 (1968).
205. Inglestam, L.: A note on Laplace transforms and strict reality in Banach algebras. Math. Z. **102**, 163—165 (1967).
206. Jacobson, N.: The radical and semi-simplicity for arbitrary rings. Amer. J. Math. **67**, 300—320 (1945).
207. Jacobson, N.: A topology for the set of primitive ideals in an arbitrary ring. Proc. Nat. Acad. Sci. U.S.A. **31**, 333—338 (1945).

208. Johnson, B. E.: Continuity of homomorphisms of topological algebras. Proc. Cambridge Philos. Soc. **60**, 171—172 (1964).
209. Johnson, B. E.: An introduction to the theory of centralizers. Proc. London Math. Soc. (3) **14**, 299—320 (1964).
210. Johnson, B. E.: A commutative semi-simple Banach algebra which is not dual. Bull. Amer. Math. Soc. **73**, 407—409 (1967).
211. Johnson, B. E.: The uniqueness of the (complete) norm topology. Bull. Amer. Math. Soc. **73**, 537—539 (1967).
212. Johnson, B. E.: Continuity of homomorphisms of algebras of operators. J. London Math. Soc. **42**, 537—541 (1967).
213. Johnson, B. E.: Centralizers and operators reduced by maximal ideals. J. London Math. Soc. **43**, 231—233 (1968).
214. Johnson, B. E.: The Wedderburn decomposition of Banach algebras with finite dimensional radical. Amer. J. Math. **90**, 866—876 (1968).
215. Johnson, B. E.: Continuity of derivations on commutative Banach algebras. Amer. J. Math. **91**, 1—10 (1969).
216. Johnson, B. E.: Continuity of homomorphisms of algebras of operators II. J. London Math. Soc. (2) **1**, 81—84 (1969).
217. Johnson, B. E.: Cohomology in Banach algebras. Mem. Amer. Math. Soc. **127**, (1972).
218. Johnson, B. E.: Approximate diagonals and cohomology of certain annihilator Banach algebras. Amer. J. Math. **94**, 685—698 (1972).
219. Johnson, B. E., Sinclair, A. M.: Continuity of derivations and a problem of Kaplansky. Amer. J. Math. **90**, 1067—1073 (1968).
220. Kaashoek, M. A.: Locally compact semi-algebras and spectral theory. Nieuw Arch. Wisk. **17**, 8—16 (1969).
221. Kaashoek, M. A.: On the peripheral spectrum of an element in a strict closed semi-algebra. Coll. Math. Soc. J. Bolyai 5. Tihany (Hungary) 1970, 319—332.
222. Kaashoek, M. A., West, T. T.: Locally compact monothetic semi-algebras. Proc. London Math. Soc. (3) **18**, 428—438 (1968).
223. Kaashoek, M. A., West, T. T.: Semi-simple locally compact monothetic semi-algebras. Proc. Edinburgh Math. Soc. (2) **16**, 215—219 (1969).
224. Kadison, R. V.: Irreducible operator algebras. Proc. Nat. Acad. Sci. U.S.A. **43**, 273—276 (1957).
225. Kadison, R. V., Ringrose, J. R.: Cohomology of operator algebras I. Type I von Neumann algebras. Acta Math. **126**, 227—243 (1971).
226. Kahane, J. P., Żelazko, W.: A characterization of maximal ideals in commutative Banach algebras. Studia Math. **29**, 339—343 (1968).
227. Kamowitz, H.: Cohomology groups of commutative Banach algebras. Trans. Amer. Math. Soc. **102**, 352—372 (1962).
228. Kamowitz, H., Scheinberg, S.: The spectrum of automorphisms of Banach algebras. J. Functional Analysis **4**, 268—276 (1969).
229. Kaplansky, I.: Topological rings. Amer. J. Math. **69**, 153—183 (1947).
230. Kaplansky, I.: Topological rings. Bull. Amer. Math. Soc. **54**, 809—826 (1948).
231. Kaplansky, I.: Regular Banach algebras. J. Indian Math. Soc. **12**, 57—62 (1948).
232. Kaplansky, I.: Dual rings. Ann. of Math. **49**, 689—701 (1948).
233. Kaplansky, I.: Normed algebras. Duke Math. J. **16**, 399—418 (1949).
234. Kaplansky, I.: Topological representations of algebras, II. Trans. Amer. Math. Soc. **68**, 62—75 (1950).
235. Kaplansky, I.: The structure of certain operator algebras. Trans. Amer. Math. Soc. **70**, 219—255 (1951).

236. Kaplansky, I.: Symmetry of Banach algebras. Proc. Amer. Math. Soc. **3**, 396—399 (1952).
237. Kaplansky, I.: Ring isomorphisms of Banach algebras. Canad. J. Math. **6**, 374—381 (1954).
238. Katznelson, Y.: Algèbres caracterisées par les fonctions que opèrant sur elles. C. R. Acad. Sci. Paris Sér. A–B **247**, 903—905 (1958).
239. Katznelson, Y.: Sur les algèbres dont les éléments non-négatifs admettent des racines. Ann. Sci. École Norm. Sup. (3) **77**, 167—174 (1960).
240. Katznelson, Y.: A characterization of the algebra of all continuous functions on a compact Hausdorff space. Bull. Amer. Math. Soc. **66**, 313—315 (1960).
241. Kelley, J. L., Namioka, I.: Linear topological spaces. New York: Van Nostrand 1963.
242. Kelley, J. L., Vaught, R. L.: The positive cone in Banach algebras. Trans. Amer. Math. Soc. **74**, 44—45 (1953).
243. Kellogg, C. N.: Centralizers and H^*-algebras. Pacific J. Math. **17**, 121—129 (1966).
244. Keown, E. R.: Reflexive Banach algebras. Proc. Amer. Math. Soc. **6**, 252—259 (1955).
245. Keown, E. R.: Some new Hilbert algebras. Trans. Amer. Math. Soc. **128**, 71—87 (1967).
246. Kleinecke, D. C.: On operator commutators. Proc. Amer. Math. Soc. **8**, 535—536 (1957).
247. Kleinecke, D. C.: Almost-finite, compact, and inessential operators. Proc. Amer. Math. Soc. **14**, 863—868 (1963).
248. Koopman, B. O.: Exponential limiting products in Banach algebras. Trans. Amer. Math. Soc. **70**, 256—276 (1951).
249. Kurepa, S.: On ergodic elements in Banach algebras. Glasnik Mat. Ser. III **18**, 43—47 (1963).
250. Kurepa, S.: On roots of an element of a Banach algebra. Publ. Inst. Math. (Beograd) **15**, 5—10 (1961).
251. Kuzmin, E. N.: An estimate in the theory of normed rings. Sibirsk Mat. Ž. **9**, 727—728 (1968).
252. Lance, E. C.: Quadratic forms on Banach spaces. Proc. London Math. Soc. (3) **25**, 341—357 (1972).
253. Lardy, L. J., Lindberg, J. A.: On maximal regular ideals and identities in the tensor product of commutative Banach algebras. Canad. J. Math. **21**, 639—647 (1969).
254. Laursen, R.: Annihilator ideals in Banach algebras. Monatsh. Math. **73**, 400—405 (1969).
255. Laursen, K. B.: Tensor products of Banach algebras with involution. Trans. Amer. Math. Soc. **136**, 467—487 (1969).
256. Lebow, A.: Maximal ideals in tensor products of Banach algebras. Bull. Amer. Math. Soc. **74**, 1020—1022 (1968).
257. LePage, C.: Sur quelques conditions entraînant la commutativité dans les algèbres de Banach. C. R. Acad. Sci. Paris Sér. A–B **265**, 235—237 (1967).
258. Lindberg, J. A.: Factorization of polynomials over Banach algebras. Trans. Amer. Math. Soc. **112**, 356—368 (1964).
259. Lindberg, J. A.: Algebraic extensions of commutative Banach algebras. Pacific J. Math. **14**, 559—583 (1964).
260. Lindberg, J. A.: A class of commutative Banach algebras with unique complete norm topology and continuous derivations. Proc. Amer. Math. Soc. **29**, 516—520 (1971).
261. Ljubić, Ju. I.: Operator norms of matrices. Uspehi Mat. Nauk **18**, 161—164 (1963).
262. Loomis, L. H.: An introduction to abstract harmonic analysis. New York: Van Nostrand 1953.

263. Lorch, E. R.: The theory of analytic functions in normed abelian vector rings. Trans. Amer. Math. Soc. **54**, 414—425 (1943).

264. Lorch, E. R.: The structure of normed abelian rings. Bull. Amer. Math. Soc. **50**, 447—463 (1944).

265. Lorch, E. R.: Normed rings—the first decade. Proc. Sym. Spectral theory and diff. problems, Stillwater 1951, 249—258.

266. Loy, R. J.: Maximal ideal spaces of Banach algebras of derivable elements. J. Austral. Math. Soc. **11**, 310—312 (1970).

267. Loy, R. J.: Uniqueness of complete norm topology and continuity of derivations on Banach algebras. Tôhoku Math. J. **22**, 371—378 (1970).

268. Luchins, E. H.: On strictly semi-simple Banach algebras. Pacific J. Math. **9**, 551—554 (1959).

269. Luchins, E. H.: On radicals and continuity of homomorphisms into Banach algebras. Pacific J. Math. **9**, 755—758 (1959).

270. Lumer, G.: Fine structure and continuity of spectra in Banach algebras. An. Acad. Brasil Ci. **26**, 229—233 (1954).

271. Lumer, G.: Semi-inner-product spaces. Trans. Amer. Math. Soc. **100**, 29—43 (1961).

272. McCarthy, C. A.: On open mappings in Banach algebras. J. Math. Mech. **8**, 415—418 (1959).

273. McCarthy, C. A.: On open mappings in Banach algebras II. Bull. Amer. Math. Soc. **65**, 66 (1959).

274. Mackey, G. W.: Isomorphisms of normed linear spaces. Ann. of Math. **43**, 244—260 (1942).

275. McKilligan, S. A., White, A. J.: Representations of L-algebras. Proc. London Math. Soc. **(3) 25**, 655—674 (1972).

276. Maltese, G.: Multiplicative extensions of multiplicative functionals in Banach algebras. Arch. Math. (Basel) **21**, 502—505 (1970).

277. Mate, L.: The Arens product and multiplier operators. Studia Math. **28**, 227—234 (1966/67).

278. Matsushita, S.: Positive linear functions on self-adjoint Banach algebras. Proc. Japan Acad. **29**, 427—430 (1953).

279. Mazur, S.: Sur les anneaux lineaires. C. R. Acad. Sci. Paris Sér. A–B **207**, 1025—1027 (1938).

280. Milne, H.: Banach space properties of uniform algebras. Bull. London Math. Soc. **4**, 323—326 (1972).

281. Mirkil, H.: The work of Šilov on commutative semi-simple Banach algebras. Notas Mat., Rio de Janeiro No. 20, 1966.

282. Moore, R. T.: Hermitian functionals on Banach algebras and duality characterizations of C^*-algebras. Trans. Amer. Math. Soc. **162**, 253—266 (1971).

283. Murphy, I. S.: Continuity of positive linear functionals on Banach *-algebras. Bull. London Math. Soc. **1**, 171—173 (1969).

284. Naimark, M. A.: Rings with involution. Uspehi Mat. Nauk **3**, 52—145 (1948).

285. Naimark, M. A.: On a continuous analogue of Schur's lemma. Dokl. Acad. Nauk SSSR **98**, 185—188 (1954).

286. Naimark, M. A.: Normed rings. Groningen: Noordhoff 1964.

287. Newburgh, J. D.: The variation of spectra. Duke Math. J. **18**, 165—176 (1951).

288. Newman, D. J.: A radical algebra without derivations. Proc. Amer. Math. Soc. **10**, 584—586 (1959).

289. Oka, K.: Sur les fonctions analytiques de plusieurs variables I–IX. Collected reprints of nine papers from various journals. Tokyo: Iwanami Shoten 1961.

290. Oliver, H. W.: Non-complex methods in real Banach algebras. J. Functional Analysis **6**, 401—411 (1970).

291. Olubummo, A.: Left completely continuous $B^{\#}$-algebras. J. London Math. Soc. **32**, 270—276 (1957).

292. Olubummo, A.: $B^{\#}$-algebras with a certain set of left completely continuous elements. J. London Math. Soc. **34**, 367—369 (1959).

293. Olubummo, A.: On the existence of an absolute minimal norm in a Banach algebra. Proc. Amer. Math. Soc. **11**, 718—722 (1960).

294. Olubummo, A.: Operators of finite rank in a reflexive Banach space. Pacific J. Math. **12**, 1023—1027 (1962).

295. Olubummo, A.: Weakly compact $B^{\#}$-algebras. Proc. Amer. Math. Soc. **14**, 905—908 (1963).

296. Olubummo, A.: Complemented Banach algebras. Canad. J. Math. **16**, 149—150 (1964).

297. Ono, T.: Note on a B^{*}-algebra. J. Math. Soc. Japan **11**, 146—158 (1959).

298. Ono,T.: A real analogue of the Gelfand-Naimark theorem. Proc. Amer. Math. Soc. **25**, 159—160 (1970).

299. Palmer, T. W.: Characterizations of C^{*}-algebras. Bull. Amer. Math. Soc. **74**, 538—540 (1968).

300. Palmer, T. W.: Characterizations of C^{*}-algebras II. Trans. Amer. Math. Soc. **148**, 577—588 (1970).

301. Palmer, T. W.: Real C^{*}-algebras. Pacific J. Math. **35**, 195—204 (1970).

302. Palmer, T.W.: The Gelfand-Naimark pseudo-norm on Banach *-algebras. J. London Math. Soc. (**2**) **3**, 59—66 (1971).

303. Paterson, A. L. T.: Solutions of equations involving analytic functions defined on subsets of Banach algebras. Proc. London Math. Soc. (**3**) **22**, 325—338 (1971).

304. Pickford, J. G.: Topics in Banach algebra theory. Ph. D. thesis. Swansea 1972.

305. Porta, H.: Two-sided ideals of operators. Bull. Amer. Math. Soc. **75**, 599—602 (1969).

306. Potapov, V. P.: The multiplicative structure of J-contractive matrix functions. Amer. Math. Soc. Transl. **15** (**2**), 131—243 (1960).

307. Pryce, J. D.: On type F semi-algebras of continuous functions. Quart. J. Math. Oxford Ser. (**2**) **16**, 65—71 (1965).

308. Przeworska-Rolewicz, D., Rolewicz, S.: On quasi-Fredholm ideals. Studia Math. **26**, 67—71 (1965).

309. Pták, V.: On the spectral radius in Banach algebras with involution. Bull. London Math. Soc. **2**, 327—334 (1970).

310. Pták, V.: Banach algebras with involution. Manu. Math. **6**, 245—290 (1972).

311. Pym, J. S.: Convolution and the second dual of a Banach algebra. Proc. Cambridge Philos. Soc. **65**, 597—599 (1969).

312. Rainwater, J.: A remark on regular Banach algebras. Proc. Amer. Math. Soc. **18**, 255—256 (1967).

313. Read, T. T.: The powers of a maximal ideal in a Banach algebra and analytic structure. Trans. Amer. Math. Soc. **161**, 235—248 (1971).

314. Rennison, J. F.: A note on subalgebras of codimension 1. Proc. Cambridge Philos. Soc. **68**, 673—674 (1970).

315. Rickart, C. E.: Banach algebras with an adjoint operation. Ann. of Math. **47**, 528—550 (1946).

316. Rickart, C. E.: The singular elements of a Banach algebra. Duke Math. J. **14**, 1063—1077 (1947).

317. Rickart, C. E.: The uniqueness of norm problem in Banach algebras. Ann. of Math. **51**, 615—628 (1950).

318. Rickart, C. E.: Representation of certain Banach algebras on Hilbert space. Duke Math. J. **18**, 27—39 (1951).

319. Rickart, C. E.: Spectral permanence for certain Banach algebras. Proc. Amer. Math. Soc. **4**, 191—196 (1953).
320. Rickart, C. E.: An elementary proof of a fundamental theorem in the theory of Banach algebras. Michigan Math. J. **5**, 75—78 (1958).
321. Rickart, C. E.: General theory of Banach algebras. New York: Van Nostrand 1960.
322. Rossi, H.: The local maximum modulus principle. Ann. of Math. **72**, 1—11 (1960).
323. Rota, G. C., Strang, W. G.: A note on the joint spectral radius. Indag. Math. **22**, 379—381 (1960).
324. Rudin, W.: Fourier analysis on groups. New York: Wiley 1962.
325. Rudin, W.: Real and complex analysis. New York: McGraw-Hill 1966.
326. Russo, B., Dye, H. A.: A note on unitary operators in C^*-algebras. Duke Math. J. **33**, 413—416 (1966).
327. Sakai, S.: C^*-algebras and W^*-algebras. Erg. d. Math. Band 60. Berlin–Heidelberg–New York: Springer 1971.
328. Sawon, Z., Warzecha, A.: On the general form of subalgebras of codimension 1 of Banach algebras with unit. Studia Math. **29**, 249—260 (1968).
329. Saworotnow, P. P.: On a generalization of the notion of H^*-algebra. Proc. Amer. Math. Soc. **8**, 49—55 (1957).
330. Saworotnow, P. P.: On the embedding of a right complemented algebra into Ambrose's H^*-algebra. Proc. Amer. Math. Soc. **8**, 56—62 (1957).
331. Saworotnow, P. P.: On the realization of a complemented algebra. Proc. Amer. Math. Soc. **15**, 964—966 (1964).
332. Saworotnow, P. P.: On two-sided H-algebras. Pacific J. Math. **16**, 365—370 (1966).
333. Schatten, R.: A theory of cross-spaces. Ann. of Math. Studies 26, Princeton 1950.
334. Schatten, R.: Norm ideals of completely continuous operators. Berlin–Göttingen–Heidelberg: Springer 1960.
335. Schatz, J. A.: Representation of Banach algebras with an involution. Canad. J. Math. **9**, 435—442 (1957).
336. Schauder, J.: Über lineare, vollstetige Funktionaloperationen. Studia Math. **2**, 183—196 (1930).
337. Sebestyen, Z.: Continuous involutions in Banach algebras. Mat. Lapok **20**, 133—136 (1969).
338. Segal, I.: Decomposition of operator algebras I, II. Mem. Amer. Math. Soc. **9**, New York, 1951.
339. Sherbert, D. R.: Banach algebras of Lipschitz functions. Pacific J. Math. **13**, 1387—1399 (1963).
340. Sherbert, D. R.: The structure of ideals and point derivations in Banach algebras of Lipschitz functions. Trans. Amer. Math. Soc. **111**, 240—272 (1964).
341. Sherman, S.: The second adjoint of a C^*-algebra. Proc. Int. Cong. Math., Cambridge **1**, 470 (1950).
342. Sherman, S.: Non-negative observables are squares. Proc. Amer. Math. Soc. **2**, 31—33 (1951).
343. Shilov, G.: On the extension of maximal ideals. Dokl. Akad. Sci. SSSR **29**, 83—84 (1940).
344. Shilov, G.: On normed rings possessing one generator. Mat. Sb. **21**, 25—47 (1947).
345. Shilov, G.: On regular normed rings. Trav. Inst. Math. Stekloff 21, Moskow 1947.
346. Shilov, G.: On decomposition of a commutative normed ring in a direct sum of ideals. Mat. Sb. **32**, 353—364 (1954).
347. Shirali, S.: Symmetry in complex involutory Banach algebras. Duke Math. J. **34**, 741—745 (1967).
348. Shirali, S.: Symmetry of Banach *-algebras without identity. J. London Math. Soc. **(2) 3**, 143—144 (1971).

349. Shirali, S.: Representability of positive functionals. J. London Math. Soc. (2) 3, 145—150 (1971).

350. Shirali, S., Ford, J. W. M.: Symmetry in complex involutory Banach algebras II. Duke Math. J. 37, 275—280 (1970).

351. Shirokov, F. V.: Proof of a conjective of Kaplansky. Uspehi Mat. Nauk 11, 167—168 (1956).

352. Siddiqi, J. A.: On a characterization of maximal ideals. Canad. M. Bull. 13, 219—220 (1970).

353. Sinclair, A. M.: Continuous derivations on Banach algebras. Proc. Amer. Math. Soc. 20, 166—170 (1969).

354. Sinclair, A. M.: Jordan homomorphisms and derivations on semi-simple Banach algebras. Proc. Amer. Math. Soc. 24, 209—214 (1970).

355. Sinclair, A. M.: Jordan homomorphisms on a semi-simple Banach algebra. Proc. Amer. Math. Soc. 25, 526—528 (1970).

356. Sinclair, A. M.: The norm of a Hermitian element in a Banach algebra. Proc. Amer. Math. Soc. 28, 446—450 (1971).

357. Sinclair, A. M.: The states of a Banach algebra generate the dual. Proc. Edinburgh Math. Soc. (2) 17, 341—344 (1971).

358. Sinclair, A. M.: The Banach algebra generated by a Hermitian operator. Proc. London Math. Soc. (3) 24, 681—691 (1972).

359. Sinclair, A. M.: Annihilator ideals in the cohomology of Banach algebras. Proc. Amer. Math. Soc. 33, 361—366 (1972).

360. Singer, I. M., Wermer, J.: Derivations on commutative normed algebras. Math. Ann. 129, 260—264 (1955).

361. Sluis, A. v. d.: The order of growth of the exponential function in a Banach algebra. Nieuw Arch. Wisk (3) 7, 95—101 (1959).

362. Smiley, M. F.: Right H^*-algebras. Proc. Amer. Math. Soc. 4, 1—4 (1953).

363. Smiley, M. F.: Right annihilator algebras. Proc. Amer. Math. Soc. 6, 698—701 (1955).

364. Smiley, M. F.: Real Hilbert algebras with identity. Proc. Amer. Math. Soc. 16, 440—441 (1965).

365. Smithies, F.: Extensions of ideals in associative algebras. Proc. Cambridge Philos. Soc. 55, 277—281 (1959).

366. Spatz, I. N.: Smooth Banach algebras. Proc. Amer. Math. Soc. 22, 328—329 (1969).

367. Stampfli, J. G.: An extreme point theorem for inverses in a Banach algebra with identity. Proc. Cambridge Philos. Soc. 63, 993—994 (1967).

368. Stampfli, J. G., Williams, J. P.: Growth conditions and the numerical range in a Banach algebra. Tôhoku Math. J. 20, 417—424 (1968).

369. Stein, J. D.: Homomorphisms of semi-simple algebras. Pacific J. Math. 26, 589—594 (1968).

370. Stinespring, W. F.: Positive functions on C^*-algebras. Proc. Amer. Math. Soc. 6, 211—216 (1955).

371. Stone, M.H.: On the theorem of Gelfand-Mazur. Ann. Polon. Math. 25, 238—240 (1953).

372. Stout, E. L.: The theory of uniform algebras. Tarrytown-on-Hudson: Bogden & Quigley 1971.

373. Strzelecki, E.: Metric properties of normed algebras. Studia Math. 23, 41—51 (1963).

374. Strzelecki, E.: Algebras under a minimal norm. Colloqu. Math. 11, 41—52 (1964).

375. Suciu, I.: Eine natürliche Erweiterung der kommutativen Banachalgebren. Rev. Roumaine Math. Pures Appl. 7, 483—491 (1962).

376. Suciu, I.: Bruchalgebren der Banachalgebren. Rev. Roumaine Math. Pures Appl. 8, 313—316 (1963).

377. Suzuki, N.: Representation of certain Banach *-algebras. Proc. Japan Acad. 45, 696—699 (1969).

378. Takeda, Z.: Conjugate space of operator algebras. Proc. Japan Acad. **30**, 90—95 (1954).
379. Taylor, D. C.: A characterization of Banach algebras with approximate unit. Bull. Amer. Math. Soc. **74**, 761—766 (1968).
380. Taylor, J. L.: The analytic functional calculus for several commuting operators. Acta Math. **125**, 1—38 (1970).
381. Taylor, J. L.: Functions of several non-commuting variables. Bull. Amer. Math. Soc. **79**, 1—34 (1973).
382. Ting, W. L.: On cohomology groups of Banach algebras. Proc. Amer. Math. Soc. **21**, 175—178 (1969).
383. Titchmarsh, E. C.: The zeros of certain integral functions. Proc. London Math. Soc. (2) **25**, 283—302 (1926).
384. Titchmarsh, E. C.: Theory of functions, second edition. London: Oxford University Press 1939.
385. Tomita, M.: The second dual of a C^*-algebra. Man. Fac. Sci. Kyushu Univ. **21**, 185—193 (1967).
386. Tomiuk, B. J.: Structure theory of complemented Banach algebras. Canad. J. Math. **14**, 651—659 (1962).
387. Tomiuk, B. J.: Left completely continuous semi-algebras. Fund. Math. **64**, 123—145 (1969).
388. Tomiuk, B. J., Wong, P. K.: The Arens product and duality in B^*-algebras. Proc. Amer. Math. Soc. **25**, 529—535 (1970).
389. Tomiyama, J.: Tensor products of commutative Banach algebras. Tôhoku Math. J. (2) **12**, 143—154 (1960).
390. Tornheim, L.: Normed fields over the real and complex fields. Michigan Math. J. **1**, 61—68 (1952).
391. Vala, K.: On compact sets of compact operators. Ann. Acad. Sci. Fenn. Ser. A 1 **351** (1964).
392. Vala, K.: Sur les éléments compacts d'une algèbre normée. Ann. Acad. Sci. Fenn. Ser. A 1 **407** (1967).
393. Varopoulos, N. Th.: Continuity des formes linéaire positives sur une algebre de Banach avec involution. C.R. Acad. Sci. Paris Sér. A–B **258**, 1121—1124 (1964).
394. Varopoulos, N. Th.: Sur les formes positive d'une algèbre de Banach. C. R. Acad. Sci. Paris Sér. A–B **258**, 2465—2467 (1964).
395. Varopoulos, N. Th.: A note on the abstract Wiener-Pitt phenomenon. Proc. Cambridge Philos. Soc. **61**, 297—298 (1965).
396. Varopoulos, N. Th.: Tensor algebras and harmonic analysis. Acta Math. **119**, 51—112 (1967).
397. Vidav, I.: Über eine Vermutung von Kaplansky. Math. Z. **62**, 330 (1955).
398. Vidav, I.: Eine metrische Kennzeichnung der selbstadjungierten Operatoren. Math. Z. **66**, 121—128 (1956).
399. Vidav, I.: Über die Darstellung der positiven Funktionale. Math. Z. **68**, 362—366 (1958).
400. Vidav, I.: Sur un système d'axiomes caracterisant les algèbres C^*. Glasnik Mat. Ser. III **16**, 189—193 (1961).
401. Vowden, B. J.: On the Gelfand-Naimark theorem. J. London Math. Soc. **42**, 725—731 (1967).
402. Waelbroeck, L.: Le calcul symbolique dans les algèbres commutatives. J. Math. Pures Appl. **33**, 147—186 (1954).
403. Wallen, L. J.: On the magnitude of x^n-1 in a normed algebra. Proc. Amer. Math. Soc. **18**, 956 (1967).

404. Walsh, B.: Banach algebras of scalar type elements. Proc. Amer. Math. Soc. **16**, 1167—1170 (1965).

405. Wang, J. K.: Multipliers of commutative Banach algebras. Pacific J. Math. **11**, 1131—1149 (1961).

406. Warzecha, A.: The general form of subalgebras with unity of finite codimension of Banach algebras with unit. Bull. Acad. Polon Sci. Sér. Sci. Math. Astronom. Phys. **17**, 237—242 (1969).

407. Wendel, J. G.: On isometric isomorphisms of group algebras. Pacific J. Math. **1**, 305—311 (1951).

408. Wendel, J. G.: Left centralizers and isomorphisms of group algebras. Pacific J. Math. **2**, 251—261 (1952).

409. Wermer, J.: On algebras of continuous functions. Proc. Amer. Math. Soc. **4**, 866—869 (1953).

410. Wermer, J.: Banach algebras and several complex variables. Chicago: Markham 1971.

411. Wermer, J.: Quotient algebras of uniform algebras. Symp. on functions algebras and rational approximation. University of Michigan 1969.

412. Wilansky, A., Zeller, K.: Banach algebra and summability. Illinois J. Math. **2**, 378—385 (1958).

413. Willcox, A. B.: Some structure theorems for a class of Banach algebras. Pacific J. Math. **6**, 177—192 (1956).

414. Wong, P. K.: On the Arens product and annihilator algebras. Proc. Amer. Math. Soc. **30**, 79—83 (1971).

415. Wong, P. K.: The Arens product and duality in B^*-algebras II. Proc. Amer. Math. Soc. **27**, 535—538 (1971).

416. Wong, P. K.: Modular annihilator A^*-algebras. Pacific J. Math. **37**, 825—834 (1971).

417. Ylinen, K.: Compact and finite-dimensional elements of normed algebras. Ann. Acad. Sci. Fenn. Ser. A 1 **428** (1968).

418. Yood, B.: Transformations between Banach spaces in the uniform topology. Ann. of Math. **50**, 486—503 (1949).

419. Yood, B.: Additive groups and linear manifolds of linear transformations between Banach spaces. Amer. J. Math. **71**, 663—677 (1949).

420. Yood, B.: Banach algebras of bounded functions. Duke Math. J. **16**, 151—163 (1949).

421. Yood, B.: Topological properties of homomorphisms between Banach algebras. Amer. J. Math. **76**, 155—167 (1954).

422. Yood, B.: Difference algebras of linear transformations on a Banach space. Pacific J. Math. **4**, 615—636 (1954).

423. Yood, B.: Periodic mappings on Banach algebras. Amer. J. Math. **77**, 17—28 (1955).

424. Yood, B.: Homomorphisms on normed algebras. Pacific J. Math. **8**, 373—381 (1958).

425. Yood, B.: Faithful *-representations of normed algebras. Pacific J. Math. **10**, 345—363 (1960).

426. Yood, B.: On the extension of modular maximal ideals. Proc. Amer. Math. Soc. **14**, 615—620 (1963).

427. Yood, B.: Noncommutative Banach algebras and almost periodic functions. Illinois J. Math. **7**, 305—321 (1963).

428. Yood, B.: Faithful *-representations of normed algebras II. Pacific J. Math. **14**, 1475—1487 (1964).

429. Yood, B.: On axioms for B^*-algebras. Bull. Amer. Math. Soc. **76**, 80—82 (1970).

430. Yood, B.: On non-negative spectrum in Banach algebras. Proc. Edinburgh Math. Soc. (to appear).

431. Żelazko, W.: A characterization of multiplicative linear functionals in complex Banach algebras. Studia Math. **30**, 83—85 (1968).

432. Żelazko, W.: Concerning extensions of multiplicative linear functionals in Banach algebras. Studia Math. **31**, 495—499 (1968).
433. Żelazko, W.: Concerning non-commutative Banach algebras of type ES. Colloqu. Math. **20**, 121—126 (1969).
434. Zeller-Meier, G.: Sur les automorphismes des algèbres de Banach. C. R. Acad. Sci. Paris Sér. A–B **264**, 1131—1132 (1967).

Additional Bibliography

435. Alexander, F. E.: On annihilator and dual A^*-algebras (to appear).
436. Alexander, F.E.: The dual and bi-dual of certain A^*-algebras. Proc. Amer. Math. Soc. **38**, 571—576 (1973).
437. Alexander, F. E.: Some counter-examples on annihilator, dual and complemented A^*-algebras (to appear in J. London Math. Soc.).
438. Bachelis, G. F.: Homomorphisms of Banach algebras with minimal ideals. Pacific J. Math. **41**, 307—312 (1972).
439. Barnes, B. A.: Banach algebras which are ideals in a Banach algebra. Pacific J. Math. **38**, 1—7 (1971).
440. Bernard, A.: Quotients of operator algebras. Lecture at University of Aberdeen 1973.
441. Brown, G., Pryce, J. D.: Stability theorems for wedges. Proc. Edinburgh Math. Soc. (2) **17**, 201—214 (1970/71).
442. Carleson, L.: Multiplicative functionals on Banach algebras. Seminars on analytic functions. Vol. 2. Princeton: Princeton University Press 1958.
443. Craw, I. G.: Axiomatic cohomology for Banach modules. Proc. Amer. Math. Soc. **38**, 68—74 (1973).
444. Dietrich, W. E.: On the ideal structure of Banach algebras. Trans. Amer. Math. Soc. **169**, 59—74 (1972).
445. Dowson, H. R.: On an unstarred operator algebra. J. London Math. Soc. (2) **5**, 489—492 (1972).
446. Feldman, C.: The Wedderburn principal theorem in Banach algebras. Proc. Amer. Math. Soc. **2**, 771—777 (1951).
447. Foiaş, C.: On a commutative extension of a commutative Banach algebra. Pacific J. Math. **8**, 407—410 (1958).
448. Gelbaum, B. R.: Banach algebra bundles. Pacific J. Math. **28**, 337—349 (1969).
449. Gelbaum, B. R.: Q-uniform Banach algebras. Proc. Amer. Math. Soc. **24**, 344—353 (1970).
450. Harte, R.: The spectral mapping theorem in several variables. Bull. Amer. Math. Soc. **78**, 871—875 (1972).
451. Harte, R.: Relatively invariant systems and the spectral mapping theorem. Bull. Amer. Math. Soc. **79**, 138—142 (1973).
452. Holub, J. R.: Bounded approximate identities and tensor products. B. Austral. Math. Soc. **7**, 443—445 (1972).
453. Inglestam, L.: Hilbert algebras with identity. Bull. Amer. Math. Soc. **69**, 794—796 (1963).
454. Inglestam, L.: On semigroups generated by topological nilpotent elements. Illinois J. Math. **13**, 172—175 (1969).
455. Jacobson, N.: Structure of rings, third edition. Amer. Math. Soc. Coll. Publ. 37, Providence 1968.

456. Kadison, R. V.: A representation theory for commutative topological algebras. Mem. Amer. Math. Soc. **7** (1951).

457. Leptin, H.: Die symmetrische Algebra eines Banachschen Raumes. J. Reine Angew. Math. **239/240**, 163—168 (1969).

458. Loy, R. J.: Identities in tensor products of Banach algebras. B. Austral. Math. Soc. **2**, 253—260 (1970).

459. Lumer, G.: Etats, algèbres quotients et sous-espaces invariants. C. R. Acad. Sci. Paris Sér. A–B **274**, 1308—1311 (1972).

460. McKilligan, S. A.: On the representation of the multiplier algebra of some Banach algebras. J. London Math. Soc. (2) **6**, 399—402 (1973).

461. McKilligan, S. A.: Duality in B^*-algebras. Proc. Amer. Math. Soc. **38**, 86—88 (1973).

462. Mizioɫek, J. K., Müldner, T., Rek, A.: On topologically nilpotent algebras. Studia Math. **43**, 41—50 (1972).

463. Ogasawara, T., Yoshinga, K.: Weakly completely continuous Banach *-algebras. J. Sci. Hiroshima Univ. Soc. A–I **18**, 15—36 (1954).

464. Paschke, W. L.: A factorable Banach algebra without bounded approximate unit (to appear in Pacific J. Math.).

465. Paterson, A. L. T., Sinclair, A. M.: Characterizations of isometries between C^*-algebras. J. London Math. Soc. (2) **5**, 755—761 (1972).

466. Pták, V.: Un théorème de factorisation. C. R. Acad. Sci. Paris Sér. A–B **275**, 1297—1299 (1972).

467. Robbins, D. A.: Existence of a bounded approximate identity in a tensor product. B. Austral. Math. Soc. **6**, 443—445 (1972).

468. Scheinberg, S.: The spectrum of an automorphism. Bull. Amer. Math. Soc. **78**, 621—623 (1972).

469. Seid, H. A.: A corollary to the Gelfand-Mazur theorem. Amer. Math. Monthly **77**, 282—283 (1970).

470. Shirali, S.: On the Jordan structure of complex Banach *-algebras. Pacific J. Math. **27**, 397—404 (1968). Correction: ibid. **31**, 834 (1969).

471. Sinclair, A. M.: Annihilator ideals in the cohomology of Banach algebras. Proc. Amer. Math. Soc. **33**, 361—366 (1972).

472. Spicer, D. Z.: A commutativity theorem for Banach algebras. Colloqu. Math. **27**, 107—108 (1973).

473. Tomiuk, B. J.: Duality and the existence of weakly completely continuous elements in a B^*-algebra. Glasgow Math. J. **13**, 56—60 (1972).

474. Tomiuk, B. J.: Modular annihilator A^*-algebras (to appear in Canad. Math. Bull.).

475. Tomiuk, B. J., Wong, P. K.: Weakly semi-completely continuous A^*-algebras. Illinois J. Math. **16**, 653—662 (1972).

476. Tomiuk, B. J., Wong, P. K.: Annihilator and complemented Banach *-algebras. J. Austral. Math. Soc. **6**, 20—47 (1972).

477. Tomiyama, J.: Primitive ideals in tensor products of Banach algebras. Math. Scand. **30**, 257—262 (1972).

478. Torrance, E.: Maximal C^*-sub-algebras of Banach algebras. Proc. Amer. Math. Soc. **25**, 622—624 (1970).

479. Tsuji, M.: Potential theory in modern function theory. Tokyo: Maruzen 1959.

480. Varopoulos, N. Th.: Sur les quotients des algèbres uniformes. C. R. Acad. Sci. Paris Sér. A–B **274**, 1344—1346 (1972).

481. Varopoulos, N. Th.: Sur le produit tensoriel des algèbres normées. C. R. Acad. Sci. Paris Sér. A–B **276**, 1193—1195 (1973).

482. Varopoulos, N. Th.: Some remarks on Q-algebras (to appear in Ann. Inst. Fourier (Grenoble)).

483. Wilansky, A.: Topological divisors of zero and Tauberian theorems. Trans. Amer. Math. Soc. **113**, 240—251 (1964).
484. Wong, P. K.: On the Arens product and annihilator algebras. Proc. Amer. Math. Soc. **30**, 79—83 (1971).
485. Yood, B.: Corrections to "Periodic mappings on a Banach algebra". Amer. J. Math. **78**, 222—223 (1956).
486. Yood, B.: Continuity for linear maps on Banach algebras. Studia Math. **31**, 263—266 (1968).
487. Żelazko, W.: On a certain class of non-removable ideals in Banach algebras. Studia Math. **44**, 87—92 (1972).
488. Zygmund, A.: Trigonometric series, second edition. 2 volumes. Cambridge 1959.

Index

The page number of definitions is indicated by bold face figures.

Index of Symbols

Ergebnisse der Mathematik und ihrer Grenzgebiete

Prices are subject to change without notice